BEHAVIOR
OF THE
LOWER ORGANISMS

BEHAVIOR
OF THE
LOWER ORGANISMS

by H. S. JENNINGS *with a new*
foreword by CHING KUNG

INDIANA UNIVERSITY PRESS
Bloomington   London

Published in Canada by Fitzhenry & Whiteside Limited, Don Mills, Ontario

Manufactured in the United States of America

ISBN 0–253–31135–7 cl   LC62–16164
ISBN 0–253–31136–5 pa

PREFACE

THE objective processes exhibited in the behavior of the lower organisms, particularly the lower animals, form the subject of the present volume. The conscious aspect of behavior is undoubtedly most interesting. But we are unable to deal directly with this by the methods of observation and experiment which form the basis for the present work. Assertions regarding consciousness in animals, whether affirmative or negative, are not susceptible of verification. This does not deprive the subject of consciousness of its interest, but renders it expedient to separate carefully this matter from those which can be controlled by observation and experiment. For those primarily interested in the conscious aspects of behavior, a presentation of the objective facts is a necessary preliminary to an intelligent discussion of the matter.

But apart from their relation to the problem of consciousness and its development, the objective processes in behavior are of the highest interest in themselves. By behavior we mean the general bodily movements of organisms. These are not sharply distinguishable from the internal physiological processes; this will come forth clearly in the present work. But behavior is a collective name for the most striking and evident of the activities performed by organisms. Its treatment as subsidiary to the problems of consciousness has tended to obscure the fact that in behavior we have the most marked and in some respects the most easily studied of the organic processes. Such treatment has made us inclined to look upon these processes as something totally different from the remainder of those taking place in organisms. In behavior we are dealing with actual objective processes (whether accompanied by consciousness or not), and we need a knowledge of the laws controlling them, of the same sort as our knowledge of the laws of metabolism. In many respects behavior presents an exceptionally favorable field for the study of some of the chief problems of life. The processes of behavior are regulatory in a high degree. Owing to their striking character, the way in which regulation occurs becomes more evident than in most other fields, so that they present a most favorable opportunity for study of this matter. To the regulatory aspect of behavior special attention is paid in the following pages.

The modifiability of the characteristics of organisms has always been a subject of the greatest importance in biological science. In most fields the study of this matter is beset with great difficulties, for the modifications require long periods and their progress is not easily detectible. In the processes of behavior we have characteristics that are modifiable with absolute ease. In the ordinary course of behavior variations of action are continually occurring, as a result of many internal and external causes. We see quickly and in the gross the changes produced by the environment, so that we have the best possible opportunity for the study of the principles according to which such changes occur. Permanent modifications of the methods of action are easily produced in the behavior of many organisms. When we limit ourselves to the subjective aspect of these, thinking only of memory, or the like, we tend to obscure the general problem involved. This problem is: What lasting changes are producible in organisms by the environment or otherwise, and what are the principles governing such modifications? Perhaps in no other field do we have so favorable an opportunity for the study of this problem, fundamental for all biology, as in behavior. There seems to be no *a priori* reason for supposing the laws of modification to be different in this field from those found elsewhere. The matter needs to be dealt with from an objective standpoint, keeping the general problem in mind.

A study of behavior from the objective standpoint will help us to realize that the activities with which we deal in other fields of physiology are occurring in a substance that is capable of all the processes of behavior, including thought and reason. This may aid us to be on our guard against superficial explanations of physiological processes.

But the chief interest of the subject of the behavior of animals undoubtedly lies, for most, in its relation to the development of psychic behavior, as shown by man. The behavior of the lowest organisms must form a fundamental part of comparative psychology.

In the special field of the behavior of the lowest organisms the foundations of our knowledge were laid by Verworn, in 1889, in his "Psycho-physiologische Protistenstudien." Binet, in his "Psychic Life of Micro-organisms" (1889), gave a most readable essay on the subject, presenting it frankly from the psychical standpoint. Lukas, in his "Psychologie der niedersten Tiere" (1905), has recently again dealt with the questions of consciousness in lower animals, the treatment of objective processes being subsidiary to this matter.

The present work was designed primarily as an objective description of the known facts of behavior in lower organisms, that might be used, not only by the general reader, but also as a companion in actual laboratory experimentation. This description, comprising Parts I and II of the present work, on the Protozoa and lower Metazoa, respectively, was made as far as possible independent of any theoretical views held by the writer ; his ideal was indeed to present an account that would include the facts required for a refutation of any of his own general views, if such refutation is possible. These designs have involved a fuller statement of details, with sometimes their repetition under new experimental conditions, than would have been necessary if the theoretical discussion had been made primary, and only such facts adduced as would serve to illustrate the views advanced. But the scientific advantages of the former method were held to outweigh the literary advantages of the latter.

As originally written, this descriptive portion of the work was more extensive, including, besides the behavior of the Protozoa and Cœlenterata, systematic accounts of behavior in Echinoderms, Rotifera, and the lower worms, together with a general chapter on the behavior of other invertebrates. The work was planned to serve as a reference manual for the behavior of the groups treated. But the exigencies of space compelled the substitution of a chapter on some important features of behavior in other invertebrates for the systematic accounts of the three groups last mentioned. The accounts of the Protozoa and of the Cœlenterata as representative of the lowest Metazoa remain essentially as originally written.

After this objective description was prepared, the need was felt for an analysis of the facts, such as would bring out the general relations involved. Part III is the result. Thus the conclusions set forth in Part III are the result of a deliberate analysis of the facts presented in a description which had been made before the conclusions had been drawn. The selection of facts set forth in the descriptive parts of the work has therefore been comparatively little affected by the general theories held by the writer. The loss of unity toward which this fact tends has perhaps its compensation in the impartiality which it helps to give the descriptions.

The writer is conscious of the necessarily provisional nature of most general conclusions at the present stage of our knowledge, and the analysis given in Part III is presented with this provisional character fully in mind. The reader should approach it in a similar attitude.

Since the book is written primarily from a zoölogical standpoint, it would be appropriate in some respects to entitle it " Behavior of the Lower Animals." But the broader title seems on the whole best, since the treatment of unicellular forms involves consideration of many organisms that are more nearly related to plants than to animals.

The figures have been drawn for the present work by my wife. Figures not credited to other authors are either new or taken from my own previous works.

The author is much indebted to the Carnegie Institution of Washington for making possible a year of uninterrupted research, devoted largely to studies preliminary to the preparation of this work and to its actual composition. He is further indebted for the use of a number of figures first published by the Carnegie Institution.

UNIVERSITY OF PENNSYLVANIA,
December 11, 1905.

FOREWORD TO THE 1976 EDITION

A renaissance in the study of cell motility and microbial behavior is taking place in the mainstream of experimental biology today. Recent developments make this classic work of Jennings more valuable than ever.

Present research is focused largely on the ultrastructures, molecular components, and biochemical and bioelectric events related to motility and behavior of microorganisms. We now know how a bacterial flagellum and its basal complex are constructed and that such a flagellum does not "beat" but rotates at its base (Berg, 1975). It has also been established that the beat of eucaryotic cilia (flagella) involves sliding of the microtubules instead of contraction (Satir, 1974). Studies of other forms of nonmuscular motility have also yielded much information. Microtubules and microfilaments are involved in a large variety of cellular and intracellular motions. A comprehensive review of cell motility can be found in Inoué and Stephens (1975) and the record of a recent conference on Cell Motility at Cold Spring Harbor (Goldman et al., in press).

Beyond the machinery of motion, considerable work has also been done on what Jennings would call the "regulation" (see the last chapter). For example, work on the bacterial responses to attractants and repellants (chemotaxis) has been extensive (Alder, 1975). A receptor protein that binds to galactose (an attractant) has been identified and studied (Adler, 1974; Boos, 1974). Some components of the regulatory mechanism, the "brain" of the bacterium, have been genetically identified (Parkinson, 1975). Jennings would also be pleased that research on his favorite *Paramecium* has advanced. The "avoiding reaction," which he so carefully described in his book, has been found to be the result of an action potential that can be understood in terms of standard metazoan physiological concepts (Eckert, 1972; Naitoh and Eckert, 1974). Furthermore, the "action system" (p. 107) of paramecium can now be genetically modified to give predictable but often bizarre behavior (Kung et al., 1975). Interests in behavioral regulation led to the recent Gordon Conference on Sensory Transductions in Microorganisms.

More comprehensive reviews on different aspects of various microbial behavior can be found in Volume XI of *Acta Protozologica, Symposium on Motile System of Cells* (1972), Pérez-Miravete (1973), Eisenstein (1975), and Carlile (1975).

Either implicitly or explicitly, most recent workers on lower organisms consider them "models" for higher organisms. Such "modeling" is not at the gross behavioral level but at the physical and chemical level. Thus, the avoiding reaction of a paramecium is not studied as a model of, say, the escape reflex of crayfish, but it is thought that the mechanisms for the action potentials of the two organisms are alike. Bacterial chemotaxis and smelling in animals may be different pieces of behavior but the molecular events in chemoreception or sensory transduction may be quite similar. Electron microscopy tells us that there is tremendous structural conservatism among eucaryotic cells. Ribosomes, mitochondria, basal bodies (centrioles), etc., from protozoan to man look much alike. Biochemistry tells us that an even deeper universality exists. Rhodopsin, actin, myosin, etc., are found in microorganisms and, at the level of molecular interactions, classification of life forms becomes practically meaningless. The diversity of morphology and behavior is thought to be built on a set of universal physical structures and events conserved through evolution.

Thus, recent development is no longer in the direction of more description and direct observation of behavior. Laboratories are occupied with finding and fitting together the nuts and bolts of the machinery in cellular architecture, metabolism, heredity, development, and, now, motility and behavior. Such is the trend of experimental biology today. Given this trend, it is not likely that a book on the *behavior* and not on the *behavioral machinery* of lower organisms will be written in the near future. This makes Jennings' volume invaluable as a source of information and inspiration.

CHING KUNG

Laboratory of Molecular Biology and Department of Genetics
University of Wisconsin-Madison
January, 1976

REFERENCES

Adler, J. 1974. In E. Sorkin and D. Platz, ed., *Antibiotics and Chemotherapy*, *vol. 19. Chemotaxis: Its Biology and Biochemistry.* White Plains, N.Y.:S. Karger.

Adler, J. 1975. *Ann. Rev. Biochemistry 44:*341–356.

Berg, H. C. 1975. *Ann. Rev. Biophys. Bioeng. 4:*119–136.

Boos, W. 1974. In E. Sorkin and D. Platz, ed., *Antibiotics and Chemotherapy*, *vol. 19. Chemotaxis: Its Biology and Biochemistry.* White Plains, N.Y.:S. Karger.

Carlile, M. J. 1975. *Primitive Sensory and Communication Systems: The Taxes, Tropisms of Micro-organisms and Cells.* New York: Academic Press.

Eckert, R. 1972. *Science 176:*473–481.

Eisenstein, E. M. 1975. *Aneural Organisms: Their Significance for Neurobiology.* New York: Plenum Press.

Goldman, R. D., T. D. Pollard, and J. L. Rosenbaum. In press. *Cold Spring Harbor Conference on Cell Proliferation, vol. 3. Cell Motility.*

Inoué, S. and R. E. Stephens. 1975. *Society of General Physiologists Series*, *vol. 30. Molecules and Cell Movement.* New York: Raven Press.

Kung, C., S. Y. Chang, Y. Satow, J. Van Houten, and H. Hansma. 1975. *Science 188:*898–904.

Naitoh, Y. and R. Eckert. 1974. In M. A. Sleigh, ed., *Cilia and Flagella.* New York: Academic Press.

Parkinson, J. S. 1975. *Cell 4:*183–188.

Pérez-Miravete, A. 1973. *Behaviour of Micro-organisms.* New York: Plenum Press.

Satir, P. 1974. In M. A. Sleigh, ed., *Cilia and Flagella.* New York: Academic Press.

Behavior of the Lower Organisms is a descriptive and experimental classic which is still cited in contemporary controversies. While not infallible (Mast and Lashley, 1916) and not representative of more recent developments (Fraenkel and Gunn, 1940; Carthy, 1958; Jensen, 1959), it continues to be a basic text for the student of animal behavior. But it is much more: it is a major work of a remarkable scientist, Herbert Spencer Jennings (1868–1947), and a work which is important in the history of experimental biology and behavioristic psychology.

Jennings, one of the giants of American biology during the first half of this century, was the son of a small-town physician who encouraged his son's naturalistic interests. After graduating from high school, Jennings taught for a time in one-room country schools, and then spent a year (1889) instructing at a small college. He then went to Michigan for his undergraduate training (1890–1893) and there finally decided on a career in biology, became familiar with rotifers and protozoa, the forms he used in nearly all of his subsequent research, and made surveys on these forms for the U. S. Commission of Fish and Fisheries. After a year of graduate study at Michigan, he went to Harvard, where he finished his graduate training (A.M., 1895; Ph.D., 1896). His thesis was a descriptive study of the development of rotifers, but he meanwhile acquired from C. B. Davenport an enthusiasm for the experimental methods which were just then spreading among biologists. In 1897 he received a traveling fellowship and spent a year in Europe, in Jena with Verworn and at the Naples biological station, where he began the program of work and developed the point of view summarized in this book. He continued his work on behavior during the following decade, which was spent in several temporary positions. A research assistantship with the Carnegie Institution of Washington enabled him to return to Naples and to prepare his first book, *Contributions to the Study of the Behavior of Lower Organisms* (1904). In 1906, the year the present volume was published, Jennings left the University of Pennsylvania and went to the Johns Hopkins University, where he remained until his retirement in 1938. While Jennings was completing his work on behavior, Mendel's laws of inheritance were being rediscovered. Soon after arriving at Johns Hopkins, Jennings began the study of inheritance in protozoa and he continued to work and write on genetics and problems of general biology for the rest of his life. While for a time he taught laboratory courses in behavior of lower organisms, even this stopped after S. O. Mast joined the faculty at Johns Hopkins in 1911.[1]

Jennings' work can be seen as a continuation of the evolutionary study of mind that began with Darwin and occupied such early workers as Romanes, Lubbock, and Preyer. These men had the principal aim of demonstrating that various mental functions had evolved just as had man's physical organs. The question of where sensations, perceptions, memory, consciousness, etc., first arose in the phylogenetic series was primary for these workers. Meanwhile other workers (Verworn, Jacques Loeb) were concerned with explaining animal behavior in terms of physio-chemical influences and without use of anthropomorphic, psychic, or mentalistic terms, i.e., those which refer to human subjective experience. Loeb modeled his explanations on Sachs' theory of tropisms of plants, in which stimulus-directed movement occurs until bilaterally symmetrical stimulation results. Jennings accepted the objective, experimental *modus operandi* of Verworn and Loeb, but denied the general utility of the tropism theory with its emphasis on external stimuli. He emphasized the complexity and variability of behavior in lower organisms and the importance of internal factors as determinants of behavior. He retained both the interest in evolutionary origins and the concern for objectivity.[2]

Jennings stopped active work with behavior around 1908, well before the behavioristic views of John B. Watson, usually considered the founder of psychological behaviorism, appeared (1912–14); the question then arises as to how Jennings influenced the behavioristic movement in American psychology. To answer that question we must consider the reaction his work evoked from certain other workers.

Jennings' books received criticism from two quarters, from the defenders of Loeb's tropism theory and from the psychologists who were offended by Jennings' language and methodology. Jennings' reply to the first group of critics, who felt his attack on Loeb's doctrines was a retreat from objectivism, is of interest because it effectively states his basic approach and his relation to the still earlier work of Loeb.

Certain authors seem to identify the "tropism theory" with the view that the behavior of organisms is to be explained by objective, experimentally determinable factors. They feel that an attack on the "tropism theory" is an attack on this view; this comes forth notably in the criticisms made by Loeb and Torrey, and it is evident in the attitude of some other writers.

There is, so far as I can see, nothing in the facts and relations which I have brought out that in any way opposes the principle that behavior is to be explained by objective, experimentally determinable factors—or indeed that bears in any way on the question. I have simply assumed throughout that it *is* to be explained in that way, and I do not see how experimental investigations can proceed on any other basis. Beginning my work in 1896, when the movement led by Loeb against the use of psychic concepts in explaining objective phenomena was in full swing, I considered that battle fought and won; I have, therefore, ever since proceeded, without discussion or ado of any sort, on that basis. Everyone must recognize the tremendous service done by Loeb in championing through thick and thin the necessity for the use of objective, experimental factors in the analysis of behavior. No convinced experimentalist, knowing the previous history of the subject, can reread, as I have just done, Loeb's early work on behavior without being filled with admiration for the clear-cut enunciation, defense, and application of the principles on which valuable experimental work has rested since that time, and on which it must continue to rest.

Any differences of opinion between Loeb and myself are then matters of detail; they concern merely the results of the application of the agreed principles of investigation. It has seemed to me that some of the experimentalists have rested content with superficial explanations; that they do not realize the complexity of the problems with which they were dealing. This has been the history of most applications of experiment to biology; the more thorough the work, the deeper are the problems seen to be.

Thus I have not hesitated to bring forth facts tending to show the inadequacy of the physio-chemical factors thus far set forth, and doubtless some have suspected that this was done with the concealed purpose of discrediting the general adequacy of such factors. This is a complete mistake; I did not till lately realize even the existence of such a suspicion. Complete confidence in the experimental method removes anxiety as to the effect of criticizing the details of its application. My objections are only to the adequacy of particular factors; they are based on experimental grounds, and the difficulties they raise are to be resolved only by experimental study. There is a vast difference between holding that behavior is fundamentally explicable on experimental grounds, and holding that we have already so explained it.

In a recent paper Loeb[3] has intimated that even if the behavior of the organisms under consideration were as complex as that of man, the same objective and experimental methods must be used in analyzing it. To this I fully agree, and the behavior of man is of course no more to be excepted from this treatment than is that of any other organism. [1908]

One of the reasons why Jennings received criticism from the followers of Loeb was that he used concepts which applied to human behavior (trial and error, perception, choice, etc.) in describing the behavior of lower organisms. Jennings continually emphasized, however, that he referred only to the "objective manifestations" of these concepts, that is, to the verifiable behavior to which the concepts refer and not to the subjective experiences of man. It is the clear enunciation of the idea that psychological concepts must be defined in terms of objectively observable, verifiable behavior that marks Jennings as a behaviorist.

Very different criticisms came from the psychological quarter. In the *Psychological Bulletin* the reviewer spiced his comments with ridicule while attacking the suggestion of continuity between the behavior of lower animals and of man. The reviewer argued that only if the behavior of lower and higher animals were equally complex would the use of a common terminology for both be justified. The use of introspective evidence by the reviewer also contrasts with behavioristic practice, as does the emphasis of the reviewer on the concept of consciousness:

Unfortunately, Jennings, while not a psychologist, has nevertheless in this volume wandered off into the green pastures of the psychologist (and has even nibbled at the stubble of the philosopher)! He was not content to allow his experimental facts to stand as facts, but must needs raise the question which stands (needlessly) as the bête noir of the student of behavior. Are the lower organisms conscious? Or, to phrase it from the objective standpoint, "Do there exist in the lower organisms objective phenomena of a character similar to those which we find in the behavior of man?" Jennings gives an affirmative answer: "So far as the objective evidence goes, there is no difference in kind, but a *complete continuity between the behavior of lower and of higher organisms*" (italics ours). Jennings then goes on to say that "no statement concerning consciousness in animals is open to verification or refutation by observation and experiment." ". . . All that experiment and observation can do is to show us whether the behavior of lower organisms is objectively similar to the behavior that in man is accompanied by consciousness. If this question is answered in the affirmative, as the facts seem to require . . . then it may perhaps be said that objective investigation is as favorable to the view of the general distribution of consciousness as it could well be." It is at this point that we must raise the question which is fundamental to our science. Have we any other criterion than that of behavior for assuming that our neighbor is conscious? And do we not determine this by the complexity of his reactions (including language under behavior)? .Complexity in conscious content is always accompanied by complexity in adjustment. This idea is the basal one in functional psychology. If my monkey's adjustments were as complex as those of my human subjects in the laboratory, I would have the same reason for drawing the conclusion as regards a like complexity in the mental processes of the two. Nor would my opinion 'then be largely dominated by general philosophical considerations drawn from other fields.' My inferences would alike in the two cases be based upon observed *facts* of behavior. Jennings has not shown, nor has anyone else shown that the behavior of lower organisms is objectively similar to that in man. To make the reviewer's position clear, Jennings' statements concerning the presence of perception in lower organisms may be cited. "When we say an animal *perceives* something, or shows a perception of something, we base this statement upon the observation that it reacts in some way to this thing. On the same basis, we could make the statement that amoeba perceives all classes of stimuli which we ourselves perceive, save sound (which is, however, essentially one form of mechanical stimulation). Perception as judged from our subjective experiences means much more; how much of this may be present in animals outside of ourselves we cannot know." The flaws in Jennings' psychology are surely patent to every student of experimental psychology. Is simple reaction to a stimulus the only *objective manifestation* of perceptual behavior in man? Certainly not! There are hundreds of others beside the overt movement of the voluntary muscles which can be directly observed, such as eye movement, convergence, accommodation, changes in respiration, circulation, changes in tonus of musculature, etc., and still others which can be inferred, as concerted reaction between different cortical systems; cortical 'retention' of the modifications of past stimuli, etc. It is the task of the experimental psychologist[4] to refine upon and to add to this list of objective manifestations of the perceptual act. So far as we know, some such complexity in adjustment is *necessary to every perceptual act*. If we may be allowed to call introspection in at this point, we find that it everywhere supports our contention that where you have complexity in content you likewise have complexity in adjustment; if subjectively to the human 'experiencer' there is more than simple reaction towards a stimulus in a perception, objectively there is more too. If Jennings would show that the adjustments of the amoeba to a sensory stimulation were as complex *from the objective or behavior standpoint* as our own adjustments to a like stimulus, we would not only be willing to grant him that his amoeba *perceives* but also we would be forced to make the assumption for the very same cogent reasons that we assume our fellow man perceives.

The same lack of psychological analysis is to be found in Jennings' assertion that lower organisms behave as though they consciously discriminate, and that they react as though they had representations. From the standpoint of the contribution of facts, the book is exceedingly valuable. That portion of the book dealing with the analysis of behavior has a somewhat dubious value because of its vagueness and complexity, and its constant allusions to pleasure and pain and to other psychical processes in man. The final chapters dealing overtly with the relationship of the behavior of lower organisms to psychic behavior should be undoubtedly greatly modified when the book comes to a second edition. [1907]

In 1905 the same reviewer had criticized Jennings' use of terms:

If we grant that Professor Jennings has taken the props from under the theory of tropisms, we must, I think, admit that he in his turn has neither proved that the positive reaction is due to pleasure nor the retractive one to pain. Finally, however much we as psychologists would like to believe that even in the lowest organisms the method of reaction is trial and error, we have to admit, I think, that it is straining the point to include the movements which Professor Jennings describes as taking place in the above organisms, under that method.

Jennings' reply to his critics (1908) contains material relevant to these reviews from the *Psychological Bulletin:*

In some recent writings one finds indications of a curious dualism, as if the behavior of lower organisms were to be analyzed in the objective, experimental way, but the behavior of higher animals and man were not. This takes most often the form of objection to any comparison between the objective features of the behavior of higher and lower animals, or to the use of the same terms in speaking of them, with a tendency toward accusations of vitalism or "psychologizing," against those making such comparisons. Such accusations evidently depend on the premise that the behavior of higher animals is to be explained only by vitalism or by "psychologizing."

When one is tempted to accuse an opponent in such ways, it is worthwhile to first examine whether the tendency to read psychic or vital factors into the phenomena does not lie in the mind of the accuser, rather than in that of the accused. When one has consciously and consistently taken the ground that the behavior of all organisms, including man, is to be analyzed in the objective, experimental way, and that there is no ground for expecting a failure of this method at any point, there is less occasion for anxiety at use of similar terms for similar objective phenomena throughout the series.

For example, the "method of trial and error" is as much an objective phenomenon, to be explained by experimentally determinable factors, in the dog or man, as in the infusorian. The undoubted great differences between the exemplifications of the "method" at the two extremes are matters for experimental analysis and demonstration, if the experimental method is not to fail. They do not show that the fundamental principle involved is different, and it is this common fundamental principle to which the common name calls attention. How far we should avoid words that have ever had any psychic connotation whatever is a matter on which there may be divergence of opinion; but it is most important to realize that this is totally distinct from the question whether the psychic connotation is of any use in objective experimental analysis. If this distinction is lost sight of, a divergence in practical details is taken for a conflict in fundamental principles, to the detriment of experimental science. That it is impossible to avoid such words completely is seen when we find in the writings of such men as Loeb the frequent use of such terms as "associative memory." Of course it is to only the objective phenomena that Loeb refers; but this is precisely the case also with other experimentalists accused of similar practices! [pp. 708–709]

Who was this reviewer, the psychologist who chided Jennings for his use of "trial and error" and for defining "perception" in terms of objective behavior instead of conscious content? Who was it that was disquieted by a suggestion of continuity between lower and higher animals? Who was it that invoked introspection in argument? The reviewer was John B. Watson, who seven years later wrote as follows:

Psychology as the behaviorist views it is a purely objective experimental branch of natural science. Its theoretical goal is the prediction and control of behavior. Introspection forms no essential part of its methods, nor is the scientific value of its data dependent upon the readiness with which they lend themselves to interpretation in terms of consciousness. The behaviorist attempts to give a unitary scheme of animal response. He recognizes no dividing line between man and brute. The behavior of man, with all of its refinement and complexity, forms only a part of his total field of investigation. [1914, p. 1][5]

It seems evident that Watson's views changed radically between 1907 and 1914 and became similar to those of Jennings. The impetus for this change was not simply familiarity with Jennings' works. Watson was at Chicago when he wrote his reviews. The following year he joined the faculty of the university where Jennings was Professor of Zoology. Later, he wrote:

In the fall of 1908 I went to Hopkins—met Baldwin, Jennings, Dunlap, Lovejoy, and Dr. Adolf Meyer. The whole tenor of my life changed. I tasted freedom in work without supervision. I was lost in and happy in my work. . . . As soon as possible, I began work with Jennings, taking his courses in evolution and his lab work on the behavior of lower organisms. [Watson, 1934, p. 277]

Not only did Watson meet and study under Jennings, but he interacted with other students of Jennings. He read Mast's book in manuscript and collaborated on a number of behavioral papers with K. S. Lashley, whose dissertation (1914) was done under Jennings on the inheritance of tentacle number in hydra. It is Lashley that Watson singles out for acknowledgment of debt and appreciation in his preface to his first book, but while four other teachers and colleagues are also cited, Jennings is not. Watson's espousal of a point of view that he had originally ridiculed must certainly have galled him, and it seems reasonable that Watson would balk at public pronouncements of debt to Jennings, a biologist, particularly when he and the psychologists he wished to influence were sensitive about the independent status of psychology as a science.

In a later book, Watson comes close to acknowledging a debt to Jennings:

Behaviorism, as I tried to develop it in my lectures at Columbia in 1912 and in my earliest writings, was an attempt to do one thing—to apply to the experimental study of man the same kind of procedure and the same language of description that many research men have found useful for so many years in the study of animals lower than man. [1930, p. v.]

In this book Watson includes a long quotation from Jennings' book *The Biological Basis of Human Nature* (1930) and in the preface thanks him for the use of the quotation.

In 1904 and 1907, Watson was dissatisfied with Jennings' approach; in 1908 he moved to the same institution, took courses under Jennings, and collaborated with one of Jennings' students; by 1914 he had adopted a point of view which differed little in principle from that of Jennings, if much in style. Where Jennings was cautious and calm, Watson was vehement, but the theme was the same—objective, experimental analysis of behavior. As Jennings acknowledged, this theme was not original with him but traceable to Loeb, and Jennings' contribution was its careful extension to more complex situations than those consistent with Loeb's theory of tropisms. Watson further extended that view to still more complex problems (human language, thought, etc.). All honor to Watson for that extension, but in fairness he should not be called the father of behaviorism unless the grandfather Jennings and the great-grandfather Loeb be proclaimed as well. A less gracious interpretation might be to call Watson not the father of behaviorism, but the midwife who delivered it from the uterine environment of lower organisms, where it was conceived by Loeb and developed by Jennings, into the world of man where it could further mature.

It should be noted that Watson had received training under Loeb, at Chicago, and this early training may have laid the foundation for Watson's later behavioristic views. However, it did not prevent him from reviewing Loeb's work in much the same manner he afforded Jennings (Watson, 1907, pp. 291–293) and there is little evidence of Watson's behavioristic views until after he came into direct contact with Jennings at Johns Hopkins.

It is apparent that *Behavior of the Lower Organisms* is a volume of considerable historical importance in the development of behavioristic psychology. It is also an introduction to the behavioristic position which is all the more effective because it lacks the polemic Watson directed against the now-banished introspective dogma. As an introduction to behaviorism it may well excite and direct psychologists of the future as it did those of the recent past.[6]

DONALD D. JENSEN

Department of Psychology
Indiana University
February, 1962

NOTES

1. See Sonneborn (1948) and Geiser (1934) for additional biographical material.
2. See Mast (1911) for a detailed historical review.
3. Pfuger's Archiv, 1906, *115*, p. 581 (Jennings' footnote).
4. At the time of the review experimental psychology employed mainly the method of experimental introspection, i.e., introspection by trained observers under standard conditions.
5. Note also Watson's frequent references to lower organisms (pp. 5, 27, 33, 46, etc.), his definition of sensation in terms of differential response (p. 13), the positive tone of his references to Jennings (pp. 56, 430), etc.
6. Preparation of the foreword was supported by grant G-17501 of the National Science Foundation.

REFERENCES

Carthy, J. D. 1958. *An Introduction to the Behavior of Invertebrates.*

Fraenkel, G. and Gunn, D. L. 1940. *The Orientation of Animals.*

Geiser, S. W. 1934. *Bios 5:*3–18.

Jennings, H. S. 1908. *Science 27:*698–710.

Jensen, D. D. 1959. *Behaviour 15:*82–122.

Mast, S. O. 1911. *Light and the Behavior of Organisms.*

Mast, S. O. and Lashley, K. S. 1916. *J. Exp. Zool. 21:*281–293.

Sonneborn, T. M. 1948. *Genetics 33:*1–4.

Watson, J. B. 1905. *Psychol. Bull. 2:*144–147.

Watson, J. B. 1907. *Psychol. Bull. 4:*288–291, 291–293.

Watson, J. B. 1914. *Behavior: An Introduction to Comparative Psychology.*

Watson, J. B. 1930. *Behaviorism.*

Watson, J. B. 1934. In Murchison, ed., *A History of Psychology in Autobiography: Vol. III.*

CONTENTS

PART I

BEHAVIOR OF UNICELLULAR ORGANISMS

CHAPTER I

BEHAVIOR OF AMŒBA

CHAPTER II

BEHAVIOR OF BACTERIA

CHAPTER III

BEHAVIOR OF INFUSORIA; PARAMECIUM

Structure ; Movements ; Method of Reaction to Stimuli

CHAPTER IV

BEHAVIOR OF PARAMECIUM (CONTINUED)

Special Features of the Reactions to a Number of Different Classes of Stimuli

CHAPTER V

BEHAVIOR OF PARAMECIUM (CONTINUED)

Reactions to Electricity and Special Reactions

CHAPTER VI

BEHAVIOR OF PARAMECIUM (CONTINUED)

Behavior under Two or More Stimuli ; Variability of Behavior ; Fission and Conjugation ; Daily Life ; General Features of the Behavior

CHAPTER VII

BEHAVIOR OF OTHER INFUSORIA

Action Systems. Reactions to Contact, to Chemicals, to Heat and Cold

CHAPTER VIII

REACTIONS OF INFUSORIA TO LIGHT AND TO GRAVITY

CHAPTER IX

REACTIONS OF INFUSORIA TO THE ELECTRIC CURRENT

CHAPTER X

MODIFIABILITY OF BEHAVIOR IN INFUSORIA, AND BEHAVIOR UNDER NATURAL CONDITIONS. FOOD HABITS

PART II

BEHAVIOR OF THE LOWER METAZOA

CHAPTER XI

INTRODUCTION AND BEHAVIOR OF CŒLENTERATA

CHAPTER XII

GENERAL FEATURES OF BEHAVIOR IN OTHER LOWER METAZOA

PART III

*ANALYSIS OF BEHAVIOR IN LOWER ORGANISMS, WITH A
DISCUSSION OF THEORIES*

CHAPTER XIII

CHAPTER XIV

CHAPTER XV

CHAPTER XVI

ANALYSIS OF BEHAVIOR IN LOWER ORGANISMS

CHAPTER XVII

ANALYSIS OF BEHAVIOR (CONTINUED)

CHAPTER XVIII

ANALYSIS OF BEHAVIOR (CONTINUED)

CHAPTER XIX

CHAPTER XX

CHAPTER XXI

BEHAVIOR AS REGULATION, AND REGULATION IN OTHER FIELDS

BEHAVIOR
OF THE
LOWER ORGANISMS

PART I

THE BEHAVIOR OF UNICELLULAR ORGANISMS

CHAPTER I

THE BEHAVIOR OF AMŒBA

1. Structure and Movements of Amœba

The typical Amœba (Fig. 1) is a shapeless bit of jellylike protoplasm, continually changing as it moves about at the bottom of a pool amid the débris of decayed vegetation. From the main protoplasmic mass there are sent out, usually in the direction of locomotion, a number of

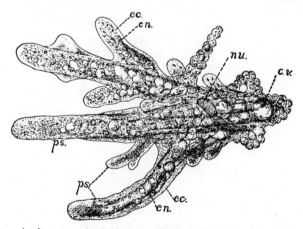

Fig. 1. — *Amœba proteus*, after Leidy (1879) (slightly modified). *c.v.*, contractile vacuole; *ec.*, ectosarc; *en.*, endosarc; *nu.*, nucleus; *ps.*, pseudopodia.

lobelike or pointed projections, the pseudopodia (Fig. 1, *ps.*). These are withdrawn at intervals and replaced by others. Within the mass of protoplasm certain differentiations are observable. Covering the outer surface there is usually, though not always, a transparent layer containing no granules; this is called the ectosarc (Fig. 1, *ec.*). Within this the protoplasm is granular, and contains bits of substance taken as food, vacuoles filled with water, and certain other structures. This

granular protoplasm is known as the endosarc (Fig. 1, *en.*). Within the fluidlike endosarc we find two well-defined structures. One is a disk-like or rounded, more solid body, known as the nucleus (Fig. 1, *nu.*).

FIG. 2. — *Amœba limax*, after Leidy (1879).

The other is a spherical globule of water, which at intervals collapses, emptying the contained water to the outside. This is the contractile vacuole (Fig. 1, *c.v.*).

There are many different kinds of Amœbæ, varying in their appearance and structure. For our purposes it will be sufficient to distinguish three main types. In one type the form is very irregular and changeable, and there are many pseudopodia. Of this type *Amœba proteus* (Fig. 1) is the commonest species. In a second type the animal usually moves forward rapidly as a single elongated mass, the protoplasm seeming very fluid. *Amœba limax* (Fig. 2) is a representative of this group. A third type consists of slowly

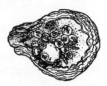

FIG. 3. — *Amœba verrucosa*, after Leidy (1879).

moving Amœbæ, of nearly constant form, usually having wrinkles on the surface, and with the thick ectosarc much stiffened, so that it does not appear fluid in character. The commonest representative of this type is *Amœba verrucosa* (Fig. 3).

In its usual locomotion the movement of Amœba is in many respects comparable to rolling, the upper surface continually passing forward

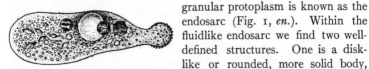

and rolling under at the anterior end, so as to form the lower surface. This may best be seen by mingling soot with water containing many Amœbæ. Fine granules of soot cling readily to the surface of Amœbæ of the *verrucosa* type, and more rarely to Amœbæ of other types. Such particles which are clinging to the upper surface move steadily forward till they reach the anterior edge. Here they are rolled over and come in contact with the substratum. They then remain quiet till the Amœba has passed across them. Then they pass upward again at the posterior end, and forward once more to the anterior edge. This

FIG. 4. — Paths of two particles attached to the outer surface of Amœba. That portion of the paths that is on the lower surface is represented by broken lines. The two particles were seen to complete the circuit of the animal five or six times in the paths shown. (The Amœba was of course progressing; no attempt is made to represent this in the figure.)

is repeated as long as the particles cling to the surface.   Single particles have been seen to pass thus many times around the body of the animal. Diagrams of the movements of the particles clinging to the surface are shown in Figs. 4 and 5.

It is not only the outermost layer of the ectosarc that thus moves forward.   On the contrary, the whole substance of the Amœba, from the

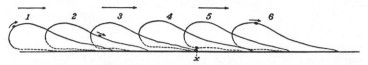

Fig. 5. — Diagram of the movements of a particle attached to the outer surface of *Amœba verrucosa*, in side view. In position 1 the particle is at the posterior end; as the Amœba progresses, it moves forward, as shown at 2, and when the Amœba has reached the position 3, the particle is at its anterior edge, at *x*. Here it is rolled under and remains in position, so that when the Amœba has reached the position 4, the particle is still at *x*, at the middle of its lower surface. In position 5 the particle is still at the same place *x*, save that it is lifted upward a little as the posterior end of the animal becomes free from the substratum. Now as the Amœba passes forward, the particle is carried to the upper surface, as shown at 6. Thence it continues forward, and again passes beneath the Amœba.

outer surface to the interior of the endosarc, moves steadily forward as a single stream, only the part in contact with the substratum being at rest.   At times small particles are at first attached to the outer surface, then gradually sink through the ectosarc into the endosarc.   Throughout the entire process of sinking inward the movement is steadily forward.

It is clear, then, that Amœba rolls, the upper surface continually passing across the anterior end to form the lower surface.   The anterior edge is thin and flat and is attached to the substratum, while the posterior

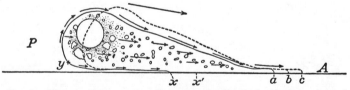

Fig. 6. — Diagram of the movements in a progressing Amœba in side view. *A*, anterior end; *P*, posterior end. The large arrow above shows the direction of locomotion; the other arrows show the direction of the protoplasmic currents, the longer ones representing more rapid currents. From *a* to *x* the surface is attached and at rest. From *x* to *y* the protoplasm is not attached and is slowly contracting, on the lower surface as well as above. *a*, *b*, *c*, successive positions occupied by the anterior edge. As the animal rolls forward, it comes later to occupy the position shown by the broken outline.

end is high and rounded, and is not attached to the substratum.   A very good idea of the character of the movements of Amœba may be obtained in the following way: Pin two edges of a handkerchief together, so as to make a flat cylinder. Within this place some heavy objects that will fill part of the cylinder, and lay the whole on a flat surface. Now pull forward the upper surface of the cloth near the anterior edge, a little at a time, bringing it in contact with the substratum. If this process is continued, the handkerchief rolls slowly forward with thin anterior edge and high posterior portion, — the weight within dragging behind. The lower surface is at rest while the upper surface moves forward. In all these respects the movement is like the locomotion of Amœba. A diagram of the movement of Amœba as it would appear in side view is given in Fig. 6.

While typically all the currents are forward in a progressing Amœba, any portion of the protoplasm may be excluded temporarily from the currents. This is especially common at the posterior end or tail, which is often composed of quiet protoplasm, covered with wrinkles or papillæ But the substance of the tail is in the course of time drawn into the currents and passes forward.

In the formation of pseudopodia the movement is much like that at the anterior end of the body. If the pseudopodium is in contact with the substratum, the upper surface moves forward while the lower surface is at rest. If the pseudopodium is sent forth freely into the water, its entire surface moves outward, in the same direction as the tip. These movements have been determined by observing the motion of particles attached to the outer surface of extending pseudopodia.

In some Amœbæ, according to Rhumbler (1898, 1905), the external protoplasmic currents turn backward at the sides of the anterior end, so that there is produced a fountainlike arrangement, an internal current forward, external currents backward. Such currents resemble those due to local decrease of surface tension in a drop of inorganic fluid.

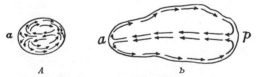

A   b

FIG. 7.—Currents in a drop of fluid when the surface tension is decreased on one side. *A*, the currents in a suspended drop, when the surface tension is decreased at *a*. After Berthold. *B*, axial and surface currents in a drop of clove oil in which the surface tension is decreased at the side *a*. The drop elongates and moves in the direction of *a*, so that an anterior (*a*) and a posterior (*p*) end are distinguishable.

Through such a local decrease the tension or *pulling* along the surface is lessened in a given region, so that the remainder of the surface film is pulling more strongly; it therefore drags the surface layer of the drop away from the point of lowered tension. The result is that currents pass on the surface in all directions away from this point (Fig. 7).

At the same time the inward pressure is decreased in the region of lowered tension, while elsewhere the pressure remains the same. Hence the internal fluid of the drop is pressed out toward the region where the film is weakened; a current flows in the central part of the drop toward this point. This current may produce a projection at the point of lowered tension, provided the surface currents do not carry the fluid back as fast as it is brought forward.

It was long supposed that the movements of all sorts of Amœbæ were of this character. As a natural conclusion, it was commonly held that locomotion and the formation of pseudopodia in Amœba are due to a local decrease in surface tension at the region of forward movement. As our account shows, most Amœbæ do not move at all as do liquid drops whose movements are produced through changes in surface tension.[1] Rolling movements with all currents forward cannot be produced experimentally through local changes in the surface tension of a drop of fluid. It is necessary, therefore, to abandon the surface tension theory for those Amœbæ that move in the way shown in Fig. 6. If the theory is still maintained for the Amœbæ with backward currents, this involves holding that the movements are due to fundamentally different causes in different Amœbæ; this is the view maintained by Rhumbler (1905).

While most Amœbæ roll as they progress, different species differ greatly in special features of their movements. The species of the *verrucosa* type (Fig. 3) move slowly and change form very little, not sending out pseudopodia. Those of the *limax* type (Fig. 2) move more rapidly and change form more frequently, but they rarely send out pseudopodia. Finally, in the *proteus* type (Fig. 1) the form is excessively changeable, many pseudopodia extending and retracting. Many Amœbæ show what might be called specialized habits in their usual movements.

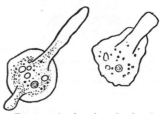

FIG. 8. — *Amœba velata*, showing the antennalike anterior pseudopodium projecting freely into the water. After Penard (1902).

For example, *Amœba angulata* and *Amœba velata* usually send forth at the anterior edge a pseudopodium which extends freely into the water and waves back and forth, serving as a feeler or antenna

[1] According to Rhumbler (1905), such movements are most readily seen in a species of Amœba living parasitically in the intestine of the cockroach. Whether the currents on the upper surface are actually backward, where the interior currents are forward, as is required if the movements are to be explained by local decrease of surface tension, has not been shown.

(Fig. 8). Many other special peculiarities of movement are described in the great work of Penard (1902) on these organisms.

2. Reactions of Amœba to Stimuli

The conditions under which Amœba lives are not always the same, and as the conditions change, the behavior of Amœba changes also. Such changes in behavior are usually called *reactions*, while the external agents that induce them are called *stimuli*.

A. Reaction to Contact with Solids

One of the commonest stimuli is that due to contact with a solid object. If a solid body strikes strongly against one side or one end of a

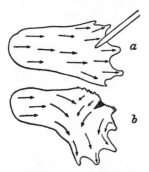

moving Amœba, the part affected contracts and releases its hold on the substratum, and the internal currents start away from it. The Amœba changes its course and moves in another direction. We may call this a negative reaction, since it takes the animal away from the source of stimulation. This reaction can be produced experimentally by touching the animal, under the microscope, with the tip of a glass rod drawn to a minute point (Fig. 9). The animal does not, as a rule, move directly away from the side touched, but merely in some other direction than toward this side. If we touch it at the anterior edge, the part touched stops and contracts, while the current turns to one side at this point, so that the animal moves at an angle with its former course (Fig. 9). Often the course is

Fig. 9. — Negative reaction to mechanical stimulation in Amœba. An Amœba advancing in the direction shown by the arrows is stimulated with the tip of a glass rod at its anterior edge (*a*). Thereupon this part is contracted, the currents are changed, and a new pseudopodium sent out (*b*).

altered only a little in this way. But if all of one side or one end is strongly stimulated, then a pseudopodium may be sent out on the side opposite, so that the animal moves almost directly away from the stimulated region (Fig. 10).

By repeatedly stimulating Amœba it is possible to drive it in any desired direction. The advancing edge is touched with the rod; it thereupon withdraws. A new pseudopodium is sent out elsewhere. If this does not lead in the direction desired, it is touched, causing retraction, whereupon the Amœba tries a new direction. This continues till a pseudopodium is sent out in the direction desired by the experimenter. The animal may now be compelled to follow a definite straight course, by stimulating any pseudopodium which tends to diverge from this course.

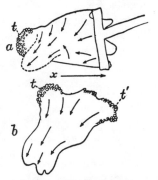

If the posterior end of a moving Amœba is stimulated, the animal continues to move forward, usually hastening its course a little. The posterior end is of course already contracted, and the new stimulation merely causes it to contract a little more.

The negative reaction is of course the method by which Amœba avoids obstacles. If an Amœba in creeping comes against a small solid body, the reaction is often less sharply defined than in the cases which we have thus far described. A typical example is shown in Fig. 11. A progressing Amœba came in contact at the middle of its anterior edge with the end of a dead alga filament. Thereupon the protoplasm ceased

FIG. 10. — Negative reaction to a mechanical stimulus when the entire anterior end is strongly stimulated. *a* and *b*, successive stages. The arrow *x* shows the original direction of motion; the arrows in *a* show the currents immediately after stimulation. In *b* a new tail (*t′*) has been formed from the former anterior end, uniting with the old tail (*t*).

to flow forward at the point of contact *c*, while on each side of this point the motion continued as before. In a short time, therefore, the animal had the form and position shown by the broken outline in Fig. 11; the filament projected deeply into a notch at the anterior edge. Motion continued in this manner would have divided the Amœba into two parts. But soon motion ceased on one side (*x*), while it continued on the side *y*. The currents in *x* became reversed

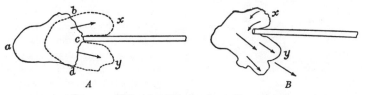

FIG. 11. — Method by which Amœba avoids an obstacle.

and flowed around the end of the filament into *y*, as shown at *B*, Fig. 11. Thus the animal had avoided the obstacle by reversing a part of the current and flowing in another direction.

But not all mechanical stimuli cause a negative reaction. Sometimes Amœba, on coming in contact with a solid body, turns and moves toward it, — responding thus by a positive reaction. At times an Amœba which is moving along on the glass slip used in microscopic work comes in contact by its upper surface with the under surface of the cover-glass. Thereupon it sometimes pushes forth a pseudopodium

FIG. 12. — *Amœba velata* passing from the slide to the cover-glass, side view. After Penard (1902). At *a* the animal is creeping in the usual way, with the tentaclelike pseudopodium projecting into the water. At *b* the pseudopodium has reached the cover-glass and attached itself. At *c* the animal has released its hold on the slide, and is now attached to the cover alone.

on this under surface; the pseudopodium attaches itself; the Amœba releases its hold on the slide, and now continues its course on the under side of the cover-glass. Penard (1902) has observed this in *Amœba velata*, when the long, tentaclelike anterior pseudopodium of this animal comes during its feeling movement in contact with the cover-glass. The process is represented in Fig. 12. In a similar manner Amœbæ frequently pass to the under side of the surface film of water, creeping on this as if it were a solid body.

Under certain circumstances Amœba seems especially disposed toward this positive reaction. Sometimes an Amœba is left suspended in the water, not in contact with anything solid. Under such circumstances the animal is as nearly completely unstimulated as it is possible for an Amœba to be; it is contact only with the water, and that uniformly on all sides. But such a condition is most unfavorable for its normal activities; it cannot move from place to place, and has no opportunity to obtain food. Amœba has a method of behavior by which it meets these unfavorable conditions. It usually sends out long, slender pseudopodia in all directions, as illustrated in Fig. 13.

FIG. 13. — *Amœba proteus* suspended in the water, showing the long pseudopodia extended in all directions. After Leidy (1879).

The body may become reduced to little more than a meeting point for these pseudopodia. It is evident that the sending out of these long arms greatly increases the chances of coming in contact with a solid body, and it is equally evident that contact with a solid is under the circumstances exactly what will be most advantageous to the animal. As soon as the tip of one of the pseudopodia does come in contact with something solid, the behavior changes (Fig. 14). The tip of the pseudopodium

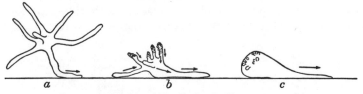

FIG. 14. — Method by which a floating Amœba passes to a solid.

spreads out on the surface of the solid and clings to it. Currents of protoplasm begin to flow in the direction of the attached tip. The other pseudopodia are slowly withdrawn into the body, while the body itself passes to the surface of the solid. After a short time the Amœba, which had been composed merely of a number of long arms radiating in all directions from a centre, has formed a collected flat mass, creeping along a surface in the usual way. This entire reaction seems a remarkable one in its adaptiveness to the peculiar circumstances under which the organism has been placed.

Positive reactions toward solid bodies are particularly common in the process of obtaining food. In our account of the food reactions we shall give examples of striking and long-continued reactions of this sort.

B. Reactions to Chemicals, Heat, Light, and Electricity

Reactions to Chemicals. — If a strong chemical in solution diffuses against one side or end of the body, the Amœba contracts the part affected, releasing it from the substratum, while the protoplasmic currents start in some other direction. The animal has thus changed its course. The reaction to chemicals can best be shown in the following way. The tip of a capillary glass rod is moistened, then dipped in some powdered chemical, preferably a colored one, such as methyline blue. This tip is then, under the microscope, brought close to one side of an Amœba in an uncovered drop of water. As soon as the diffusing chemical comes in contact with one side of the body, the reaction occurs. Chemicals that are fluid may be drawn into an excessively fine capillary tube and the tip of this held near the Amœba. Some of the variations in the reactions to chemicals are shown in Fig. 15.

Such experiments show that Amœba is very sensitive to changes in the chemical composition of the water surrounding it, and is inclined to move away whenever it comes to a region in which the water differs even slightly from that to which it is accustomed. It has been shown to react negatively when the following substances come in contact with one side of its body: methyline blue, methyl green, sodium chloride, sodium carbonate, potassium nitrate, potassium hydroxide, acetic acid, hydrochloric acid, cane sugar, distilled water, tap water, and water

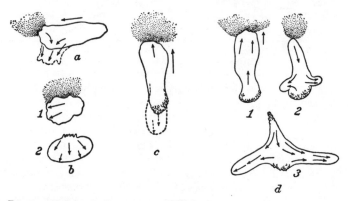

Fig. 15. — Variations in the reactions of Amœba to chemicals. The dotted area represents in each case the diffusing chemical. The arrows show the direction of the protoplasmic currents.

a. A little methyl green diffuses against the anterior end of an Amœba. The latter reacts by sending out a new pseudopodium at one side of the anterior end and moving in the direction so indicated.

b. A solution of NaCl diffuses against the right side of a moving Amœba (1). The side affected contracts and wrinkles strongly, while the opposite side spreads out (2), the currents flowing as shown by the arrows.

c. A solution of NaCl diffuses against the anterior end of an advancing Amœba. A broad pseudopodium, shown by the dotted outline, pushes out from the posterior region, above the end, and the course is reversed.

d. A solution of methyline blue diffuses against the anterior end of an Amœba (1). Thereupon a pseudopodium is sent out on each side of the posterior end at right angles with the original course (2). Into these the entire substance of the animal is drawn (3).

from other cultures than that in which the Amœba under experimentation lives.

Reaction to Heat. — If one side of an Amœba is heated, it reacts in the same negative way as to chemicals or to a mechanical shock. The reaction to heat may be observed as follows: An Amœba creeping on the under surface of the cover-glass is chosen for the experiment. The point of a needle is heated in a flame and placed against the cover-glass in front of the Amœba, or a little to one side of it. If the needle is not brought too close so as to affect the whole body instead of only one side, the animal responds by contracting the part affected and moving in some other direction.

Reactions to Light.—Light has a peculiar effect on Amœba. In general its functions seem better performed in the dark; strong light interferes with them seriously. Rhumbler (1898) observed that if Amœbæ are suddenly subjected to light while busy feeding on Oscillaria filaments, they cease to feed, and even give out the partly ingested filaments. Harrington and Leaming (1900) found that ordinary white light thrown on a moving Amœba causes it to come to rest at once. Blue light acts in the same way, while in red light the movements are as free as in darkness. Other colors have intermediate effects. Engelmann (1879) found that sudden illumination causes an extended Pelomyxa (which is merely a very large Amœba) to contract suddenly. It is well known that exposure to strong light is destructive to most lower organisms.

In correspondence with the fact that light interferes with its activities, we find that Amœba moves away from a source of strong light. If the sun is allowed to shine on it from one side, it moves, as Davenport (1897) shows, in the opposite direction. It thus moves in a general way in the same direction as the rays of light (Fig. 16). It is a peculiar fact that experiments so far have not shown a negative reaction to occur when light is thrown from directly above or below on one side or end of an Amœba. The fact that the whole body contracts when illuminated, as shown by the work of Engelmann (1879) on Pelomyxa, would lead us to expect that when a portion of the body is illuminated, this would contract, producing thus a negative reaction. But this has not been demonstrated. The experimental difficulties are great, and this may account for the lack of positive results. If future work substantiates the fact that light falling obliquely on one side causes

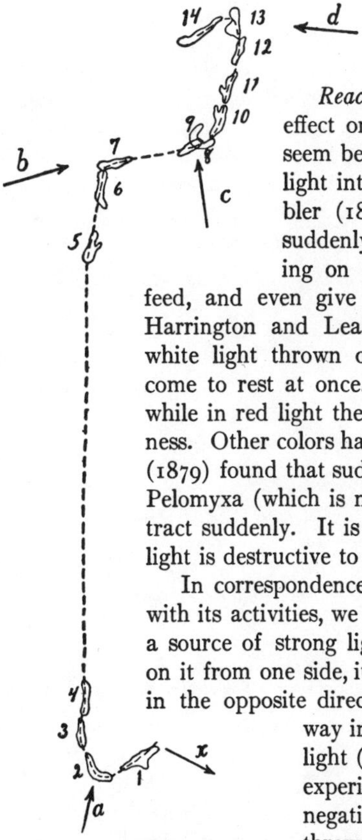

FIG. 16.—Reaction of Amœba to light, after Davenport (1897). The Amœba was first moving in the direction indicated by the arrow *x*. Light coming from the direction shown by the arrow *a* was then thrown upon it. It changed its course, occupying successively the positions 1, 2, 3, 4. The direction of the light was successively changed as indicated by the arrows *b, c, d*; the numbers 5–14 show the successive positions occupied by the animal. It will be observed that in every case as soon as the direction of the light is changed, the Amœba changes its course in a corresponding way, so as to retreat steadily from the source of light.

a reaction, while light falling from above or below on one side causes none, this would seem to indicate that the direction of the rays in passing through the body has something to do with determining the direction of locomotion. But in the myxomycete plasmodium, which resembles Amœba in its movements and in many other respects, light falling from above or below on a part of the body does produce a negative reaction, — the withdrawal of the part affected. Probably further experimentation will show the same thing to be true in Amœba.

Reaction to Electricity. — Electric currents probably form no part of the normal environment of Amœba, yet the animal reacts in a very definite way when a continuous current is passed through the water containing it. That side of the body which is directed toward the positive pole or anode contracts as if the animal were strongly stimulated here. Then a pseudopodium starts out somewhere on the side directed

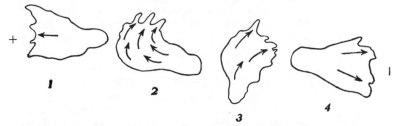

Fig. 17. — Reaction of Amœba to the electric current. The arrows show the direction of the protoplasmic currents; at 1 the direction of movement before the current acts is shown. 2, 3, 4, successive positions after the current is passed through the preparation.

toward the negative pole or cathode, and the Amœba creeps in that direction (Fig. 17). The reaction takes place throughout as if the Amœba were strongly stimulated on the anode side. If the electric current is made very strong, the anode side contracts still more powerfully, and the Amœba bursts open on the opposite side. The current is thus very injurious.

C. How Amœba gets Food

In the water in which Amœba lives are found many other minute animals and plants. Upon these Amœba preys, taking indifferently an animal or a vegetable diet. Its behavior while engaged in obtaining food is very remarkable for so simple an animal.

Spherical cysts of Euglena are a common food with *Amœba proteus.* These cysts are smooth and spherical, easily rolling when touched, so that they present considerable difficulties to an Amœba attempting to ingest them. One or two concrete cases will illustrate the behavior of Amœba when presented with the problem of obtaining such an object as food.

A spherical Euglena cyst lay in the path of an advancing *Amœba proteus.* The latter came against the cyst and pushed it ahead a short distance. The cyst did not cling to the protoplasm, but rolled away as soon as it was touched, and this rolling away continued as long as the animal moved forward. Now that part of the Amœba that was immediately behind the cyst stopped moving, so that the cyst was no longer pushed forward. At the same time a pseudopodium was sent out on each side of the cyst (Fig. 18), so that the latter was enclosed in a little bay. Meanwhile, a thin sheet of protoplasm passed from the upper surface of the Amœba over the cyst (Fig. 18, 2). The two lateral pseudopodia became bent together at their free ends; the cyst was thus held so that it could not roll away. The pseudopodia and the overlying sheet of protoplasm fused at their free ends, so that the

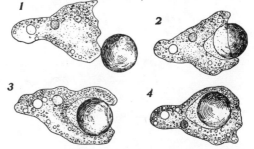

FIG. 18. — Amœba ingesting a Euglena cyst. 1, 2, 3, 4, successive stages in the process.

cyst was completely enclosed, together with a quantity of water. It was then carried away by the animal.

Amœba does not always succeed in obtaining its food so easily as in the case described. Often the cyst rolls away so lightly that the animal fails to grasp and enclose it. In such a case Amœba may continue its efforts a long time.

Thus, in a case observed by the author, an *Amœba proteus* was moving toward a Euglena cyst (Fig. 19). When the anterior edge of the Amœba came in contact with it, the cyst rolled forward a little and slipped to the left. The Amœba followed. When it reached the cyst again, the latter was again pushed forward and to the left. The Amœba continued to follow. This process was continued till the two had traversed about one-fourth the circumference of a circle. Then (at 3) the cyst when pushed forward rolled to the left, quite out of contact with the animal. The latter then continued straight forward, with broad anterior edge, in a direction which would have taken it away from the food. But a small pseudopodium on the left side came in contact with the cyst, whereupon the Amœba turned and again followed the rolling ball. At times the animal sent out two pseudopodia, one on each side the cyst (as at 4), as if trying to enclose the latter, but the spherical cyst rolled so easily that this did not succeed. At other times a single, long, slender pseudopodium was sent out, only its tip remaining in contact with the cyst (Fig. 19, 5); then the body was brought up from the rear, and the food pushed farther. Thus the chase continued until the rolling cyst and the following Amœba had described almost a complete

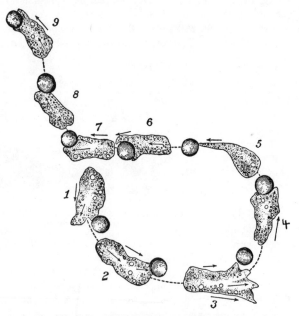

FIG. 19. — Amœba following a rolling Euglena cyst. The figures 1–9 show successive positions occupied by Amœba and cyst.

circle, returning nearly to the point where the Amœba had first come in contact with the cyst. At this point the cyst rolled to the right as it was pushed forward (7). The Amœba followed (8, 9). This new path was continued for some time. The direction in which the ball was rolling would soon have brought it against an obstacle, so that it seemed probable that the Amœba would finally secure it. But at this point, after the chase had lasted ten or fifteen minutes, a ciliate infusorian whisked the ball away in its ciliary vortex.

Such behavior makes a striking impression on the observer who sees it for the first time. The Amœba conducts itself in its efforts to obtain food in much the same way as animals far higher in the scale. In cultures containing many Amœbæ and many Euglena cysts it is not at all rare to find specimens thus engaged in following a rolling ball of food. Sometimes the chase is finally successful; sometimes it is not. Many of the cysts are attached to the substratum. Amœba often attempts to take such cysts as food, sending pseudopodia on each side of and above them, in the usual way, then covering them completely with its body. But it finally gives up the attempt and passes on.

Sometimes when a single pseudopodium comes in contact with a cyst, this pseudopodium alone reacts, stretching out and pushing the cyst ahead of it and keeping in contact with it as long as possible. Meanwhile the remainder of the Amœba moves in some other direction (Fig. 20). Finally the pseudopodium is pulled by the rest of the body away

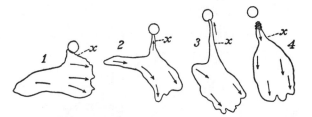

FIG. 20. — A single pseudopodium (*x*) reacts positively to a Euglena cyst, its protoplasm flowing in the direction of the cyst and pushing it forward, while the remainder of the Amœba moves in another direction. 1–4, successive forms taken. At 4 the reacting pseudopodium is pulled away from the cyst, whereupon it contracts.

from the cyst. Again, two pseudopodia on opposite sides of the body may each come in contact with a cyst. Each then stretches out, pulling a portion of the body with it, and follows its cyst. Soon the body comes to form two halves connected only by a narrow isthmus. Finally one half succeeds in pulling the other away from its attachment to the bottom. The latter half then contracts, and the entire Amœba follows the victorious pseudopodium.

Amœbæ frequently prey upon each other. Sometimes the prey is contracted and does not move; then there is no difficulty in ingesting it. Such a case has been described and figured by Leidy (1879, p. 94, and Pl. 7, Figs. 12-19). But the victim does not always conduct itself so passively as in this case, and sometimes finally escapes from its pursuer. This may be illustrated by a case observed by the present writer (Fig. 21).

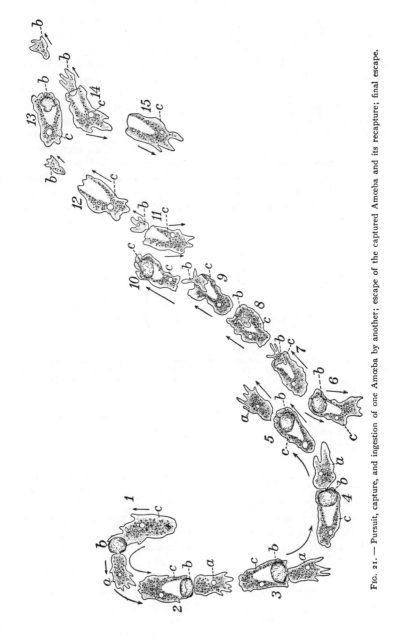

Fig. 21. — Pursuit, capture, and ingestion of one Amœba by another; escape of the captured Amœba and its recapture; final escape.

I had attempted to cut an Amœba in two with the tip of a fine glass rod. The posterior third of the animal, in the form of a wrinkled ball, remained attached to the rest of the body by only a slender cord, — the remains of the ectosarc. The Amœba began to creep away, dragging with it this ball. This Amœba may be called *a*, while the ball will be designated *b* (see Fig. 21). A larger Amœba (*c*) approached, moving at right angles to the path of the first specimen. Its path accidentally brought it in contact with the ball *b*, which was dragging past its front. Amœba *c* thereupon turned, followed Amœba *a*, and began to engulf the ball *b*. A cavity was formed in the anterior part of Amœba *c*, reaching back nearly or quite to its middle, and much more than sufficient to contain the ball *b*. Amœba *a* now turned into a new path; Amœba *c* followed (Fig. 21, at 4). After the pursuit had lasted for some time the ball *b* had become completely enveloped by Amœba *c*. The cord connecting the ball with Amœba *a* broke, and the latter went on its way, disappearing from our account. Now the anterior opening of the cavity in Amœba *c* became partly closed, leaving only a slender canal (5). The ball *b* was thus completely enclosed, together with a quantity of water. There was no adhesion between the protoplasm of *b* and *c*; on the contrary, as the sequel will show clearly, both remained independent, *c* merely enclosing *b*.

Now the large Amœba *c* stopped, then began to move in another direction (Fig. 21, at 5-6), carrying with it its meal. But the meal — the ball *b* — now began to show signs of life, sent out pseudopodia, and became very active; we shall therefore speak of it henceforth as Amœba *b*. It began to creep out through the still open canal, sending forth its pseudopodia to the outside (7). Thereupon Amœba *c* sent forth its pseudopodia in the same direction, and after creeping in that direction several times its own length, again enclosed *b* (7, 8). The latter again partly escaped (9), and was again engulfed completely (10). Amœba *c* now started again in the opposite direction (11), whereupon Amœba *b*, by a few rapid movements, escaped from the posterior end of Amœba *c*, and was free, — being completely separated from *c* (11, 12). Thereupon *c* reversed its course (12), overtook *b*, engulfed it completely again (13), and started away. Amœba *b* now contracted into a ball and remained quiet for a time. Apparently the drama was over. Amœba *c* went on its way for about five minutes without any sign of life in *b*. In the movements of *c* the ball became gradually transferred to its posterior end, until there was only a thin layer of protoplasm between *b* and the outer water. Now *b* began to move again, sent pseudopodia through the thin wall to the outside, and then passed bodily out into the water (14). This time Amœba *c* did not return and recapture *b*. The two

Amœbæ moved in opposite directions and became completely separated. The whole performance occupied about fifteen minutes. Such behavior is evidently complex. An analysis into simple reactions to simple stimuli is difficult if possible at all. We shall return to this matter later.

The method of food-taking illustrated in the behavior described is characteristic for Amœbæ of the *proteus* and *limax* types. It is sometimes said that these Amœbæ take food at the wrinkled posterior end. This, if true at all, is certainly rare; the author has never observed it, though he has seen food taken in dozens of cases. The essential features of the food reaction seem to be the movement of the Amœba toward the food body (long continued, in some cases), the hollowing out of the anterior end of the Amœba, the sending forth of pseudopodia on each side of and above the food, and the fusion of the free ends of the pseudopodia, thus enclosing the food, with a quantity of water. The reaction is thus complex; at times, as we have seen, extremely so.

In the process of taking food which we have just described there is no adherence between the protoplasm and the food body. But in *Amœba verrucosa* and its relatives foreign objects do adhere to the surface of the body, and this adherence is of much assistance in obtaining food. It partly compensates for the lack of pseudopodia in these species. But it is not alone food substances that cling to the surface of the body. Particles of soot and bits of débris of all sorts become attached in the same way. Not all these substances are taken into the body as food, so that adhesion to the surface does not account for food-taking. For this an additional reaction is necessary.

Food-taking in *Amœba verrucosa* often occurs as follows: The animal in its progress comes in contact with a small food body, such as a Euglena cyst. This adheres to the surface, and may pass forward on the upper surface of the body to the anterior edge, in the way described on a previous page. At the same time it begins to sink slowly into the body, surrounded by a layer of ectosarc. When it has rounded the anterior edge, the Amœba passes over it; then the food body passes upward again at the posterior end and forward on the upper surface. It is now sunk still more deeply into the protoplasm, and by the time it reaches the anterior edge again it has usually passed completely into the endosarc, together with the layer of ectosarc enveloping it. In this way the author has seen *Amœba verrucosa* ingest various algæ, small flagellates, Euglena cysts, and a small Amœba of the *proteus* type. Indifferent particles, such as bits of soot, which are attached to the surface at the same time, are not taken in.

Sometimes the taking of food is in *Amœba verrucosa* a much more complicated process than that just described. Rhumbler (1898) has given a very interesting account of the way in which this species feeds upon filaments of algæ many times its own length (Fig. 22). The animal settles upon the middle of an Oscillaria filament, envelopes it, and lengthens out along it (*a*). Then one end bends over (*b*), so that a loop is formed in the filament (*c*). The Amœba then stretches out on the filament again, bends it over anew, and the process is repeated until the fila-

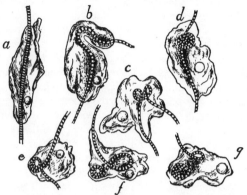

FIG. 22. — *Amœba verrucosa* coiling up and ingesting a filament of Oscillaria. After Rhumbler (1898). The letters *a* to *g* show successive stages in the process.

ment forms a close coil within the Amœba (*c* to *g*, Fig. 22). Leidy (1879, p. 86) has given a similar account of the method of feeding on filaments of algæ in Dinamœba.

Filaments that have been partly coiled up are often ejected when light is thrown upon the animal (Rhumbler, 1898).

3. FEATURES OF GENERAL SIGNIFICANCE IN THE BEHAVIOR OF AMŒBA

We find that the simple naked mass of protoplasm reacts to all classes of stimuli to which higher animals react (if we consider auditory stimulation merely a special case of mechanical stimulation). Mechanical stimuli, chemical stimuli, temperature differences, light, and electricity control the direction of movement, as they do in higher animals. In other words, Amœba has some method of responding to all the chief classes of life conditions which it meets.

The cause of a reaction — that is, of a change of movement — is in most cases some change in the environment, due either to an actual alteration of the conditions, or to the movement of the animal into new conditions. This is notably true of the reactions to mechanical, chemical, and thermal stimuli. In the reactions to light and the electric current this is not so evident at first view.   The Amœba reacts even though the light or current remains constant.   But if, as appears to be true, the stimulation occurs primarily on that side on which the light shines, or on the anode side in the reaction to electricity, then it is true that even in these cases the reacting protoplasm is subjected to changes of conditions. Since the movement of the Amœba is of a rolling character, the protoplasm of the anterior end and that of the posterior end continually interchange positions.   In an Amœba moving toward the cathode the extended protoplasm at the cathode end is gradually transferred to the anode end, and as this change takes place it contracts.   In the reaction to light the protoplasm of the anterior end directed away from the light is gradually transferred in the rolling movement to the lighted side; it then contracts.   It is therefore possible that in these cases also it is the change from one condition to another that causes reaction.

It is notable that changes from one condition to another often cause reaction when neither the first condition nor the second would, if acting continuously, produce any such effect.   Thus, Amœbæ react negatively to tap water or to water from a foreign culture, but after transference to such water they behave normally.   Harrington and Leaming (1900) show that when white light is thrown on an Amœba it ceases to move, but if this light continues, the animal resumes movement.   To constant conditions Amœba tends to become acclimatized.

But even constant conditions may induce reaction if they interfere seriously with the life activities of the animal.   Under great heat or strong chemicals the protoplasm contracts irregularly and remains thus contracted till death follows.   A different example of the production of a reaction by constant conditions is shown in the behavior of Amœbæ suspended in the water.   Under these conditions, as we have seen, the animal sends pseudopodia in all directions, taking a starlike form.   It is evident that the general condition of the organism, as well as an external change, may determine a reaction.

The fact that the nature of the behavior depends on the general condition of the organism is illustrated in another way by the observation of Rhumbler, that Amœbæ may begin to take food, then suddenly reject it.   This rejection occurs especially after subjection to light.   Apparently the light changes the condition of the animal in such a way that it no longer reacts to food as it did.

In Amœba, as in higher animals, the localization of the stimulation partially determines the reaction.   The result of stimulation on the right side is to cause movement in a direction different from that produced by stimulation on the left side.   In Amœba the relation of the movement to the localization of the stimulus is very simply determined, through the fact that it is primarily the part stimulated that responds. This part contracts or extends, thus partly determining the direction of movement. But the localization of the external stimulus is not the only factor in determining the direction of locomotion. Especially in the negative reactions certain other factors are evident, which are of much importance for understanding the behavior. After stimulation at one side or end, the new pseudopodium is as a rule not sent out in a direction exactly opposite that from which the stimulation comes. It usually appears, as we have seen, on some part of the original anterior end of the body, and at first alters the course only slightly. This is evidently connected with the fact that only the anterior end is attached to the substratum, and without such attachment locomotion cannot occur. If the pseudopodium were sent out from the unattached posterior part of the body, it would have to overcome the resistance of the contraction existing there, and would have to find the substratum and become attached to it. The new pseudopodium thus starts out from the region of least resistance, and in such a way that the new movement forms a continuation of the original one, though in a different direction. If the new direction still leaves the anterior part of the body exposed to the action of the stimulus, then a new pseudopodium is sent out in the same way, still further altering the course. This may continue till the original direction of locomotion is squarely reversed.

This is the method of changing the course that is usually seen in the reactions to mechanical (Fig. 9), chemical (Fig. 15), thermal, and electric (Fig. 17) stimuli. From Davenport's figures (Fig. 16) it appears to be likewise the method in the reactions to light.

From these facts it is clear that the direction of movement in a negative reaction is not determined entirely by the position of the stimulating agent or the part of the body on which it acts. The moving Amœba is temporarily differentiated, having two ends of opposite character, while the two sides differ from the ends. These internal factors play a large part in determining the direction of movement; the present action of Amœba, even when responding to stimuli, depends, as a result of these temporary differentiations, partly on its past action. The new pseudopodium will be sent out under most circumstances from some part of the anterior end, only under special conditions from a side, and still more rarely from the posterior end. We have here the first traces of relations which play large parts in the behavior of animals higher than Amœba. Structural differentiations have become permanent in most animals, and as such play a most important rôle in determining the direction of movement. Further, in practically all animals the past actions are, as in Amœba, important factors in determining reactions to present stimuli. In Amœba we see in the simplest way the effects of past stimuli and past reactions in determining present behavior.

As a result of this interplay of external and internal factors in determining movement, the avoidance of a stimulating agent usually occurs in Amœba by a process which we should call in higher animals one of trial. If the movement were directly and unequivocally determined by the localization of the stimulus, there would be nothing involved that could be compared to a trial. The direct withdrawal of the part stimulated is a factor due immediately to the localization of the external agent. But the sending forth of a pseudopodium in a new direction is not forced by the external agent, but is an outflow of the internal energy of the organism, and the position of this new pseudopodium is, as we have seen, determined by internal conditions. The latter factors are those which correspond to the activities that we call trial in higher animals. If the new direction of movement leads to further stimulation, a new trial is made. Such trials are repeated till either there is no further stimulation, or if it is not possible to escape completely, until the stimulation falls on the posterior end, and the animal is retreating directly from the source of stimulation.

The entire reaction method may be summed up as follows: The stimulus induces movement in various directions (as defined by internal causes). One of these directions is then selected through the fact that by subjecting the animal to new conditions, it relieves it from stimulation. This is our first example of "selection from among the conditions produced by varied movements," — a phenomenon playing a large part, as we shall see, in the behavior of organisms.

The method of reaction above described gives, with different stimuli, two somewhat differing classes of results. In the reactions to mechanical, chemical, and thermal stimuli, different directions are "tried" until the organism is moving in such a direction that it is no longer subjected to the stimulating agent; in this direction it continues to move. But in the reactions to light and to electricity new directions are tried merely until the stimulation falls upon the posterior end, and the organism is retreating directly from the source of stimulation. There is no possibility of escaping the stimulating agent completely. In the reactions to the two stimuli last mentioned the long axis of the animal must after a time take up a definite orientation with respect to the direction from which the stimulus comes, while in the reactions to other stimuli there is usually no such orientation. This difference is due, not to any essentially different method of reacting in the two cases, but merely to the peculiar distribution of the stimulating agents; light and electricity act continuously, and always affect a certain side of the organism, while this is not true of the other agents.

If an intense stimulus acts on the entire surface of Amœba at once, the animal contracts irregularly and ceases to move. If the acting agent is very powerful, the Amœba may remain contracted till it dies; otherwise it usually soon begins locomotion again.

We may classify the various changes in behavior due to stimulation into three main types, which may be called the positive reaction, the negative reaction, and the food reaction; these have already been described in detail. These types are not stereotyped; each varies much in details under different conditions. The movements in these reactions are clearly not the direct results of the simple physical action of the agents inducing them (see Jennings, 1904 *g*). As in higher animals, so in Amœba, the reactions are indirect. The effect of external agents is to cause internal alterations, and these determine the movements. It is therefore not possible to predict the movements of the organism from a knowledge of the direct physical changes produced in its substance by the agent in question.

What decides whether the reaction to a given stimulus shall be positive or negative? This question touches the fundamental problem of behavior. The nature of the physical or chemical action of an agent does not alone determine the reaction, for to the same agent opposite reactions may be given, depending on its intensity, or upon various attendant circumstances. If we should make a chemical or physical classification of the agents affecting movement in Amœba, this would not coincide with a classification based on the reactions given. But the agents which produce a negative reaction are in general those which injure the organism in one way or another, while those inducing the positive reaction are beneficial. Any agent which directly injures the animal, such as strong chemicals, heat, mechanical impact, produces the negative reaction. The positive reaction is known to be produced only by agents which are beneficial to the organism. It aids the animal to find solid objects on which it can move, and is the chief factor in obtaining food. Thus the behavior of Amœba is directly adaptive; it tends to preserve the life of the animal and to aid it in carrying on its normal activities.

It may perhaps be maintained that certain reactions are not adaptive; for example, that to the electric current. The reaction in this case does not tend to remove the organism from the action of the stimulating agent. But it is instructive to imagine in such a case an organism with possibilities of high intelligence — say even a human being — placed under similar conditions, with similar limitations of sense and of locomotive power. Would it give a more adaptive reaction than Amœba? Evidently, the conditions are such that it is impossible for the animal to escape by any means from the current. Since the stimulation apparently comes most strongly from the anode side, it is natural to move in the opposite direction. The method of the negative reaction is that of a trial of certain directions of movement. This method is in essence an adaptive one, and if it fails in the present case, certainly no better course of action can be suggested.

Can the behavior of Amœba be resolved throughout into direct unvarying reactions to simple stimuli, — into elements comparable to simple reflexes?

For most of the behavior described in the preceding pages the stimuli can be recognized in simple chemical or physical changes in the environment. Yet there are certain trains of action for which such a resolution into unvarying reactions to simple stimuli seems unsatisfactory. This is notably true for some of the food reactions. In watching an Amœba following a rolling food ball, as in Fig. 19, one seems to see the animal, after failing to secure the food in one way, try another. Again, in the pursuit of one Amœba by another, it is difficult to conceive each phase of action of the pursuer to be completely determined by a simple present stimulus. For example, in Fig. 21, after Amœba b has escaped completely and is quite separate from Amœba c, the latter reverses its course and recaptures b (at 11–13). What determines the behavior of c at this point? If we can imagine all the external physical and chemical conditions to remain the same, with the two Amœbæ in the same relative positions, but suppose at the same time that Amœba c has never had the experience of possessing b, — would its action be the same? Would it reverse its movement, take in b, then return on its former course? One who sees the behavior as it occurs can hardly resist the conviction that the action at this point is partly determined by the changes in c due to the former possession of b, so that the behavior is not purely reflex.

Of less interest than the case just mentioned are modifications in behavior due to acclimatization, and to the interference of stimuli. Amœba may become accustomed to certain things, so as to cease reacting after a time, though the condition remains the same. Thus Verworn (1889 b) found that Amœbæ which at first react to a weak electric current may after a time continue their usual movements, without regard to the current. Harrington and Leaming (1900), as we have seen, found that white or blue light thrown on Amœba causes it to cease moving, but if the light is continued, the movements begin again after a time. Indeed, we have recognized above the general fact that change is the chief factor in causing reaction, so that such acclimatization is a constant, normal factor in the behavior. A change in reaction due to a different cause is seen in Rhumbler's observation of the fact that Amœba after beginning to ingest food may reject it when subjected to light.

Beyond facts of this character, little is known as to the modifiability of reactions in Amœba.

LITERATURE I

(Works are cited here by giving the author's name followed by the date of publication. The full title will be found in the alphabetical list at the end of the volume. Only the important works are mentioned.)

A. General account of the behavior of Amœba, giving details of the observations on which the foregoing account is mainly based : JENNINGS, 1904 *e*.

B. Attempted physical explanations of the activities of Amœba: RHUMBLER, 1898; BÜTSCHLI, 1892; BERNSTEIN, 1900; JENSEN, 1901, 1902; VERWORN, 1892; JENNINGS, 1902 *a*, 1904 *g*; RHUMBLER, 1905.

C. General works on Amœba and its relatives: BÜTSCHLI, 1880; PENARD, 1902; LEIDY, 1879.

D. Reactions to unlocalized stimuli, and to localized heat: VERWORN, 1889.

E. Reaction to electricity: VERWORN, 1889 *b*, 1896 *a*; JENNINGS, 1904 *e*.

F. Reactions to light: DAVENPORT, 1897; HARRINGTON AND LEAMING, 1900; ENGELMANN, 1879.

CHAPTER II

THE BEHAVIOR OF BACTERIA

1. STRUCTURE AND MOVEMENTS

BACTERIA are perhaps the lowest organisms having a definite form and special organs for locomotion. In these characteristics they are less simple than Amœba and resemble higher animals, though in other ways the bacteria are among the simplest of organisms. Whether they are more nearly related to animals or to plants is a question of little importance for our purposes; they are usually considered as nearer to plants.

Bacteria are minute organisms living in immense numbers in decaying organic matter, and found in smaller numbers almost everywhere.

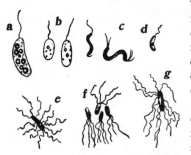

FIG. 23. — Different species of bacteria, showing the distribution of the flagella. *a, Chromatium okeni*, after Zopf ; *b, Chromatium photometricum*, after Engelmann ; *c, Spirillum undula*, after Migula ; *d, Vibrio choleræ*, after Fischer ; *e*, Bacilli of typhus, after Fischer ; *f, Bacillus syncyaneus*, after Fischer ; *g, Clostridium butyricum*, after Fischer.

They have characteristic definite forms (Fig. 23); some are straight cylindrical rods; some are curved rods; some are spiral in form; others are spherical, oval, or of other shapes. The individuals are often united together in chains.

While some bacteria are quiet, others move about rapidly. The movements are produced by the swinging of whiplike protoplasmic processes, the flagella or cilia. The flagella may be borne singly or in numbers at one end of the body, or may be scattered over the entire surface. Figure 23 shows the distribution of flagella in a number of species.

In most bacteria we can distinguish a permanent longitudinal axis, and along this axis movement takes place. Thus both the form, and in correspondence with it, the movement, are more definite than in Amœba. If the bacterium is quiet, we can predict that when it moves it will move in the direction of this axis; for Amœba such a prediction cannot be

26

made. In some bacteria the two ends are similar, and movement may take place in either direction. In others the two ends differ, one bearing flagella, while the other does not. In these species the movement is still further determined; the end bearing the flagella is anterior in the usual locomotion. In none of the bacteria can we distinguish upper and lower surfaces or right and left sides. As the bacterium swims, it revolves continually on its long axis; the significance of this revolution will be considered in our account of behavior in the infusoria.

2. REACTIONS TO STIMULI

The movements of the bacteria are not unordered, but are of such a character as to bring about certain general results, some of which at least are conducive to the welfare of the organism. If a bacterium swimming in a certain direction comes against a solid object, it does not remain obstinately pressing its anterior end against the object, but moves in some other direction. If some strong chemical is diffusing in a certain region, the bacteria keep out of this region (Fig. 24). They often collect about bubbles of air, and about masses of decaying animal or plant material. Often they gather about small green plants (Fig. 25), and in some cases a large number of bacteria gather to form a well-defined group without evident external cause.

How are such results brought about? To answer this question, we will examine carefully the behavior of the large and favorable form, Spirillum[1] (Fig. 23, *c*). Spirillum is a spiral rod, bearing a bunch of flagella at one end. In a thriving culture a large proportion of the individuals bear flagella at both ends and can swim indifferently in either direction. It is said by good authorities that such specimens are preparing to divide.

When Spirillum comes against an obstacle, it responds by the simplest possible reaction, — by a reversal of the direction of movement. In specimens with flagella at each end the new direction is continued till a new stimulation causes a new reversal. In bacteria with flagella at only one end, the movement backward is continued only a short time, then the forward movement is resumed. Usually when the forward movement is renewed, the path followed is not the same as the original path, but forms an angle with it; the bacterium has thus turned to one side. Whether this turning is due to currents in the water or other

---

[1] There are several species of Spirillum found in decaying organic matter. The species have not been clearly determined in most of the work on behavior, and this is not of great importance, as the behavior is essentially the same in character throughout.

accidental conditions, or, as is more probable, is determined in some way by the structure of the organisms, has not been discovered. In the infusoria, as we shall see, the latter is the case.

The reversal of movement of course carries the organism away from the agent causing it. We find that the same reaction is produced when the bacterium comes to a region where some repellent chemical is diffusing in the water. This is well shown when a drop of ⅕ per cent NaCl is introduced with a capillary pipette beneath the cover-glass of a preparation swarming with actively moving Spirilla. The bacteria at first keep up their movement in all directions, but on coming to the edge of the drop of salt solution the movement is reversed. Hence none of the bacteria enter the drop, and it remains empty, like the chemicals in Fig. 24.

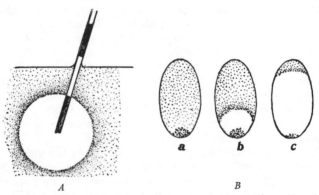

A   *B*

FIG. 24. — Repulsion of bacteria by chemicals. *A*. Repulsion of *Chromatium weissii* by malic acid diffusing from a capillary tube. After Miyoshi (1897). *B*, Repulsion of Spirilla by crystals of NaCl. *a*, Condition immediately after adding the crystals; *b* and *c*, later stages in the reaction. After Massart (1891).

They react in this way toward solutions of most acids and alkalies, as well as toward many salts and other chemicals. A drop of these chemicals remains entirely empty when introduced into a preparation of Spirilla.

This simple reversal of movement is the method by which avoidance of any agent takes place; in other words, it is the method of the negative reactions in bacteria. Bacteria also collect in certain regions, as we have seen, — about air bubbles, green plants, food, etc.; they have thus what are called "positive reactions" as well as negative ones. What is the behavior in the formation of such collections?

One finds, rather unexpectedly, that the positive reaction is produced in essentially the same way as the negative one, — by a simple reversal of movement under certain conditions. If we place water containing many Spirilla on a slide, allowing some small air bubbles to remain beneath the cover-glass, we find after a time that the bacteria are collecting about the bubbles. The course of events in forming the collections is seen to be as follows: At first the Spirilla are scattered uniformly, swimming in all directions. They pass close to the air bubble without change in the movements. But gradually the oxygen throughout the preparation becomes used up, while from the air bubble oxygen diffuses into the water. After a time therefore the bubble must be conceived as surrounded by a zone of water impregnated with oxygen. Now the bacteria begin to collect about the bubble. They do not change their direction of movement and swim straight toward the center of diffusion of the oxygen. On the contrary the movement continues in all directions as before. A Spirillum swimming close to the bubble into the oxygenated zone does not at first change its movement in the least. It swims across the zone until it reaches the other side, where it would again pass out into the water containing no oxygen. Here the reaction occurs; the organism reverses its movement and swims in the opposite direction. If the specimen has flagella at each end, it continues its reversed movement until the opposite side of the area containing the oxygen is reached; then the movement is reversed again. This is continued, the direction of movement being reversed as often as the organism comes to the outer boundary of the zone of oxygen within which it is swimming.[1] Thus the bacterium oscillates back and forth across the area of oxygen. Specimens having flagella at but one end swim backward only a short distance after reaching the boundary of the area, then start forward again.

As a result of this way of acting the bacterium of course remains in the oxygenated area. The latter thus retains every bacterium that enters it. Many bacteria, swimming at random, enter the area in the way described, react at the outer boundary, and remain; thus in the course of time the area of oxygen swarms with the organisms, while the surrounding regions are almost free from them. The finding of the oxygen then depends upon the usual movements of the bacteria, — not upon movements specially set in operation or directed by the oxygen.

Thus the positive and negative reactions of the bacteria are produced in the same way; both take place through the reversal of the movement when stimulated. The stimulus is some change in the nature of the surrounding medium. In the negative reaction the change is from ordinary water to water containing some chemical; in the posi-

[1] The bacterium may of course come against the bubble itself; the movement is then reversed in the same way.

tive reaction it is the change from water containing oxygen to water containing none.

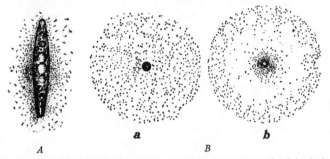

FIG. 25. — Collections of bacteria about algæ, due to the oxygen produced by the latter. *A*, Spirilla collected about a diatom. After Verworn. *B*, Bacteria gathered about a spherical green alga cell in the light. *a* shows the condition immediately after placing the bacteria and alga on a slide; no collection has yet formed. *b*, Condition two minutes later; part of the bacteria have gathered closely about the cell. After Engelmann (1894).

Spirilla collect in the way above described about any source of oxygen. Green plants give off oxygen in the light, so that the bacteria collect about desmids, diatoms, and other microscopic plants, in a lighted preparation, in the same way as about air bubbles (Fig. 25). Many other bacteria react in the same way to oxygen; notably the ordinary bacterium of decaying vegetable infusions, *Bacterium termo*. Bacteria react to exceedingly minute quantities of oxygen, so that it is possible to use them as tests for the presence of small amounts of this substance. Engelmann calculates that a bacterium may react to one one-hundred-billionth of a milligram of oxygen. By means of such reactions he has carried on investigations to determine whether various green or colorless organisms do or do not give off oxygen; results may be attained in this way that could scarcely be reached otherwise (Fig. 26). Spirillum (especially *S. tenue*) is so remarkably sensitive to oxygen that many individuals may react to the oxygen produced by a single specimen of another smaller bacterium (Engelmann).

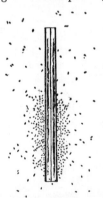

FIG. 26. — An experiment of Engelmann (1894), showing that when a diatom is partly lighted, only the part exposed to the light produces oxygen. The upper half of the diatom was in the shade, the lower half in the light. The bacteria have gathered only about the lighted half of the diatom.

When bacteria collect about bubbles or near the edge of the cover-glass as a reaction to oxygen, certain differences are to be observed in different species. Spirilla usually gather in a narrow zone a short distance from the air surface, while *Bacterium termo* and most other species collect in another zone, a little closer to the air. These relations are illustrated for Spirilla and certain infusoria in Fig. 27. In such cases it is found that the reversal of movement is brought about in two different regions.

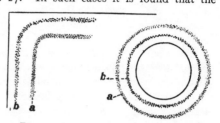

Passage from the zone in which the quantity of oxygen is adapted to the particular species, to a region having less oxygen, causes the reversal; passage to a region having more oxygen (next to the air surface) causes the reversal with even greater precision. As a re-

FIG. 27.—Collections of Spirilla, *a*, and a ciliate infusorian Anophrys, *b*, at the corner of the cover-glass, and about a bubble. Each remains in a narrow zone a certain distance from the air surface, the bacteria farther away than the infusoria. After Massart.

sult, each species remains swimming about within the narrow zone adapted to it, at a short distance from the air.

Thus any given species is adapted to a certain concentration of oxygen, which may be called its optimum. Passage from the optimum in either direction — toward more oxygen or less oxygen — causes the reversal of movement, so that the bacteria remain in the optimum.

Oxygen is of course necessary, or at least useful, to these bacteria; most of them become immobilized soon if oxygen is excluded from the water. The reversal of movement on passing to a region of less oxygen is thus an adaptive reaction. It is probable that the concentration in which each species tends to remain is that most favorable to its life activities. Some bacteria (the so-called *anaërobic* species) do not require oxygen, and these bacteria do not collect in an oxygen-

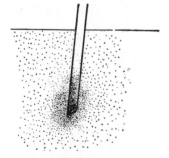

FIG. 28. —Collection of *Chromatium weissii* in and about a capillary tube containing 0.3 per cent ammonium nitrate. After Miyoshi (1897).

ated area. One of these, *Amylobacter*, is known to avoid oxygen in all effective concentrations; that is, it reverses its movement on coming to a region containing oxygen (Rothert, 1901).

Many bacteria collect in various other chemicals in the same manner as in solutions of oxygen (see Fig. 28). Such collections are usually formed in food substances; meat extract, for example, is an agent which produces such collections in most species of bacteria. Pfeffer (1884) found that *Bacterium termo* forms collections in meat extract, asparagine, peptone, white of egg, conglutin, grass extract, leucin, urea, and various other substances which might serve as nourishment. The so-called *sulphur bacteria* use hydrogen sulphide in their nutritive processes, and are found to collect in solutions of this substance (Miyoshi, 1897).

Many bacteria collect also in solutions of chemicals which probably do not serve directly as food.[1] *Bacterium termo* collects markedly in weak solutions of potassium carbonate, so that this is a favorable substance for demonstrating the collections. It collects also in most salts of potassium, and in a less marked way in many other inorganic chemicals. Indeed, this species may be said to gather in weak solutions of most inorganic chemicals, save in those of the powerful acids and alkalies. This bacterium lives on decaying vegetation, from which many chemicals diffuse into the surrounding water; potassium salts especially are given off in this manner. The tendency of the organisms to collect in such salts therefore keeps them in proximity to the decaying vegetation which serves them as nourishment; these reactions are thus indirectly adaptive. But *Bacterium termo* collects in certain chemicals that are not thus given off by decaying vegetation. Pfeffer (1888) found that they gather in salts of rubidium, cæsium, lithium, strontium, and barium, with which under natural conditions they never come in contact. It has been suggested that this may be explained as due to a similarity in the effect of these chemicals to the effects of others which they do meet under natural conditions. The organisms react thus in the same way to similar stimulation, without regard to its diverse source in different cases.

Many other bacteria resemble *Bacterium termo* in collecting in solutions of a great variety of chemicals. Miyoshi (1897) found that the sulphur bacterium *Chromatium weissii* forms collections in weak solutions of hydrogen sulphide, potassium nitrate, ammonium nitrate (Fig. 28), calcium nitrate, sodium-potassium tartrate, ammonium phosphate, monosodium phosphate, sodium chloride, cane sugar, grape sugar, asparagine, and peptone.

Some reactions can hardly be considered in any way adaptive. Rothert (1901) found that Amylobacter and another bacterium collect

---

[1] The method of testing the reaction to chemicals has usually been as follows. A capillary glass tube is filled with the solution to be tested, and one end is sealed. The open end is then brought into the fluid containing bacteria; these then enter the tube (Fig. 28) or leave it empty (Fig. 24, *A*), depending on their reaction to the chemical.

in weak solutions of ether. From the method by which the gatherings are produced, it is, of course, evident that collection in any agent signifies merely that the organisms are less repelled by this agent than by the surrounding conditions. All such collections are doubtless to be conceived as brought about by a reversal of the movement on passing from the dilute chemical to water containing none of the chemical. In many cases this has been determined by direct observations;[1] in other cases the observations have not been made.

If the chemical is stronger, the reversal of movement is produced when the bacteria come in contact with it, so that strong chemicals as a rule remain empty. Thus the same chemicals that, when dilute, produce a "positive reaction" cause, when stronger, a negative reaction. All substances in dilute solutions of which Spirillum gathers are avoided if stronger solutions are used. Miyoshi found this to be true also for *Chromatium weissii;* and it is indeed a general rule for bacteria.

Why should the bacteria avoid strong solutions of the very substances that when weak are "attractive"? It is, of course, well known that strong solutions are as a rule injurious; the negative reaction is therefore distinctly adaptive under these conditions. Even when we can see no use for the positive reaction, as in the case of the collecting of Amylobacter in a solution of ether, we find that the reaction becomes negative as soon as the solution becomes injurious. Amylobacter keeps out of stronger solutions of ether.

Yet the bacteria are no more infallible in detecting injurious substances than are higher organisms. If a poisonous chemical is mixed with a solution in which the bacteria naturally collect, the organisms may continue to enter a drop of the solution, where they are killed. So Pfeffer (1888, p. 628) found that if to an attractive solution of 0.019 per cent potassium chloride be added 0.01 per cent mercuric chloride, *Bacterium termo* and *Spirillum undula* continue to pass into the solution, though they are there immediately killed. *Bacterium termo* swarms into solutions of morphine (*morphium chloride*), where after ten minutes to an hour all motion ceases.

To just what action of the strong solution is the repellent effect, when it occurs, due? Strong solutions may be injurious from two different classes of causes. The specific properties of the given chemical may cause injuries when acting intensely, and this might induce the negative reaction. But farther, in any strong solution the osmotic pressure is high, and this produces injury in organisms by withdrawing the

---

[1] The reversal of motion under these circumstances has been described especially by Pfeffer (1884), Rothert (1901), and Jennings and Crosby (1901).

water from the protoplasm (*plasmolysis*).   The reaction of bacteria might then be due to this physical effect of strong solutions.

If the repellent effects of strong chemicals are due to their osmotic pressure, then all solutions having equal osmotic pressure must be equally repellent.   This gives a method of testing the matter.   Bacteria have been subjected to the action of many chemicals in solutions of equivalent osmotic pressure, with the following results.   There are many strong chemicals which cause reaction when the osmotic pressure is very low, — much lower than in the weakest solutions required to produce reaction in other substances.   Such are, as a rule, the strong mineral acids and alkalies (Pfeffer); such are potassium cyanide, potassium oxalate, sodium carbonate, sodium sulphite, and potassium nitrate in the experiments of Massart (1889).   The reactions produced by these substances can be due then only to their chemical effects, without regard to the osmotic pressure.   On the other hand, Massart has shown that in two species of bacteria — *Spirillum undula* and *Bacterium megatherium* — the repellent power of a large number of chemicals is proportional to the osmotic pressure of the solutions.   It appears probable therefore that the osmotic pressure is the cause of the reaction.[1]   In certain other bacteria it has been demonstrated that there is no such sensitiveness to osmotic pressure.   *Bacterium termo* enters the strongest solutions of attractive salts.   This is supposed to be because its protoplasm is permeable to the salts in question.   Taken all together, the experimental results demonstrate that in many cases the negative reaction is due to the chemical properties of the substance, and they render it probable that in some other cases the reaction is due to the osmotic pressure.

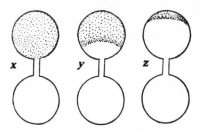

FIG. 29. — Repulsion of Spirilla of sea water by distilled water. The upper drop consists of sea water containing Spirilla; the lower of distilled water.   At *x* these have just been united by a narrow neck.   At *y* and *z* the bacteria are driven back before the advancing distilled water.   After Massart (1891).

It is not always more concentrated solutions that cause the reversal of movement.   Bacteria that live in sea water keep out of areas of dis-

[1] This conclusion is weakened by the fact that the bacteria are much less repelled by several substances — glycerine, asparagine, dextrose, and saccharose — even when they are so concentrated as to have higher osmotic pressure than the repellent solutions of the substances above mentioned (Massart, 1889).   This is explicable only by making certain special, unproved assumptions for each case.   The matter needs further investigation.

tilled water in the same way (Fig. 29). This result may be due to the fact that the osmotic pressure of the distilled water is less than that of the sea water. On the other hand, it is possible that it is due merely to the cessation of the chemical action of certain components of the sea water. The case would then be comparable to the reaction induced when bacteria come to a region containing no oxygen, as described in the preceding pages.

Most bacteria do not react to light. But there are certain bacteria for whose successful development light is required, and in these species we find that reaction to light occurs in the same manner as the reaction to oxygen in others. The species which react to light belong chiefly to the group of sulphur bacteria. They contain a purple coloring matter (*bacterio-purpurin*), which acts in a manner analogous to the chlorophyl of higher plants. By its aid, through the agency of light, these bacteria break up and assimilate carbon dioxide, giving off oxygen.

Engelmann (1882 *a*, 1888) made a thorough study of the relations to light in one of these bacteria, *Chromatium photometricum* (Fig. 23, *b*). This organism moves actively and develops well in diffuse light, but in the dark movement soon ceases and development stops. Only in the light does it assimilate carbon dioxide and give off oxygen. In correspondence with this, *Chromatium photometricum* collects in lighted areas. This takes place in the same manner as the collection of bacteria in oxygen. Engelmann placed the bacteria on a glass slide, in the usual way, then illuminated a certain spot from below, while light was cut off from the remainder of the preparation. He found that the bacteria do not react on entering the lighted area. But when once within this area, on coming to the outer boundary they suddenly reverse their movement and swim backward a distance. Then they start forward again; on coming anew to the boundary they react as before, and this happens every time they reach the confines of the lighted area. Thus none leave the light; all those that enter the lighted area remain, and a dense collection is soon formed here. In every detail the phenomena are parallel to those found in the reactions of other bacteria to oxygen, as described in previous pages.

A sudden decrease of light causes the same backward movement that is observed when the bacteria come to the edge of the lighted area. If the light is suddenly decreased by closing the diaphragm of the microscope, all the bacteria at once swim backward a distance, — often ten to twenty times their length. This shows that the reaction is not due to the difference in illumination of two ends or two sides of the organism, but only to the sudden decrease in light. This is shown also by the fact that the bacteria may swim completely across the boundary of the lighted region into the dark before reacting; the reaction then carries them back into the light. With the smaller bacteria the reaction usually occurs in this manner, while in larger species (*Monas okeni; Ophido-monas sanguinea*) the reversal of movement occurs when only one end has passed into the dark. A sudden increase of light merely causes the organisms to swim forward a little more rapidly.

The purple bacteria are sensitive in different degrees to lights of different colors, tending to gather in certain colors more than in others. This is shown in a most striking way when a spectrum is thrown on a preparation of *Chromatium photometricum* (Fig. 30). The largest num-

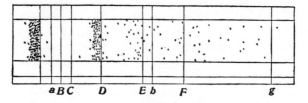

ᵃB C   D   E b   F   g

Fig. 30. — Distribution of bacteria in a microscopic spectrum. The largest group is in the ultra-red, to the left; the next largest group in the yellow-orange, close to the line *D*. After Engelmann.

ber of the bacteria collect in the ultra-red rays, which do not affect the human eye at all. There is another collection in yellow-orange, while a few are scattered through the green and blue. None are found in the red, the violet, or ultra-violet. These collections arise in the same manner as those in the white light. Bacteria swimming from blue toward yellow-orange, or from red toward ultra-red, do not react at all, but continue their course. But specimens swimming in the opposite direction react in the usual way, by leaping back, when they come to the outer boundary of the ultra-red or the orange-yellow. Hence, in the course of time, if the bacteria continue moving, almost all of them will be found in the two regions last named.

It is a most interesting fact that the colors in which the bacteria collect are exactly those which are most absorbed by them, and are also those which are most favorable to their metabolic processes. Engelmann showed that most oxygen is given off, and hence that most carbon dioxide is assimilated, in the ultra-red rays, while next to the ultra-red the orange-yellow are most favorable to these processes. The reactions of these bacteria to light are therefore adapted with remarkable precision to bringing them into regions which offer the best conditions for their development. This is the more remarkable when we consider that under natural conditions the bacteria rarely if ever have opportunity to react to the separated spectral colors.

Besides the purple bacteria, a green form, *Bacterium chlorinum*, is known to assimilate carbon dioxide and to collect in light, in the same manner as do the purple species.

The precise method by which bacteria react to heat and cold has been little studied. Mast (1903) has shown that Spirilla do not react at all to changes in temperature. If a portion of the preparation containing them is heated, they continue to pass into this region just as before, though they may be at once killed by the heat. They may pass also into a cold region, where motion gradually ceases.

The reaction to the electric current, like that to heat and cold, is in need of a thorough examination. Verworn found that when subjected to a continuous current some bacteria pass to the anode, others to the cathode.

When placed in a vertical tube, some kinds of bacteria pass upward to the top, in opposition to the force of gravity, while others gather at the lower end (Massart, 1891). The factors on which this reaction to gravity depends, and the precise way in which the reaction takes place, are unknown.

Bacteria often react to contact with solids by settling down and becoming quiet on the surface of the solid, which is usually some food body. *Bacterium termo* thus forms dense collections on the surface of such an object as a fly's leg.

3. General Features in the Behavior of Bacteria

We find that the chief reactions of bacteria, so far as they have been precisely determined, take place through a single movement, — a temporary reversal of the direction of swimming. This reaction is so simple as to be comparable to a reflex action as we find it in an isolated muscle. Whether the bacteria collect in a certain region or avoid it depends on what it is that produces this reversal of movement. The reaction is caused as a rule by a change in the environment of the organism. This change is usually brought about by the movement of the bacterium into a region differing from that which it previously occupied, but it may be due to an active alteration of the environment, as when light is suddenly cut off. For the reaction to occur with the result of a general movement of the organisms into a certain region, it is not necessary that different parts of the body should be differently stimulated, as we found to be the case in Amœba. The only requirement for producing a general movement of the organisms in a certain direction is that movement in any other direction shall result in such a change as will produce the reversal of movement. Not every change in the environment produces a reaction. A change leading toward a certain optimum condition produces no reaction, while a change of opposite character causes the reversal of movement. A negative change in the environment — the decrease or cessation of action of a certain agent — may be as effective a stimulus as is a positive change due to the entrance of a new agent into action. This is well illustrated in the reactions to light and oxygen. All these relations we shall meet again, more fully illustrated, in the behavior of infusoria.

The strength of the change necessary to cause a reaction has been found by Pfeffer to vary in accordance with Weber's law. This as usually formulated expresses certain relations between sensation and stimulus in man. According to this law, it is the relative change in the environment, not the absolute change, that causes a perceptible difference in sensation. Thus if a certain perceptible weight x is pressing on the skin of certain parts of the body, it requires an additional weight of about $\frac{1}{3}x$ to produce a noticeable difference in the sensation; if the original weight is $2x$, then an additional weight of $\frac{2}{3}x$ is required. In general the additional weight must be about one-third the original one before a noticeable difference in sensation is produced. In the bacteria we know nothing about sensations, but if we substitute reaction for sensation, similar relations are found to hold good. Pfeffer found that if *Bacterium termo* is cultivated in 0.01 per cent meat extract, they collect noticeably in capillary tubes containing 0.05 per cent meat extract, but not in a weaker solution. For producing reaction the inner fluid must therefore be five times as strong as the outer. If now the culture fluid is raised to a strength of 0.1 per cent meat extract, then five times this strength — namely, 0.5 per cent — is required to induce the bacteria to collect. If the culture fluid is 1 per cent, the fluid in the capillary tube must be 5 per cent in order to produce the usual reaction. The fluid in which the bacteria collect must be always five times as strong as that in which they live. It is the relative change, not the absolute change, that induces reaction. This agreement between the relation of sensation to stimulus in man and that of reaction to stimulus in these low organisms is of great interest.

There is a considerable amount of variation in the reactions among different individuals of the same species. Thus, Rothert found that specimens of Amylobacter from a certain culture were markedly negative to oxygen and positive to ether, while in specimens from another culture these reactions were hardly observable. Even among individuals of the same culture there is variation. Engelmann found that when the light falling on a group of individuals of Chromatium was suddenly decreased, a few react to even very slight changes, a larger number to more considerable changes, while some hardly react at all. "Nervous" and "apathetic" individuals, Engelmann says, can be distinguished in any group. Even in the same individual the reaction may vary. Engelmann found that if the light was suddenly decreased, then restored, and at once decreased again, the bacteria usually do not react to the second decrease, though they did to the first.

Among different kinds of bacteria there are, as we have seen, certain constant differences in the reactions. A relation of great significance becomes evident on examining the facts; *behavior under stimulation depends on the nature of the normal life processes,* — especially the metabolic processes. Bacteria that require oxygen in their metabolism collect in water containing oxygen; bacteria to which oxygen is useless or harmful avoid oxygen. Bacteria that use hydrogen sulphide in their metabolism gather in that substance. Bacteria that require light for the proper performance of their metabolic processes gather in light, while others do not. When one color is more favorable than others to the metabolic processes the bacteria gather in that color, even though they may under natural conditions have no experience with separated spectral colors. Keeping in mind that all these collections are formed through the fact that the organisms reverse their movement at passing out of the favorable conditions, these relations can be summed up as follows: Behavior that results in interference with the normal metabolic processes is changed, the movement being reversed, while behavior that does not result in interference or that favors the metabolic processes is continued.

This statement doubtless does not express the behavior completely, yet the general fact which it sets forth is on the whole clearly evident. The result of this method of action is to make the behavior regulatory, or adaptive. Through it, the bacteria, like higher organisms, avoid injurious conditions and collect in beneficial ones. There are some exceptions to this; the adaptiveness is not perfect, as nothing is perfect under all conditions. The exceptions are perhaps not more numerous in these lowest organisms than in the highest ones.

Putting all together, the behavior of the bacteria may be summed up as follows: They swim about in a direction determined by the position of the body axis, until the movement subjects them to an unfavorable change; thereupon they reverse and swim in some other direction. With rapid movements and much sensitiveness to unfavorable influences, this soon results in their finding and remaining in the favorable regions. In the presence of a localized region of favorable conditions (food or oxygen, for example) the organisms do not show movement in a single direction, adapted to reaching these favorable conditions. On the contrary, they show movements in all sorts of directions; one of these is finally continued or selected by its success. We find again behavior based on the "selection from among the conditions produced by varied movements."

LITERATURE II

(On the behavior of Bacteria)

ENGELMANN, 1881, 1882 a, 1888, 1894; JENNINGS AND CROSBY, 1901; MASSART, 1889, 1891, 1891 a; MAST, 1903; MIYOSHI, 1897; PFEFFER, 1884, 1888; ROTHERT, 1901, 1903.

CHAPTER III

THE BEHAVIOR OF INFUSORIA; PARAMECIUM

Structure; Movements; Method of Reaction to Stimuli

INTRODUCTORY

THE name *Infusoria* is applied to those unicellular organisms (aside from bacteria) that swim by means of cilia or flagella, as well as to a few others. The organs of locomotion are protoplasmic processes on the body surface. Where these are short and numerous, they are called cilia; where they are long and the organism bears but one or a small number, they are called flagella. The organisms bearing cilia are classed together as Ciliata; those with flagella are the Flagellata. Figure 31 shows a number of characteristic forms of the Ciliata. Along with the infusoria we shall take up other unicellular organisms or developmental stages that swim by means of such protoplasmic processes, — for example, spermatozoa and swarm spores.

The infusoria are commonly found, as the name implies, in infusions of decaying animal and vegetable matter. One of the commonest and best known of the infusoria is Paramecium, found in water containing decaying marsh plants, or in hay infusion with which some marsh or pond water has been mixed. The behavior of Paramecium has been studied more than that of any other infusorian, so that we shall take this up first as a representative of the group. The behavior of other species will be then examined to discover how far the relations in Paramecium are typical, and to bring out differences — especially points for which Paramecium is not a favorable object of study.

1. Behavior of Paramecium; Structure

Paramecium (Fig. 32) is a whitish, cigar-shaped animal, living in immense numbers in decaying vegetable infusions, and visible to the naked eye as a minute, elongated particle. The anterior part of the body is slender but blunt, the posterior part thicker, but more pointed. Thus the two ends differ, as in some bacteria, and there is a further

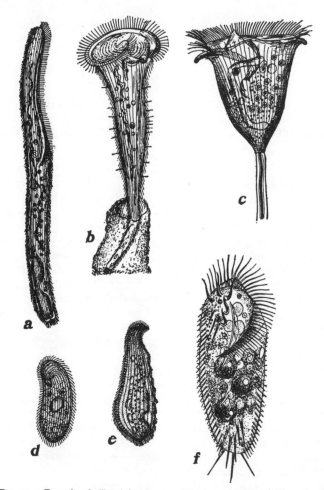

FIG. 31. — Examples of ciliate infusoria.   *a, Spirostomum ambiguum* Ehr., after Stein. *b, Stentor ræselii* Ehr., after Stein.   *c, Vorticella nebulifera* O. F. M., after Bütschli.   *d, Colpidium colpoda* Ehr., after Schewiakoff, from Bütschli.   *e, Loxophyllum meleagris* O. F. M., after Bütschli.   *f, Stylonychia mytilus* Ehr., after Engelmann.

differentiation of the lateral surfaces.   One side, the oral surface, bears a broad, oblique groove, known as the oral groove, or peristome, extending from the mouth in the middle of the body forward to the anterior end. When the animal is placed with the oral surface below, the groove extends from the right behind toward the left in front (see Fig. 32). The animal is thus not bilaterally symmetrical, but slightly spiral in form. The surface opposite the oral groove is marked by the presence near it of two large contractile vacuoles; this may be called the aboral surface. By considering the oral surface as ventral we may distinguish for convenience right and left sides. The entire body is covered with fine cilia, set in oblique rows. Those at the posterior end are a little longer than the others.

As to internal structure, we may distinguish an outer firm layer known as the ectosarc, enclosing an inner fluid portion, the endosarc. The ectosarc is covered by a thin outer cuticle; below this it is thickly set with rodlike sacs, placed perpendicular to the surface and known as trichocysts; the contents of these may be discharged as fine threads. The endosarc contains two nuclei, the large macronucleus and the minute micronucleus, together with numerous masses of food, most of them enclosed in vacuoles of water. The endosarc

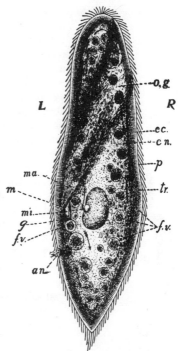

FIG. 32. — Paramecium, viewed from the oral surface. *L*, left side; *R*, right side. *an.*, anus; *ec.*, ectosarc; *en.*, endosarc; *f.v.*, food vacuoles; *g*, gullet; *m*, mouth; *ma.*, macronucleus; *mi.*, micronucleus; *o.g.*, oral groove; *P.*, pellicle; *tr.*, trichocyst layer. The arrows show the direction of movement of the food vacuoles.

is in continual movement, rotating lengthwise of the body, in the direction shown by the arrows in Fig. 32. Between endosarc and ectosarc, but attached to the latter, are the two contractile vacuoles, which at intervals collapse, emptying their contents to the outside. From the mouth (*m*) a passageway the gullet (*g*), leads through the ectosarc into the endosarc.

2. Movements

Paramecium swims by the beating of its cilia. These are usually inclined backward, and their stroke then drives the animal forward. They may at times be directed forward; their stroke then drives the animal backward. The direction of their effective stroke may indeed be varied in many ways, as we shall see later. The stroke of the cilia is always somewhat oblique, so that in addition to its forward or backward movement Paramecium rotates on its long axis. This rotation is over to the left (Fig. 33), both when the animal is swimming forward, and when it is swimming backward. The revolution on the long axis is not due to the oblique position of the oral groove, as might be supposed, for if the animal is cut in two, the posterior half, which has no oral groove, continues to revolve.

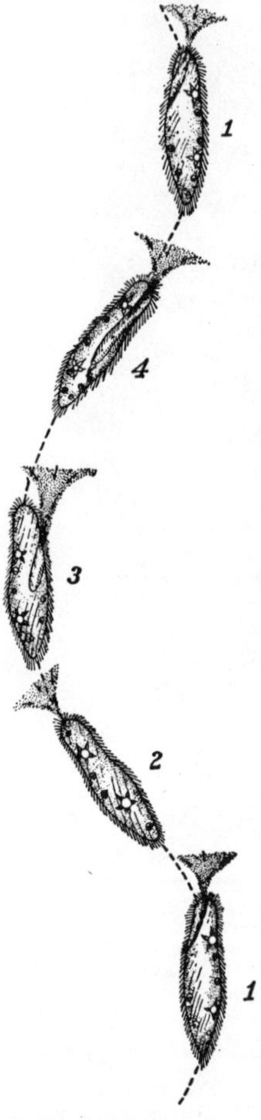

The cilia in the oral groove beat more effectively than those elsewhere. The result is to turn the anterior end continually away from the oral side, just as happens in a boat that is rowed on one side more strongly than on the other. As a result the animal would swim in circles, turning continually toward the aboral side, but for the fact that it rotates on its long axis. Through the rotation the forward movement and the swerving to one side are combined to produce a spiral course (Fig. 33). The swerving when the oral side is to the left is to the right; when the oral side is above, the body swerves downward; when the oral side is to the right the body swerves to the left, etc. Hence the swerving in any given direction is compensated by an equal swerving in the opposite direction; the resultant is a spiral path having a straight axis.

Fig. 33. — Spiral path of Paramecium. The figures 1, 2, 3, 4, etc., show the successive positions occupied. The dotted areas with small arrows show the currents of water drawn from in front.

The spiral course plays so important a part in the behavior of Paramecium that we must analyze it farther. The spiral swimming is evidently the resultant of three factors, — the forward movement, the rotation on the long axis, and the swerving toward the aboral side. Each of these factors is due to a certain peculiarity in the stroke of the cilia. The first results from the fact that the cilia strike chiefly backward. The second is due to the fact that the cilia strike, not directly backward, but obliquely to the right, causing the animal to roll over to the left. The third factor — the swerving toward the aboral side — is due largely to the greater power of the stroke of the oral cilia, and the fact that they strike more nearly directly backward. It seems partly due however to a peculiarity in the stroke of the body cilia, by which on the whole they strike more strongly toward the oral groove than away from it, thus driving the body in the opposite direction.

Each of these factors may vary in effectiveness, and the result is a change in the movements. The forward course may cease completely, or be transformed into a backward course, while the rotation and the swerving continue. Or the rotation may become slower, while the swerving to the aboral side continues or increases; then the spiral becomes much wider. This result is brought about by a change in the direction of the beating of the cilia to the left of the oral groove; they beat now to the left (toward the oral groove) instead of to the right (Fig. 34). The result of this is, as the figure shows, to oppose the rotation to the left, but to increase the swerving toward the aboral side. The width of the spiral, or the final complete cessation of the rota-

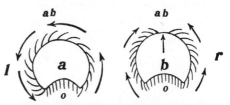

FIG. 34. — Diagrams of transverse sections of Paramecium, viewed from the posterior end, showing the change in the beat of the cilia of the left side. *a*, Stroke of the cilia in the usual forward movement. All the cilia strike toward the right side (*r*), rotating the organism to the left (*l*), as shown by the arrows. *b*, Stroke of the cilia after stimulation. The cilia of the left side strike to the left, opposing the lateral effect of the cilia of the right side. This causes the animal to cease revolving, and to swerve toward the aboral side (*ab*). *o*, Oral groove.

tion on the long axis which sometimes occurs, depends on the number and effectiveness of these cilia of the left side that beat toward the oral groove instead of away from it. A large part of the behavior of Paramecium depends, as we shall see, on the variations in the three factors which produce the spiral course.

3. ADAPTIVENESS OF THE MOVEMENTS

How does Paramecium meet the conditions of the environment? Under the answer to this question must be included certain aspects of the spiral movement, described in the foregoing paragraphs. The problem solved by the spiral path is as follows: How is an unsymmetrical organism, without eyes or other sense organs that may guide it by the position of objects at a distance, to maintain a definite course through the trackless water, where it may vary from the path to the right or to the left, or up or down, or in any intermediate direction? It is well known that man does not succeed in maintaining a course under similar but simpler conditions. On the trackless snow-covered prairie the traveller wanders in circles, try hard as he may to maintain a straight course, — though it is possible to err only to the right or left, not up or down, as in the water. Paramecium meets this difficulty most effectively by revolution on the axis of progression, so that the wandering from the course in any given direction is exactly compensated by an equal wandering in the opposite direction. Rotation on the long axis is a device which we find very generally among the smaller water organisms for enabling an unsymmetrical animal to follow a straight course. The device is marvellously effective, since it compensates with absolute precision for any tendency or combination of tendencies to deviate from a straight course in any direction whatsoever.

The normal movements of Paramecium are adaptive in another respect. The same movements of the cilia which carry the animal through the water also bring it its food. The oral cilia cause a current of water to flow rapidly along the oral groove (Fig. 33). In the water are the bacteria upon which Paramecium feeds; they are carried by this current directly to the mouth. In the gullet is a vibrating membrane which carries particles inward; the bacteria which reach the mouth are thus carried through the gullet to the endosarc, where they form food vacuoles and are digested.

Not only food, but also other substances, may be brought to Paramecium by the currents due to the movements of the cilia. It is important for understanding the behavior of this animal to realize that not only does it move forward to meet the environment, but the environment, so far as that is possible, also streams backward to meet it. If there is a chemical diffusing in the water in front of it, or if the water is warmer or colder, or differs in any other way,

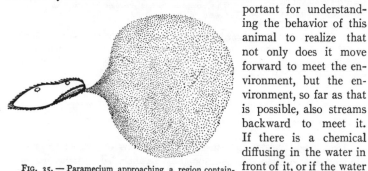

FIG. 35. — Paramecium approaching a region containing India ink (shown by the dots). The India ink is drawn out to the anterior end and oral groove of the animal.

a sample of this differentiated region is pulled backward in the form of a cone, and as a result of the stronger beating of the oral cilia, passes as a stream down the oral groove to the mouth (Fig. 33). This may best be seen by bringing near the anterior end of a resting Paramecium, by means of a capillary pipette, some colored solution, such as methyline blue, or by using in the same way water containing India ink. Or if a cloud of India ink, with a definite boundary, is produced in the water containing swimming Paramecia, a cone of the ink is seen to move out to meet the advancing animals (Fig. 35). Thus Paramecium is continually receiving "samples" of the water in front of it. Since in its spiral course the organism is successively pointed in many different directions, the samples of water it receives likewise come successively from many directions (Fig. 33). Thus the animal is given opportunity to "try" the various different conditions supplied by the neighboring environment. Paramecium does not passively wait for the environment to act upon it, as Amœba may be said, in comparison, to do. On the contrary, it actively intervenes, determining for itself what portion of the environment shall act upon it, and in what part of its body it shall be primarily affected by the varying conditions of the surrounding water. By thus receiving samples of the environment for a certain distance in advance, it is enabled to react with reference to any new condition which it is approaching, before it has actually entered these conditions.

4. Reactions to Stimuli

Let us suppose that as Paramecium swims forward in the way just described, it receives from in front a sample that acts as a stimulus, — that is perhaps injurious. The ciliary current brings to its anterior end water that is hotter or colder than usual, or that contains some strong chemical in solution, or holds large solid bodies in suspension, or the infusorian strikes with its anterior end against a solid object. What is to be done?

Paramecium has a simple reaction method for meeting all such conditions. It first swims backward, at the same time necessarily reversing the ciliary current. It thus gets rid of the stimulating agent, — itself backing out of the region where this agent is found, while it drives away the stimulus in its reversed ciliary current. It then turns to one side and swims forward in a new direction. The reaction is illustrated in Fig. 36. The animal may thus avoid the stimulating agent. If, however, the new path leads again toward the region from which the stimulus comes, the animal reacts in the same way as at first, till it finally becomes directed elsewhere. We may for convenience call this reaction, by which the animal avoids all sorts of agents, the "avoiding reaction."

In the foregoing paragraph we have given only a general outline of the behavior. The avoiding reaction has certain additional features, which add greatly to its effectiveness. After getting rid of the stimulus by swimming backward a distance there must be some way of deter-

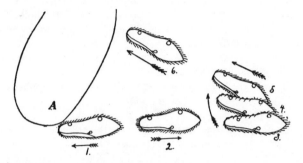

Fig. 36. — Diagram of the avoiding reaction of Paramecium. *A* is a solid object or other source of stimulation. 1–6, successive positions occupied by the animal. (The rotation on the long axis is not shown.)

mining the new direction in which the animal is to swim forward. It is evident that some method of testing the conditions in various different directions in advance would be the most effective way of accomplishing this. The infusorian now moves in precisely such a way as to make such tests. It will be recalled that in its usual course the animal is revolving on the long axis and swerving a little toward the aboral side (Fig. 33), so that it swims in a narrow spiral. After swimming backward a certain distance in response to stimulation, the revolution on the long axis becomes slower, while the swerving toward the aboral side is increased. As a result the anterior end swings about in a large circle; the animal becomes pointed successively in many different directions, as illustrated in Figs. 37 and 38. From each of these directions it receives in its ciliary vortex a "sample" of the water from immediately in advance, as the figures show. As long as the samples contain the stimulating agent, — the hot or cold water, the chemical, or the like, — the animal holds back and continues to swing its anterior end in a circle — "trying" successively many different directions. When the sample from a certain direction no longer contains the stimulating agent, the animal simply resumes its forward course in that direction. Thus its path has been changed, so that it does not enter the region of the chemical or the hot or cold water. Mechanical obstacles are avoided in precisely the same way, save that of course the ciliary vortex does not bring samples of the stimulating agent, so that the infusorian is compelled to try starting forward repeatedly in various directions, before it finds one in which it can pass freely.

This method of behaving is perhaps as effective a plan for meeting all sorts of conditions as could be devised for so simple a creature. On getting into difficulties the animal retraces its course for a distance, then tries going ahead in various directions, till it finds one in which there is no further obstacle to its progress. In this direction it continues. Through systematically testing the surroundings, by swinging the anterior end in a circle, and through performing the entire reaction repeatedly, the infusorian is bound in time to find any existing egress from the difficulties, even though it be but a narrow and tortuous passageway.

The different phases of this avoiding reaction are evidently due to modifications of the three factors in the spiral course. The swimming backward is due of course to a reversal of the forward stroke of the cilia. The turning toward the aboral side is an accentuation of the swerving that takes place always; it is due to the fact that the cilia at the left side of the body strike during the reaction toward the oral groove instead of away from it. Thus the cilia of both right and left sides now tend to turn the animal toward the aboral side. The difference between the usual condition and that found during the reaction is illustrated in Fig. 34. Finally, the decrease or cessation in the revolution on the long axis is due to the same factor as the increase in swerving toward the aboral side. During the reaction the cilia of the left side oppose the usual revolution on the long axis to the left (as shown in Fig. 34), through the same change which causes them to assist in turning the body toward the aboral side.

The avoiding reaction varies greatly under different conditions, though its characteristic features are maintained throughout. But its different phases vary in intensity depending on circumstances. The backward movement may be long continued, or may last but a short time; or there may be merely a stoppage or slowing of the forward movement. The swerving toward the aboral side may be only slightly increased, while the revolution on the long axis becomes a little slower. In this case the anterior end swings about in a small circle, as in Fig. 37, so that the animal is pointed successively in a number of directions varying only a little from the original one. With a stronger stimulus the swerving toward the aboral side is more decided, while the rotation on the long axis is slower; then the anterior end swings about a larger circle, as in Fig. 38. The Paramecium thus

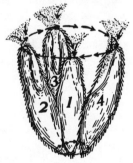

Fig. 37. — Paramecium swinging its anterior end about in a small circle, in a weak avoiding reaction. 1, 2, 3, 4, successive positions occupied.

becomes pointed successively in many directions differing much from the original one. Finally, the rotation on the long axis may completely cease, while the swerving toward the aboral side is farther increased; then the Paramecium swings its anterior end about a circle with its posterior end near the centre (Fig. 39). In this case the animal may turn directly away from the stimulating agent.

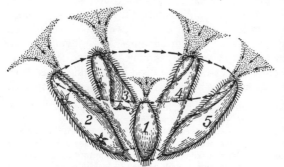

Such variations are seen when the infusoria are subjected to stimuli of different intensities. If the

FIG. 38. — More pronounced avoiding reaction. The anterior end swings about a larger circle. 1–5, successive positions occupied.

animals come in contact with any strong chemical, or with water that is very hot, they respond first by swimming a long way backward, thus removing themselves as far as possible from the source of stimulation. Then they turn directly toward the aboral side, — the rotation on the long axis completely ceasing, as in Fig. 39. In this way the animal may turn directly away from the drop and retrace its course. But often the reaction is so violent that the anterior end swings about in two or three complete circles before the animal starts forward again. Then the new path may lead it again toward the drop, when the reaction is repeated.

In marked contrast with

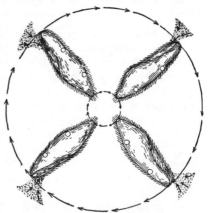

FIG. 39. — Avoiding reaction when revolution on the long axis ceases completely. The anterior end swings about a circle of which the body forms one of the radii.

this violent reaction is the behavior when the stimulus is very weak. A weak stimulus is produced for example by $\frac{1}{10}$ per cent to $\frac{1}{20}$ per cent sodium chloride, or by water only three or four degrees above the normal temperature. The Paramecium whose oral cilia bring it a sample of such water merely stops, or progresses more slowly, and begins to swing its anterior end about in a circle, as in Fig. 37, thus "trying" a number of different directions. As long as the oral cilia continue to bring it the weak salt solution or the warmed water, the animal holds back, and continues to swing its anterior end about in a circle. When the anterior end is finally pointed in a direction from which no more of the stimulating agent comes, the Paramecium swims forward. The reaction in this case is a very precise and delicate one; in a cursory view the animal seems to turn directly away from the region of the stimulus, — the revolution on the long axis and swinging of the anterior end in a circle being easily overlooked.

Between this delicate reaction and the violent one first described there exists every intermediate gradation, depending on the intensity of the stimulation.

Paramecia react to most of the different classes of stimuli which act upon them, in the way just described. Mechanical stimuli, such as solid obstacles, or disturbances in the water; chemicals of all sorts; heat and cold; light that is sufficiently powerful to be injurious; electric shocks, and certain disturbances induced by gravity and by centrifugal force, all cause the animal to respond by the avoiding reaction, so that it escapes if possible from the region or condition that acts as a stimulus. Certain peculiarities and special features in the action of the different classes of stimuli will be taken up separately in the following chapters.

Stimulating agents produce the same reaction when they act on the entire surface of the body as they do when they reach only the anterior end or oral groove. This is shown by dropping the animals directly into a $\frac{1}{2}$ per cent solution of sodium chloride, or into corresponding solutions of other chemicals; or into hot or cold water. They at once give the avoiding reaction; they swim backward, turn toward the aboral side, then swim forward, and this reaction may be repeated many times. If the stimulating agent is not so powerful as to be directly destructive, the reaction ceases after a time, and the Paramecia swim about within the solution as they did before in water.

This experiment shows clearly that the cause of the avoiding reaction does not lie in the difference in the intensity of the chemical on the two sides or two ends of the animal, as is sometimes held. For as we have just seen, the animal reacts in the same way when the entire surface of the body is subjected equally to the action of the chemical or the changed temperature. It is clear that the cause of the reaction is the *change* from one solution or temperature to another. This is evident further from the fact that the animal reacts as a rule when the change occurs, but ceases to react after the change is completed. To constant conditions Paramecium soon becomes acclimatized; it is change that causes reaction.

To this general statement there are certain exceptions. If we place the infusoria in conditions of such intense action that they are quickly destructive, — for example, in 2 per cent potassium bichromate, or in water heated to 38 degrees C.,—the animals continue to react till they die. For two or three minutes they rapidly alternate swimming backward with turning toward the aboral side and swimming forward, till death puts an end to their activity. Thus very injurious conditions may produce reaction independently of change. But as a general rule, it is some change in the conditions that causes the animal to change its behavior. The animal, having been subjected to certain conditions, becomes now subjected to others, and *it is the transition from one state to another that is the cause of reaction.* This is a fact of fundamental significance for understanding the behavior of lower organisms.

But it is not mere change, taken by itself, that causes reaction, but change *in a certain direction.* This is shown by observation of the behavior of the individuals as they pass from one set of conditions to another. If we place Paramecia on a slide in ordinary water, then introduce into the preparation, by means of a capillary pipette, a drop of $\frac{1}{4}$ per cent sodium chloride, as shown in Fig. 40, we find that the animals react at the change from the water to the salt solution, so that they do not enter the latter. If, on the other hand, the animals are first mixed with $\frac{1}{4}$ per cent salt solution, and a drop of water is introduced into the preparation (as in Fig. 40), they do not react at passing from the salt solution to the water. In the same way, Paramecia at a temperature of 30 degrees react at passing to a higher temperature, but not at passing to a lower temperature. Paramecia at 20 degrees, on the other hand, react at passing to a lower temperature, not at passing to a higher. To these relations we shall return.

A relation which is worthy of special emphasis is the following: The direction toward which the animal turns in the avoiding reaction does not depend on the side of the animal that is stimulated, but is determined by internal relations. The animal always turns toward the aboral side. It is true that with chemical stimuli the stimulation usually occurs on the oral side, so that the animal turns away from the side stimulated. But, as we have just seen, it turns in the same way when all parts of the body are equally affected by the stimulating agent. Furthermore, it is possible to apply mechanical stimuli to various parts of the body, and observe the resulting reaction. If with the tip of a fine glass point we touch the oral side of Paramecium, the infusorian turns directly away from the point touched. But if we touch the aboral side, the Paramecium turns in the same manner as before, — toward the aboral side, and hence toward the point touched. This experiment is more easily performed, and the results are more striking, with certain of the Hypotricha,[1] because these animals do not continually revolve on the long axis, as Paramecium does.

The general effect of the avoiding reaction is to cause the animals to avoid and escape from the region in which the stimulus is acting. This may be illustrated for the different classes of stimuli in the following ways.

The effects of this reaction to chemicals may best be seen by introducing a little $\frac{1}{2}$ per cent solution of sodium chloride into the water containing the animals. For this purpose water with many Paramecia is placed on a slide and covered with a long coverglass supported near its end by glass rods. A medicine dropper is drawn to a long, slender point, and with this a drop of the salt solution is introduced

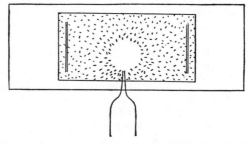

FIG. 40. — Method of introducing a chemical into a slide of infusoria.

beneath the cover-glass, as illustrated in Fig. 40. The Paramecia are swimming about in all directions, but as soon as they come to the

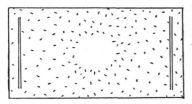

FIG. 41. — Slide of Paramecia four minutes after the introduction of a drop of $\frac{1}{2}$ per cent NaCl. The drop remains empty.

region of the salt solution, the avoiding reaction is given in the way already described, and the animals swim elsewhere. Thus the drop of salt solution remains empty (Fig. 41).

Practically all strong chemicals induce the avoiding reaction, so that Paramecia do not enter them. This has been shown for many alkalies, neutral salts, and organic substances, and for strong acids. In the case of acids the reaction differs in certain respects from the behavior under the influence of other chemicals; this will be brought out later.

The reaction to heat or cold may easily be shown by placing a drop

[1] See Chapter VII.

of hot or cold water on the cover-glass of a slide of Paramecia, or by touching the cover-glass with a hot wire, or a piece of ice. The animals respond by the avoiding reaction, just as when stimulated by a chemical, so that the hot or cold region remains vacant. The intensity of the reaction depends on the temperature, and very hot water causes a much more decided reaction than very cold water.

The avoiding reaction is seen under mechanical stimulation when a specimen in swimming comes against an obstacle. It may also be shown by touching the anterior end of the animal with a fine glass point. A slight disturbance in the water may be induced by injecting a fine stream of water against the animal with a pipette drawn to a capillary point; the animal then responds by the avoiding reaction, thus swimming elsewhere.

Special features in the reactions to various different classes of stimuli will be dealt with in the next chapter.

5. " POSITIVE REACTIONS "

The reactions thus far described have the effect of removing the animal from the source of stimulation; they might therefore be characterized as *negative*. But Paramecia are known also to collect in certain regions, giving rise to what are commonly known as positive reactions. How are these brought about?

A simple experiment throws much light on the cause of such collections. Under usual conditions the animals avoid a $\frac{1}{10}$ per cent solution of NaCl, so that when a drop of this is introduced into a slide of Paramecia, they leave it empty. But if we mix the animals with $\frac{1}{2}$ per cent NaCl, then introduce into a slide of this mixture a drop of $\frac{1}{10}$ per cent NaCl, in the way shown in Fig. 40, we find that the Paramecia quickly collect in this drop, though under ordinary circumstances they avoid it. Very soon the drop of $\frac{1}{10}$ per cent NaCl is swarming with the infusoria, as in Fig. 43, while very few remain in other parts of the preparation. The phenomena are identical with what has often been called positive chemotaxis.

Careful observation of the movements of the individuals shows, as might be expected, that the Paramecia collect in the $\frac{1}{10}$ per cent NaCl merely because they avoid the stronger solution more decidedly. Passage from the $\frac{1}{10}$ per cent solution to the $\frac{1}{2}$ per cent solution causes the avoiding reaction, while passage in the reverse direction does not. The details of the behavior are as follows: The Paramecia in the $\frac{1}{2}$ per cent NaCl are swimming rapidly in all directions, so that many of them are carried toward the drop. On reaching its boundary they do not react in any way, but swim directly into it. They continue across till they reach the farther boundary, where they come in contact again with the $\frac{1}{2}$ per cent solution. Here the reaction occurs. The animals give the avoiding reaction, swimming backward, turning toward the aboral side, and starting forward again, etc. They of course soon come in contact again with the outlying $\frac{1}{2}$ per cent NaCl, whereupon they react as before, and this continues, so that they do not leave the drop of $\frac{1}{10}$ per cent NaCl. The path of a single Paramecium in such a drop is like that shown in Fig. 44. Since all the infusoria that enter the drop of $\frac{1}{10}$ per cent NaCl remain, it soon swarms with them.

In place of NaCl, we may use pairs of solutions of other chemicals, one stronger than the other, — taking pains of course not to employ concentrations that are decidedly injurious. With any of the ordinary inorganic salts or alkalies the animals collect in the weaker solution, through the fact that they avoid the stronger one in the way described above. The same concentration of a given chemical may play opposite rôles in successive experiments, depending on whether it is associated with a weaker or a stronger solution. In the former case the Paramecia avoid it; in the latter they gather within it. If the weaker solution surrounds a drop of the stronger, the latter is left empty, and the Paramecia remain scattered through the preparation, as in Fig. 41. If the stronger solution surrounds the weaker, the latter becomes filled with the Paramecia, as in Fig. 43, while the former is left nearly empty. Thus with the same pair of substances we get either a dense aggregation (or what is often called positive chemotaxis), or a certain area left vacant ("negative chemotaxis"), depending on the relation of the two fluids to each other.

If we use pure water in place of the weaker solution, we get the same result; the Paramecia collect in the drop of water. This is easily shown by introducing a drop of water into a preparation of Paramecia that have been mixed with $\frac{1}{4}$ per cent NaCl; the water soon swarms with the infusoria. The culture water in which Paramecia live usually contains various salts, and is often alkaline in reaction. If a drop of distilled water is added (as in Fig. 40) to a preparation of infusoria in such culture water, the animals gather in the distilled water.

The same results may be obtained with water of differing temperatures. This is done by surrounding an area of water at the normal temperature with water at a temperature considerably higher or lower. The Paramecia may be placed on a slide in the usual way, with a cover-glass supported by glass rods. This slide is then placed on a bottle or other vessel containing water heated to forty-five or fifty degrees. As soon as the Paramecia begin to move about more rapidly in conse-

quence of the heat, a drop of cold water is placed on the upper surface
of the cover-glass. At once a dense collection of Paramecia is formed

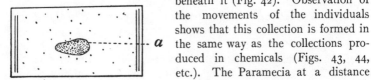

beneath it (Fig. 42). Observation of
the movements of the individuals
shows that this collection is formed in
the same way as the collections produced in chemicals (Figs. 43, 44,
etc.). The Paramecia at a distance
from the cooled region do not turn
and swim directly toward it. But the
Paramecia are swimming rapidly in
all directions, and many enter every
instant the region beneath the drop.

FIG. 42. — A slide of Paramecia is
heated to 40 or 45 degrees, then a drop of
cold water (represented by the outline *a*)
is placed on the upper surface of the
cover-glass. The animals collect beneath
this drop, as shown in the figure.

They do not react on entering, but on reaching the opposite side, where
they would pass out again into the heated water, they give the avoiding reaction. This is repeated every time they come to the other
boundary of the drop, so that
the path of an individual within
the cooled region is similar to that
shown in Fig. 44. Every Paramecium that enters the cooled
region therefore remains, and
soon a dense swarm is formed.

A collection may be formed in
the same way by resting the slide
of Paramecia on a piece of ice and

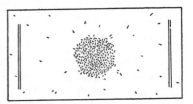

FIG. 43. — Collection of Paramecia in a
drop of $\frac{1}{30}$ per cent acetic acid.

placing a drop of warmed water on the upper surface; the Paramecia
now collect in the warmed region. But the collection is never so pronounced as in the experiment last described, because the Paramecia
when cooled move less rapidly.

Thus the Paramecia collect in certain regions because they give
the avoiding reaction when passing from certain conditions to others,
while when passing in the reverse direction they do not. Paramecia
at the normal temperature give the reaction at passing both to hotter
and to colder water; they therefore tend to gather in water at the usual
temperature. This temperature at which they gather may be spoken
of as the optimum. Passage away from the optimum induces the
avoiding reaction; passage toward the optimum does not.

In the case of the chemicals thus far considered, the animals give
the reaction at passing from the weaker to the stronger solution, not at
passing in the opposite direction, so that they collect in the weaker solution. The optimum for these substances is thus zero, and this natu-

rally results in the tendency of the animals to collect in distilled water. But there are certain chemicals of which the optimum is a certain positive concentration, so that Paramecia give the avoiding reaction at passing to weaker solutions or to water containing none of the substance in question. This is the case with acids and with oxygen. If a drop of very weak acid is introduced into a slide of Paramecia (Fig. 43) that are in ordinary water, the animals quickly gather in the drop. This may be shown by the use of about $\frac{1}{100}$ to $\frac{1}{50}$ per cent of the ordinary laboratory solutions of hydrochloric or sulphuric acid, or of $\frac{1}{50}$ to $\frac{1}{25}$ per cent acetic acid. In a short time the drop is swarming with Paramecia.[1]

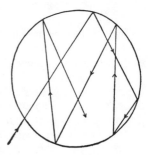

Fig. 44. — Path followed by a single Paramecium in a drop of acid.

Observation shows that the method of collecting in the acid is the same as in the cases before described. The rapid movements of the animals in all directions are what carry them into the drop. They do not react in any way at the moment of entering it, but swim across. At the point where they would pass out into the surrounding water they respond by the avoiding reaction; hence they return to the acid. This is repeated each time that they come to the boundary. Hence all that enter the acid remain till it is crowded. The path of a single Paramecium within a drop of acid is shown in Fig. 44.

In the formation of all these collections the natural roving movements play an essential part. These movements cause any given specimen in the course of a short time to cross almost any given area in the preparation, and hence bring the animals to the introduced drop. The animals do not turn and swim in radial lines toward the drop of acid. If a ring is marked on the upper surface of the cover-glass, as many Paramecia will be found to pass beneath this ring before a drop of acid is placed beneath it as after. But in the latter case all that pass beneath the ring remain, and the collection results. If we wait, before introducing the acid, till all have become nearly quiet, no collection is produced.

We may sum up the usual behavior of Paramecium under the various stimuli of the environment in the following way. The natural condition of the animal is movement. In constant external conditions (unless destructive) the movements are not changed, — that is, there

[1] In all these experiments it is assumed, of course, that the preparation contains the infusoria in very large numbers. With scattered specimens only, the results are slow and not striking.

is no reaction, — even though these conditions do not represent the optimum. But as its movements carry the animal from one region to another, the environmental conditions affecting it are of course changed, and some of these changes in condition act as stimuli, causing the animal to change its movements. If the environmental change leads toward the optimum, there is no reaction, but the existing behavior is continued. To a change leading away from the optimum (in either a plus or minus direction), Paramecium responds by the "avoiding reaction." This consists essentially in a return to a previous position, through a backward movement, then in "trying" different directions of movement till one is found which leads toward the optimum. Expressed in a purely objective way, the animal performs movements which subject it successively to many different environmental conditions. As soon as one of the conditions thus reached is of such a character as to remove the cause of stimulation, the avoiding reaction ceases and the infusorian continues in the condition now existing. This method of reacting causes the animals to collect in certain regions (as near the optimum as possible), and to avoid other regions. Thus are produced the so-called positive and negative reactions. The behavior may be characterized briefly as a selection from the environmental conditions resulting from varied movements.

Some details of the behavior under the different classes of stimuli will be given in the next chapter.

LITERATURE III

On the character of the movements and reactions of Paramecium: JENNINGS 1904 *h*, 1899, 1901.

CHAPTER IV

BEHAVIOR OF PARAMECIUM (*Continued*)

Special Features of the Reactions to a Number of Different Classes of Stimuli

In the preceding chapter the general method of the reactions of Paramecium to most classes of stimuli has been described. In the present chapter certain important details and special peculiarities of the behavior under the different classes of stimuli will be described.

I. MECHANICAL STIMULI

When Paramecium strikes in its forward course against a solid object, it responds usually by the avoiding reaction, as described in the preceding chapter. In such cases the stimulus affects the anterior end of the animal. But if mechanical stimuli affect other parts of the body, will this alter the nature of the reaction? This question may be answered by drawing a glass rod to an extremely fine point and touching various parts of the body with this point under the microscope. The first discovery that we make by this method of experimentation is that the anterior end is much more sensitive than the remainder of the body surface. If the anterior end is touched very lightly, the animal responds by a strong avoiding reaction, while the same or a more powerful stimulus on other parts of the body produces no reaction at all. There is some evidence drawn from other sources[1] that the region immediately about the mouth is likewise very sensitive.

A second fact brought out by these experiments is that a stimulus on the posterior part of the body produces a different reaction from a stimulus in front. If we touch the anterior end, or any point on the anterior portion of the body back nearly to the middle, the typical avoiding reaction is produced. But if we touch the middle or the posterior part of the body of a resting specimen, the animal, if it reacts at all, merely moves forward.

[1] See Chapter V.

On the other hand, as we have seen in the preceding chapter, the direction in which the animal turns in the avoiding reaction does not depend on the side of the body stimulated. The animal turns toward the aboral side as well when that side is touched, as when the oral side receives the stimulus.

The reactions which we have thus far described have the effect of removing the animal from the object with which it comes in contact, so that they may be called negative reactions. But under certain conditions, not very precisely definable, Paramecium does not avoid the object which it strikes against. On the contrary it stops and remains in contact with the object. This seems most likely to happen when the animal is swimming slowly, so that it does not strike the object violently. But this does not explain all cases; many individuals seem much inclined to come to rest against solids, while others do not. Often all the individuals in a culture are thus inclined to come to rest, while in another culture all remain free swimming, and give the avoiding reaction whenever they come in contact with a solid. Observing a single swimming specimen, it is often seen to react as follows. When it first strikes against an object it responds with a weak avoiding reaction, — swimming backward a short distance, turning a little toward the aboral side, then swimming forward again. Its path carries it against the object again, whereupon it stops and comes to rest against the surface.

FIG. 45. — Paramecium at rest against a cotton fibre, showing the motionless cilia in contact with the fibre.

The objects against which Paramecium strikes under normal conditions are usually pieces of decaying vegetable matter or bits of bacterial zoöglœa. Remaining in contact with these helps it to obtain food. The cilia that come in contact with the solid cease moving, and become stiff and set, seeming to hold the Paramecium against the object (Fig. 45). Often it is only the cilia of the anterior end that are thus in contact and immovable; in other cases cilia of the general surface of the body show the same condition. Meanwhile, the cilia of the oral groove continue in active motion, so that a rapid current passes from the anterior end down the groove to the mouth (Fig. 46). This cur-

FIG. 46. — Paramecium at rest with anterior end against a mass of bacterial zoöglœa (*a*), showing the currents produced by the cilia.

rent of course carries many of the bacteria found in the zoögloea or on the decaying plant tissue; these serve as food for the animal. The cilia of the remainder of the body usually strike only weakly and ineffectively, so that the currents about the Paramecium are almost all due to the movements of the oral cilia. The body cilia directly behind those in contact with the solid are usually quite at rest.

The function of this positive contact reaction is evidently, under ordinary conditions, to procure food for the animal. But Paramecium shows no precise discrimination, and often reacts in this way to objects that cannot furnish food. Thus, if we place a bit of torn filter paper in the water containing the animals, we often find that they come to rest upon this, gathering in a dense group on its surface, just as they do with bits of bacterial zoögloea (Fig. 47). The oral cilia drive a strong current of water to the mouth, as usual, but this bears no food.. To bits of thread, ravellings of cloth, pieces of sponge, or masses of pow-

dered carmine, Paramecium may react in the same way. In general it shows a tendency to come to rest against loose or fibrous material; in other words, it reacts thus to material with which it can come in contact at two or more parts of the body at once. To smooth, hard materials, such as glass, it is much less likely to react in this manner, so that it

FIG. 47. — Paramecia gathered in a dense mass about a bit of filter paper.

clearly shows a certain discrimination in this behavior. These hard substances, it is evident, are less likely to furnish food than the soft fibrous material to which Paramecium reacts readily. But under certain conditions Paramecium comes to rest even against a smooth glass surface, or against the surface film of the water. Specimens are often found at rest in this manner in the angle between the surface film of a drop of water and the glass surface to which it is attached.

Paramecia often behave in the manner just described with reference to bodies of very minute size, — to small bits of bacterial zoögloea, or to a single grain of carmine. Such objects are of course too small to restrain the movements that naturally result from the activity of the oral cilia in the contact reaction. These cilia continue to beat in the same manner as when the object is a large one, producing currents similar to those shown in Fig. 46. This ciliary motion of course tends to drive the animal forward, and since all the active cilia are on the oral side, it tends also to move the animal toward the aboral side. The resultant of these two motions at right angles is movement in the circumference of a circle. The animal moves in the lines of the water currents shown in Fig. 46, but in the opposite direction; it is, as it were, whirled about in its own whirlpool. The resulting path is shown in Fig. 48. This circular movement, with the oral side directed toward the centre of the circle, is seen only in specimens showing the contact reaction to objects of minute size.

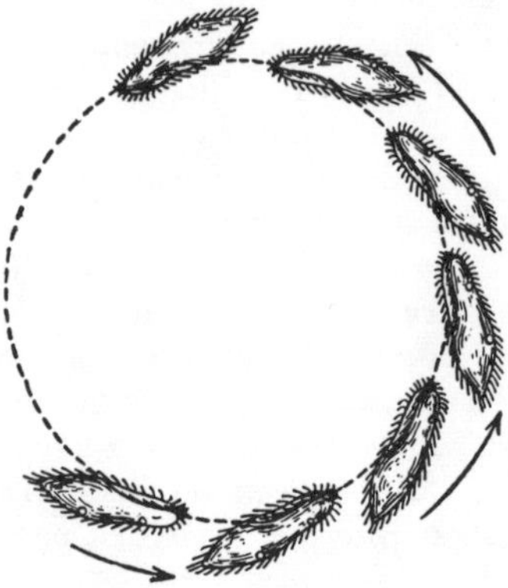

FIG. 48. — Circular path followed by Paramecium in reacting to contact with a minute particle.

The contact reaction modifies strongly the reactions to most other stimuli; this is a matter which will be taken up later.

Thus when Paramecium comes in contact with a solid object, it may react in three different ways. First, it may react either positively or negatively, this depending partly on the intensity of the stimulus, partly on the physiological condition of the Paramecium. If it reacts negatively, this reaction may take either one of two forms. If the stimulation occurs at the anterior end, the animal gives the avoiding reaction; if it occurs elsewhere, the animal merely moves forward.

2. REACTIONS TO CHEMICAL STIMULI

The reactions to chemical stimuli occur through the avoiding reaction described in the preceding chapter. As we have seen, the avoiding reaction is produced as a rule by a change from one chemical to another. With regard to this relation, there are certain facts of importance.

In all cases a certain amount of change is necessary to produce reaction; that is, the chemical must be present in a certain concentration before reaction is produced. The sensitiveness of different individuals varies greatly, and even that of given individuals changes much with changes in the conditions. It is therefore not possible to establish for any given chemical the weakest concentration that causes the avoiding reaction. But the animals when in ordinary water are very sensitive to the common inorganic chemicals, reacting to very weak solutions. Thus the weakest solutions causing reaction have been found to be for various chemicals about as follows: —

Sodium chloride, $\frac{1}{20}$ to $\frac{1}{10}$ per cent ($\frac{1}{50}$ to $\frac{1}{100}$ normal); potassium bromate, about $\frac{1}{50}$ per cent; sodium carbonate, about $\frac{1}{200}$ to $\frac{1}{300}$ per cent; copper sulphate, about $\frac{1}{800}$ per cent; potassium hydroxide, about

$\frac{1}{250}$ per cent; sodium hydroxide, $\frac{1}{500}$ per cent; sulphuric acid, $\frac{1}{800}$ per cent of an ordinary laboratory solution; hydrochloric acid, $\frac{1}{800}$ per cent of the usual solution; alcohol, 1 per cent; chloral hydrate, $\frac{2}{3}$ per cent. For the inorganic chemicals, many of these solutions are so weak as not to affect at all the sense of taste in man.

Is the reaction of Paramecium to solutions due to the chemical properties of the dissolved substance, or to its osmotic pressure? This question may be answered from the data which we possess (partly given above) as to the weakest solutions which cause reactions. If the reactions are due to osmotic pressure, then solutions having equal osmotic pressure must have equal stimulating power. The results of the experiments on the weakest solutions necessary to cause reaction show that this is not true. Thus, if the osmotic pressure of a solution of sodium chloride that will barely cause the reaction is taken as unity, the osmotic pressures of solutions of a number of other substances having the same stimulating effect are as follows: potassium bromate, $\frac{1}{9}$; sodium carbonate, $\frac{1}{22}$; copper sulphate, $\frac{1}{243}$; potassium hydroxide, $\frac{1}{16}$; sulphuric acid, $\frac{1}{52}$; ethyl alcohol, 8. The stimulating effect is not then proportional to the osmotic pressure, and must be due to the chemical properties of the substances in solution.

This is further shown by the fact that Paramecia will enter solutions of sugar and of glycerine having osmotic pressure many times as great as that of a solution of sodium chloride which they avoid. They swim into a 20 per cent solution of sugar or a 10 per cent solution of glycerine without reaction. The solutions are so concentrated that they cause plasmolysis; the Paramecia shrink into flattened plates. Just as the shrinking becomes evident to the eye of the observer, the Paramecia react in the usual way, by swimming backward and turning towards the aboral side. But this is as a rule too late to save them, and they die in the dense solution. Thus it is evident that osmotic pressure, acting by itself, produces the same "avoiding reactions" as do other stimuli, but the result is not produced till the Paramecia are already injured beyond help. The reactions to most solutions are then clearly due to their chemical properties.

Is the avoiding reaction that is produced by chemicals due directly to the injuriousness of the substance? This question may be answered by a series of experiments based on a method similar to that used in determining whether the reaction is due to osmotic pressure. If the reaction is due to the injuriousness of the chemicals, then two substances which are equally injurious must have equal powers of inducing reaction; in other words, the repelling powers of any two substances must be proportional to their injurious effects.

An extensive series of experiments has shown that this is not true (Jennings, 1899 c; Barratt, 1905). We may compare, for example, the effects of chromic acid and of potassium bichromate. The weakest solution of the former which kills the Paramecia in one minute is $\frac{1}{150}$ per cent; the weakest solution of the latter having the same effect is 1 per cent. Hence the chromic acid is 150 times as injurious as the potassium bichromate.

On the other hand, the weakest solution of chromic acid that sets in operation the avoiding reaction is still $\frac{1}{150}$ per cent, while potassium bichromate has the same effect in a $\frac{1}{20}$ per cent solution. The repellent power is thus not proportional to the injurious effects; the potassium bichromate is repellent in a strength $\frac{1}{20}$ that which is immediately injurious, which chromic acid does not repel until it has reached a strength that is already destructive. Similar relations are found for other pairs of substances. Thus the stimulating power of sodium chloride is ten times that of cane sugar, in proportion to its injuriousness.

Comparing a large number of chemicals from this point of view, it has been found that they may be divided into two classes. On the one hand are a number of substances which must be classified with potassium bichromate and sodium chloride, because their stimulating power is strong in proportion to their injurious effects. Paramecia avoid these substances markedly; if a drop of a strong solution of one of them is introduced into a preparation of the infusoria, it remains empty, and none of the infusoria are killed by it. On the other hand, there is a large number of substances which, like chromic acid and sugar, produce stimulation only where they are strong enough to be immediately injurious. When a strong solution of one of these is brought into a preparation of Paramecia, it proves very destructive, for the animals as a rule do not react until they have been injured. The following table (from Jennings, 1899 c) shows the distribution of various chemicals from this point of view : —

TABLE

1. Repellent power strong in proportion to injurious effects ; reaction protective.	2. Repellent power very weak in proportion to injurious effects ; reaction not completely protective.
LiCl, NaCl, KCl, CsCl, LiBr, NaBr, KBr, RuBr, LiI, NaI, KI, RuI, Li_2CO_3, Na_2CO_3, K_2CO_3, $LiNO_3$, $NaNO_3$, KNO_3, NaOH, KOH, NaF, KF, NH_4F, NH_4Cl, NH_4Br, NH_4I, $CaCl_2$, $SrCl_2$, $BaCl_2$, $Ca(NO_3)_2$, $Sr(NO_3)_2$, $Ba(NO_3)_2$, Potassium bromate, Potassium permanganate, Potassium bichromate, Potassium ferricyanide, Ammonium bichromate.	HF, HCl. HBr, HI, H_2SO_4, HNO_3, Acetic acid, Tannic acid, Picric acid, Chromic acid, Ammonia alum, Ammonioferric alum. Chrome alum, Potash alum, $CuSO_4$, $CuCl_2$, $ZnCl_2$, $HgCl_2$, $AlCl_3$, Copper acetate. Cane sugar, Lactose. Maltose, Dextrose, Mannite, Glycerine, Urea.

This table shows that the relative repellent power of different substances bear a somewhat definite relation to their chemical composition. All alkalies and compounds of the alkali and the earth alkali metals (save the alums, where the proportion of the metals is very small) have a relatively strong repellent effect; most other compounds have not.

While our general result is that the stimulating powers of different chemicals are not proportional to their injurious effects, yet one further fact of importance comes out clearly. All substances, whatever their nature, do produce, as soon as they become injurious, the avoiding reaction. With all the substances in the second column the avoiding reaction is produced when a strength sufficient to be injurious is reached and the reaction seems clearly due to the injuries produced. The significance of this fact will be discussed later.

In the chapter preceding the present one, we have seen that Paramecia collect in certain chemicals, owing to the fact that passage *out* of these causes the avoiding reaction. The two chief classes of chemicals in which the animals collect are acids and oxygen.

Paramecia collect in all weakly acid solutions, no matter what acid substance is present. Sulphuric, hydrochloric, nitric, hydriodic, and many other inorganic acids; acetic, formic, carbonic, propionic, and other organic acids, have been tested, and the animals have been found to gather in all. The Paramecia collect even in solutions of poisonous acid salts, such as corrosive sublimate and copper sulphate, where they are quickly killed. In all these cases they swim into the solution without reaction, but give the avoiding reaction at passing out. They give the avoiding reaction also after the injurious chemical begins to act on them, but under the circumstances this does not save them from destruction.

It seems remarkable that the animals should thus tend to gather in acids, when, as is well known, the decaying vegetable infusions in which they live are usually alkaline in character. Specimens in water that is decidedly alkaline collect even more readily in acids than do those in a neutral fluid.

A solution may contain both an acid and a repellent substance, as when $\frac{1}{25}$ per cent acetic acid is mixed with $\frac{1}{4}$ per cent sodium chloride. In this case a curious effect is produced. The Paramecia gather

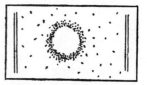

FIG. 49. — Collection of Paramecia about the periphery of a mixture of salt and acid.

in a ring about the outer edge of the solution, as in Fig. 49. They are repelled both by the inner fluid and the surrounding water. The path of a Paramecium in such a ring is similar to that shown in Fig. 50.

Strong acid solutions cause the avoiding reaction as do other chemicals. If a drop of strong acid solution is introduced into a preparation of Paramecia, the animals collect about its periphery, where the acid is diluted by the surrounding water, just as in Fig. 49. Individuals which swim against the inner strong acid respond by giving the avoiding reaction in a very pronounced way, — swimming far backward and turning toward the aboral side, for perhaps two or three or more complete turns. They react also at the outer boundary of the acid ring, so that within the ring the individual Paramecium follows such a path as is shown in Fig. 50.

FIG. 50.—Path of an individual Paramecium in such a ring as is shown in Fig. 49.

Often the reaction is not produced at the inner boundary of the ring, by the strong acid, until the Paramecium has entered far enough to be injured, or even killed. A drop of strong acid introduced into a preparation is usually soon surrounded by a zone of dead animals. Acids, as we have seen (p. 64), belong with those substances which do not produce the avoiding reaction till they have become directly injurious.

Paramecia do not, under usual conditions, collect in oxygen. If we introduce an air bubble or a bubble of oxygen into a slide preparation of Paramecia, they do not as a rule collect about it. But if the outer air is excluded from this preparation by covering its edges with vaseline, and it is allowed to stand for a long time, the behavior changes. The oxygen has of course become nearly exhausted and now the Paramecia gather about the air or the oxygen. The collections are formed in exactly the same way as are those in acids.

Thus the experiments show that all reactions to chemicals take place through the avoiding reaction, and this reaction is produced by a change in the intensity of action of the chemical in question. With some chemicals, or under certain conditions, it is a change to a greater intensity that produces the avoiding reaction; in other cases it is a change to a less intensity that produces the reaction. With acids both an increase and a decrease beyond a certain intensity produce reaction. We may express the facts for all chemicals in the following general way. For each chemical there is a certain optimum concentration in which the Paramecia are not caused to react. Passage from this optimum to regions of either greater or less concentration causes the avoiding reaction, so that the animals tend to remain in the region of the optimum, and if this region is small, to form here a dense collection. For acids and for oxygen the optimum is a certain very low concentration. For most other chemicals the optimum is zero; an increase in intensity by any effective quantity produces the avoiding reaction, while decrease in intensity has no effect. Hence the Paramecia tend to collect where none of the chemical is present.

The point needs to be brought out clearly that it is not merely passage from the absolute optimum that induces reaction, but passage in a direction leading away from the optimum. To constant conditions, even when not optimal, Paramecium becomes acclimatized; it may live for example in a $\frac{1}{10}$ per cent salt solution, though passage from water to this causes reaction. While in this salt solution, passage into conditions lying still farther from the optimum, as into $\frac{1}{2}$ per cent salt solution, causes the avoiding reaction, while passage to conditions lying nearer the optimum produces no reaction. Acclimatization to non-optimal conditions is an ever present factor in the behavior of the organisms. This is another way of stating the fact that *change* is the chief factor inducing reactions.

Acids then take a peculiar position among chemicals merely in the fact that a certain positive concentration forms the optimum, passage to a lower concentration inducing reaction. The peculiar behavior of Paramecium with respect to acids plays a large part in

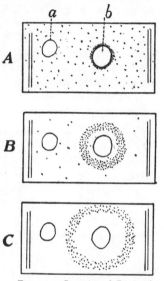

FIG. 51. — Collection of Paramecia about a bubble of CO_2. *a* is a bubble of air, *b* of CO_2. *A* shows the preparation two minutes after the introduction of the CO_2; *B*, two minutes later; *C*, eighteen minutes later.

its life under natural conditions. Paramecia produce carbon dioxide in their respiratory processes as do other organisms. This substance when dissolved in water produces an acid solution, the acidity being due to carbonic acid. In such a solution Paramecia gather as in other acids. This may be shown by introducing, by means of a capillary pipette attached to a rubber bag containing the gas, a small bubble of carbon dioxide into a slide preparation of Paramecia. The infusoria quickly gather in a dense collection about the bubble, at first pressing closely against it (Fig. 51, *A*). Later the Paramecia spread out with the diffusion of the carbon dioxide (*B*). After a time the animals are usually found chiefly about the margin of the area containing the carbon dioxide (*C*).

Now, the Paramecia gather in the solution of carbon dioxide produced by themselves, just as in that due to other causes. In this way dense spontaneous groups are formed, in which the phenomena seen in the collections about bubbles of carbon dioxide are reproduced. If a large number of Paramecia are mounted in water on a slide, they do not remain scattered, but soon gather in one or more regions (Fig. 52). Within such groups the individuals move about in all directions. On coming to an invisible outer boundary, they give the avoiding reaction in a mild form, so that they do not leave the group. The area covered by the group does not remain of the original size, but slowly enlarges, as shown in Fig. 52. It continues thus to increase in size until it covers the whole preparation.

By the use of proper indicators it can be shown that such spontaneous groups contain an acid, and this is beyond doubt due to the carbon

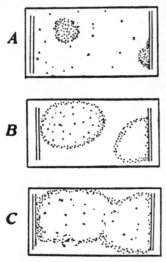

dioxide known to be produced in respiration. The groups are formed in the following way. Two or three Paramecia by chance strike against some small, loose object, a roughening of the surface of the glass, or the like, and come .to rest, in the way described in our account of the reaction to mechanical stimuli. They of course produce carbon dioxide, which diffuses into the surrounding water. Other Paramecia that swim by chance across this area of carbon dioxide of course stop and remain. They too produce carbon dioxide, so that the area grows in size; more Paramecia enter it, and finally a large and dense collection is formed. The area occupied by such a collection continually increases in size, because the Paramecia continue to produce carbon dioxide, and this continues to diffuse through the water.

FIG. 52.—Spontaneous groups formed by Paramecia. *A, B, C,* successive stages in the spreading out of such groups.

The tendency of Paramecia to gather in regions containing carbon dioxide plays a large part in their life under natural conditions, and this, together with the fact that they themselves produce carbon dioxide, explains many peculiar phenomena in their behavior. When placed in tubes or vessels of any kind, Paramecia usually show a tendency to collect into groups or clouds, having a definite boundary (Fig. 53).

This is of course a result of their reaction to carbon dioxide produced by themselves. In all experimental work on the reactions of these organisms to stimuli it is necessary to take these facts into account. For example, in order to get clear results in such work, Paramecia must not be taken with a pipette directly from a dense collection in a culture jar, and at once mounted on a slide. Such collections contain carbon dioxide, which may become unequally distributed throughout the preparation, as a result of the fact that some of the water outside the collection is likely to be taken up with the pipette at the same time.

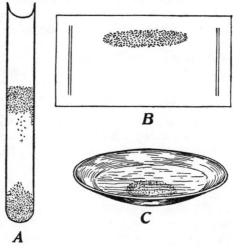

Fig. 53. — Spontaneous collections of Paramecia, due to CO_2. *A*, Collections formed in an upright tube, after Jensen. *B*, Collection formed beneath a cover-glass, when water is taken directly from a dense culture of Paramecia. *C*, Collection in the bottom of a watch-glass.

The Paramecia quickly gather in the region containing most carbon dioxide, and their reactions to other substances are inconstant and irregular, owing to the interference due to the reaction to carbon dioxide (Fig. 53, *B*). For experimental work it is always necessary before each experiment to place a few drops of the water containing the Paramecia in the bottom of a shallow watch-glass, and to aërate it thoroughly by stirring it and bringing it into contact with the air by means of the pipette. Then this aërated water and its contained Paramecia must be used for the experiments. This aëration must be repeated before each experiment, and the test for the reaction to other chemicals must be made immediately after the Paramecia are mounted, before they have had time to produce an appreciable quantity of carbon dioxide. If these precautions are neglected, the reactions of the Paramecia are inconstant, and the results of experiments are likely to be very misleading. Paramecia in a solution of carbon dioxide react to other agents in a manner entirely different from the reaction of individuals in water not containing carbon dioxide. The account of their reactions given in the present chapter assumes that the carbon dioxide has been in every case removed from the water. The experimental results described cannot be verified unless this is done.

In general it cannot be too much emphasized that in all experimental studies on the behavior of Paramecium close attention to their reactions with reference to carbon dioxide is necessary. When inconstant results are obtained, or results seeming to contradict those noted by other observers, it will often be found that inattention to the carbon dioxide produced by the animals is at the bottom of the difficulty.

3. REACTIONS TO HEAT AND COLD

As we saw in the preceding chapter, a change to a temperature decidedly above or decidedly below the optimum causes Paramecia to give the avoiding reaction, while a change leading toward the optimum does not. As a result the animals collect in temperatures as near the optimum as possible.

The effects of heat and cold differ slightly, since heat increases the rapidity of movement, while cold reduces activity. Both produce the avoiding reaction in the same way, but in heated water the reaction is continued violently till the animals escape or are killed, while in ice water the animals after a time become benumbed and sink to the bottom.

The reactions to heat and cold are seen in a striking way when the Paramecia are placed in a long tube or trough, one end of which is heated while the other is maintained at the normal temperature, or is cooled. The Paramecia then pass to the region that is nearest the optimum, forming here a collection. By changing the temperature of the ends or of the middle, the Paramecia may be driven from one end to the other or caused to gather in any part of the trough. Such experiments were devised by Mendelssohn (1895, 1902, 1902 *a, b*). He passed tubes beneath the middle and ends of the trough or slide bearing the animals, and through these tubes he conducted water of different temperatures. By changing the connections of the tubes, that end of the trough which is at first heated may later be cooled, etc., without disturbing the animals in any other way.[1]

If in this way we heat the water at one end of the trough to 38 degrees while we cool the opposite end to 10 degrees, the Paramecia collect in an intermediate region. By varying the temperatures at the two ends, the infusoria may be driven back and forth, as represented in Fig. 54, taken from Mendelssohn. By grading the temperatures properly, the sensitiveness of Paramecium to changes in temperature may be measured, and the optimum temperature determined very accurately.

[1] A simple apparatus of this sort is described and figured in Jennings, 1904.

Mendelssohn found that the optimum temperature for Paramecium lies, under ordinary conditions, between 24 and 28 degrees C., and that when there is a difference of but 3 degrees C. between the two ends of a trough 10 centimeters in length, the Paramecia gather at the end of the trough nearest the optimum (see Fig. 54). If the end *a* has a temperature of 26 degrees, the end *b* 38 degrees, the Paramecia gather at the end *a*; if now the temperature of the two ends is interchanged, the Paramecia travel from *a* toward *b*, and collect there. The same results are produced if one end has a temperature of 10 degrees, the other of 26 degrees, save that in this case the Paramecia gather at the end having the higher temperature. If Paramecia are kept for

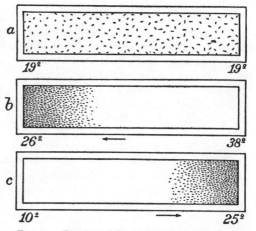

FIG. 54. — Reactions of Paramecia to heat and cold, after Mendelssohn (1902). At *a* the infusoria are placed in a trough, both ends of which have a temperature of 19 degrees. They are equally scattered. At *b* the temperature of one end is raised to 38 degrees while the other is only 26 degrees. The infusoria collect at the end having the lower temperature. At *c* one end has a temperature of 25 degrees, while the other is lowered to 10 degrees. The animals now collect at the end having the higher temperature.

some hours at a temperature of 36 to 38 degrees, the optimum becomes higher, — about 30 to 32 degrees; otherwise the phenomena remain the same.

Observation of the movement of the individuals shows that the reactions in these experiments take place in the following manner. As one end of the trough is heated above the optimum, the Paramecia in that region are seen to become more active, darting about rapidly in all directions. Those that come against the sides or end of the vessel respond by the avoiding reaction; they are thus directed elsewhere. Individuals that are swimming toward the hotter region likewise give the avoiding reaction, — at first in but a slightly marked form, stopping, swinging the anterior end about in a circle, as illustrated in Figs. 37–39, and "trying" forward movement in a number of different directions. This continues as long as they are moving toward the warmer region; but as soon as their direction of movement leads them toward the cooler region, the avoiding reaction ceases, and they continue to swim in that direction. At that end of the trough which is cooled below the optimum, similar effects are produced, save that the reaction is less rapid, and the Paramecia therefore leave this region much more slowly than they do the heated end.

Thus after a time the direction of movement of all the individuals in the hot or cold end of the trough has become changed, and all are moving, often in a well-defined group, toward the optimum region. Thus we may observe in these temperature reactions a well-defined common orientation of a large number of organisms; all are headed toward the optimum. This orientation is brought about, as we have seen, by *exclusion*. That is, movement in any other direction is stopped, through the production of the avoiding reaction, so that all finally travel in this one direction. Or, to put it more accurately, the Paramecia try every possible direction, through the avoiding reaction (Figs. 37–39), till finally they all find the only one which does not cause stimulation;· in this direction they continue to move. The method of reaction, by systematic trial of all directions, is such as to find any existing avenue of escape, no matter how narrow it may be.

4. REACTION TO LIGHT

To ordinary visible light Paramecium is not known to react in any way. If light is allowed to fall on the animals from one side only, or if one portion of the vessel containing them is strongly lighted while the rest is shaded, this has no observable effect on their movements or distribution.

But Hertel (1904) has recently shown that to powerful ultra violet light Paramecium does react. The ultra violet rays employed by Hertel came from a magnesium spectrum; they were of a wave length of 280 $\mu\mu$. When part of a drop of water containing Paramecia was subjected to this light, the animals in the lighted region at once began to move about rapidly. They therefore passed quickly into the region not lighted. Specimens moving about in this shaded region stopped at once on reaching the boundary of the lighted area, and turned away. It is evident that the reaction to light is by the usual avoiding reaction, though the details of the movement were not observed by Hertel.

When the animals were unable to escape from the light, their movement became uncoördinated, and in ten to fifty seconds it ceased. The animals were dead.

5. ORIENTING REACTIONS, TO WATER CURRENTS, TO GRAVITY, AND TO CENTRIFUGAL FORCE

In the reactions which we have thus far considered, the infusoria do not become oriented in any precise way with relation to the direction of action of the stimulating agent. But to water currents, to gravity, and to centrifugal force the animals at times react in such a way as to bring about a definite orientation, with the body axis of all the reacting individuals in line with the external force. In a water current the anterior end is directed up stream; under the influence of gravity the anterior end is directed upward, while when subjected to a centrifugal force the anterior end is directed against the action of the force.

How are these results produced, and why do the organisms take a definite axial orientation under the action of these stimuli, while they do not under most other stimuli?

In the reactions to water currents and to gravity, direct observation has shown that the orientation is produced through the movements which we have called the avoiding reaction. Under the action of a centrifugal force, observation of individuals is impossible, but beyond doubt the reaction is the same as that due to gravity.

A. *Reactions to Water Currents*

The reactions to water currents can best be studied in a tube like that shown in Fig. 55. By covering the two open ends with rubber caps filled with air, and pressing on these, the water containing the animals in the tube can be driven through

FIG. 55. — Tube used in studying the reactions to water currents.

the narrow part of the tube with any desired velocity. With a certain velocity of current most of the individuals, both those that are free swimming and those that are resting against the glass, are seen to place themselves in line with the current, with anterior end up stream. Some of the individuals usually do not react. In those that do, the reaction is brought as follows: As soon as the current begins to act, producing a disturbance in the water, the animals give the avoiding reaction in a not very pronounced form. That is, a given individual swims more slowly or stops, and swerves more strongly toward the aboral side, thus swinging the anterior end about in a circle, as in Figs. 37 and 38, "trying" various directions. It then starts forward again in one of these directions. This reaction may be repeated several times, till the infusorian finally comes into a position with anterior end directed up stream. The reaction then ceases, and the infusorian remains in this position, either swimming forward against the current, or at rest against the wall of the tube. Sometimes the reaction is a little more precise, the animal turning directly toward the aboral side till the anterior end is directed up stream. This is commonly the case with the individuals that are at rest against a solid. The reaction of the resting specimens is less easily observed, for the current easily carries them away from their attachment, when of course they behave like other free specimens.

What is the cause of the reaction to water currents? Under natural conditions the cilia of Paramecium are beating backward, driving a current of water backward over the surface, especially in the oral groove. If an external current moves in the opposite direction, or in some oblique direction, it will of course act in opposition to the cilia on that part of the body which it strikes, tending to reverse or disarrange them, and to reverse or change the direction of the usual currents. It appears not surprising that such a disturbance acts as a stimulus, causing the usual avoiding reaction until the disturbance is corrected. The correction can occur only when the animal is headed up stream; the current is then passing backward over the body in the usual direction. The reaction is essentially a response to a mechanical disturbance, comparable to that due to the touch of a solid body.

If this is the correct explanation, as seems probable, then there should be no reaction when the animal is completely immersed in a homogeneous current, — one moving at the same velocity in all parts. For as Lyon (1904) has pointed out, under these circumstances the animal is merely transferred bodily in a certain direction, along with the medium surrounding it, and at the same rate. Its relation to the enveloping fluid is the same as in quiet water; there is nothing to cause a disturbance. "Stimulation implies a change of relation between organism and environment. But if both in all their parts are moving at the same velocity, their relations do not change, and the conditions for stimulation are wanting" (Lyon, 1904, p. 150).[1] The animal should then react only when either it is in contact on one side with a solid, or when the current is moving more rapidly on one side than on the other, producing a shearing effect, with the necessarily accompanying disturbing action. Whether this is true or not is very difficult to determine, but observation seems to indicate that it is.

---

[1] This consideration, as well as the fact that individuals resting against a surface react to the current, shows the incorrectness of the theory put forward by the present author (1904 *h*), in which stimulation was supposed to be due to the variations in pressure produced through the varied movements of the animal in its spiral course.

Certain authors (Dale, 1901, Statkewitsch, 1903 *a*) have reported that Paramecia sometimes swim with the current. But in these cases irregular currents have been used, such as are produced by stirring the water containing the animals. Using a tube, the present author has found the results to be practically uniform, the animals swimming up stream. If the reverse reaction actually occurs at times, it must be due to some change of internal condition, such as results in swimming backward under certain circumstances; the direction of the current over the body would be the same in the two cases.

If the explanation of the reaction to water currents above given is correct, this reaction is clearly analogous to the compensatory movements of higher animals, as Lyon (1904) has brought out for other organisms. It is a response to unusual relations with the environment, and tends to restore the usual relations.

B. *Reactions to Gravity*

In the reaction to gravity the animals place themselves with anterior end directed upward, and as a result swim to the top of the vessel containing them, forming a collection there (Fig. 56). If the tube is inverted after the collection is formed, so that the infusoria are now at the bottom, they again direct the anterior end upward, and swim to the top. These results follow in the same way whether the upper end of the tube is open or closed, and they take place equally well when the temperature is kept uniform by immersing the tube in running water.

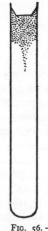

Fig. 56. — Paramecia collected at the top of a vertical tube, after Jennings (1893).

To determine the way in which the reaction occurs, it is necessary to direct the lenses of a microscope of long focus upon a region where the animals are taking up the position with long axis in the direction of gravity. This may best be accomplished by placing the animals in a U-shaped tube,

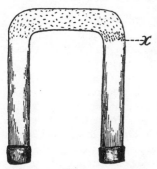

Fig. 57. — Tube used in observing the way in which Paramecium reacts to gravity.

at first with the free ends upward. After the animals have become grouped at the two free ends, the tube is inverted (Fig. 57). The Paramecia now move upward, reach the cross-piece of the U, and move across it to the opposite side. Reaching this, they at first continue the course by swimming obliquely downward, to the point x. Here the reaction occurs; the animals turn around and swim upward again. Studying the movements of the Paramecia at this point, one observes that the forward motion becomes slower, while the spiral course becomes wider. The animals swerve more strongly than usual toward the aboral side, so that the anterior end swings about in a circle, as in Figs. 37 and 38. Thus the animals are giving the avoiding reaction, "trying" successively many different positions. This is continued or repeated till after a time they come into a position with anterior end upward. The strong swerving then ceases; the animals swim upward in the usual spiral course.

The position of individuals at rest against a solid is usually quite independent of gravity. The body axis may be placed at any angle with the pull of gravity, with either end higher. The contact reaction interferes with the reaction to gravity, preventing it almost completely. Yet there is a tendency, even when in contact with a solid, to take a position with anterior end above. If Paramecia are placed in clean water in a clean, upright glass tube, in the course of time many individuals come to rest against the perpendicular walls. It will now be found, in some cases, that a considerable portion of the animals, though by no means all, are resting with the body axis nearly in line with gravity and with anterior end upward. When a swimming individual places itself in contact with the wall, it is often seen to make a sudden turn toward the aboral side, just as it comes to rest, till the anterior end is upward; then it remains in that position. The proportion thus oriented with reference to gravity is in some cultures sufficiently great, amounting perhaps to half the individuals, to show that the position is not accidental. In other cultures there may be almost no indication of any influence of gravity on the position of the attached specimens.

The precise nature of the determining factor in the reaction to gravity is very obscure. Jensen (1893) held that the reaction is due to the difference in pressure between the upper and lower portions of the organism. The cilia on the side where the pressure was greatest (the lower side) were supposed to beat more rapidly, thus turning the animal directly upward. But, as we have seen above, exact observation of the movements of the individuals shows that the reaction does not take place in this way. Moreover, the difference in pressure between the two sides of the organism is in certain reacting infusoria only one millionth of the total pressure, and this difference seems beyond question too slight to act as an effective stimulus.

Davenport (1897, p. 122) held that the reaction to gravity is due to the fact that the resistance in moving upward is greater than the resistance in moving downward, owing to the fact that the animal is heavier than water. To the changes in resistance as it swims up or down, the animal reacts. This view was accepted and elaborated by the author of the present work (Jennings, 1904 *h*). But to this can be made an objection analogous to that which is fatal to the corresponding view for the reaction to water currents. Under the uniform action of gravity, as Radl (1903, p. 139) has pointed out, it is not apparent how any such difference of resistance could be perceived by the organism. The animal would, with the same action of the cilia, and overcoming the same resistance, move somewhat more rapidly downward than upward. But it is very questionable if this slight comparative difference in rate could be perceived by the organism, — though this is of course not impossible. In any case, the fact that resting individuals may react to gravity appears fatal to the view at present under consideration.

The view having the greatest probability is perhaps that suggested by Lyon (1905). The animal contains substances of differing specific gravity; this Lyon has demonstrated. The distribution of these substances must change with the various positions taken by the animal. When the anterior end is directed downward the redistribution of internal substances thus induced acts as a stimulus, causing the usual reaction. The animal "tries" new positions till it reaches one with anterior end upward; then the reaction ceases and the animal remains in the position so reached.

Whatever the cause for the reaction to gravity, the stimulation it induces is evidently very slight, and its effect is easily annulled by the action of other agents. As we have seen, the contact reaction usually prevents the reaction to gravity. The same is true of most other stimulating agents. Almost any other stimulus that may be present produces its usual effect without interference from gravity, so that the reaction to gravity is seen clearly only in the absence of most other stimuli. Thus, if the walls of the vessel containing the animals are not clean, or if the water contains many solid particles in suspension, often no reaction to gravity can be observed.

Furthermore, the reaction to gravity becomes reversed under certain conditions. Sometimes nearly all the individuals in a given culture swim downward instead of upward. This result may be produced in cultures having originally the more usual upward tendency, in a number of different ways (Sosnowski, 1899; Moore, 1903). These will be mentioned in our section on reactions to two or more stimuli.

C. Reaction to Centrifugal Force

Conditions similar to those due to gravity may be produced by a centrifugal force, and Paramecia then react, as might be expected, in the same way as to gravity. Jensen (1893) shows that if a tube containing Paramecia is placed in a horizontal position on a centrifuge and whirled at a certain rate, the infusoria tend to swim toward that end of the tube next to the centre. In a tube 12 cm. long, with the inner end 2 cm. from the centre, the phenomena were well shown when the tube was whirled at the rate of four turns per second, for ten or fifteen minutes. In such a tube the Paramecia at the outer end, where the movement is fastest, are carried by the centrifugal force, against their active efforts, to the outer end of the tube; this is of course a purely passive phenomenon. The remainder of the Paramecia swim toward the end of the tube next the centre and collect there; this is the active part of the reaction.

This movement toward the inner end of the tube is doubtless due to the same causes, whatever they may be, that produce the upward movement in the reaction to gravity. Lyon (1905) has shown that the body contains substances of varying specific gravity, some of which collect, under strong centrifugation, at that end of the animal which is at the outer end of the tube. This redistribution is probably the cause of the reaction to centrifugal force. If the passage of such substances into the anterior end should act as a stimulus to the usual reaction, this would produce the results actually observed.

6. RELATION OF THE ORIENTATION REACTIONS TO OTHER REACTIONS

We are now in a position to define the difference between these orientation reactions and the others that we have described, and to see why the result of the avoiding reaction is to produce a certain position of the body axis in one set of cases, while it does not in the others.

In the reactions to mechanical stimuli, chemicals, osmotic pressure, heat and cold, and powerful light, the avoiding reaction is caused by the transition from one external condition to another; by a change in the intensity of action of some agent,— the change being of such a character as to lead away from the optimum. As a result, the organism tries repeated different directions of movement (in the avoiding reaction) till it hits upon one in which the transition is toward the optimum instead of away from it; in this direction it continues. This does not require the body axis to take any definite orientation, since as a rule there are various directions in which the animal can move and be on the whole approaching the optimum. Furthermore, the body axis might be in any position, provided the movement were on the whole toward the optimum.

But in the reactions to water currents, gravity and centrifugal force, it is a certain position of the body that results in stimulation; displacement of the cilia, or of certain internal constituents, occur in certain positions of the body, causing disturbances to which the animal reacts, as usual, by the avoiding reaction. This reaction consists in successively "trying," not only different directions of locomotion, but also different positions of the body axis, as a glance at Figs. 37–39 will show. As soon therefore as a position is reached in which the disturbance causing the reaction no longer exists, the reaction of course stops; the animal therefore retains this axial position.[1]

A comparison of the reactions to these two sets of agents brings out strongly the general adaptiveness and effectiveness of the reaction method of the infusorian. The avoiding reaction is of such a character as to bring about in a systematic way (1) different directions of movement; (2) different axial positions; (3) different environmental conditions (of temperature, chemicals, etc.). If any one of these puts an end to the disturbance which caused stimulation, the reaction of course stops at that point, and the animal retains the direction of movement, axial orientation, or environmental condition thus reached. If a certain axial orientation must be reached before the stimulating disturbance ceases, then the result of the reaction will be to produce this orientation. If the disturbance ceases before a common orientation of all the individuals is reached, then no common orientation will occur. In other words, the method of reaction is such as to bring about any condition whatsoever that is required in order to put an end to stimulation, — provided of course that this condition is attainable. It will therefore produce in some cases a certain direction of movement, in other cases a certain axial orientation, in other cases the retention of a certain environmental condition, just as circumstances may require.

LITERATURE IV

A. Reactions to contact with solids : PÜTTER, 1900; JENNINGS, 1897, 1899.
B. Reactions to chemicals : JENNINGS, 1897, 1899 *c* ; GREELEY, 1904 ; BARRATT, 1905.
C. Reactions to heat and cold : JENNINGS, 1904 ; MENDELSSOHN, 1895, 1902, 1902 *a*, 1902 *b*.
D. Reactions to light : HERTEL, 1904.
E. Reactions to water currents : JENNINGS, 1904 *h* : LYON, 1904, 1905.
F. Reactions to gravity and centrifugal force : LYON, 1905 ; JENSEN, 1893 ; JENNINGS, 1904 *h* ; SOSNOWSKI, 1899 ; MOORE, 1903.

---

[1] It is worthy of note that the position of orientation is not one in which a median plane of symmetry takes up a definite position with reference to the external agent, as is sometimes set forth. The infusorian when oriented continues to revolve on its long axis, so that no more can be maintained than that the longitudinal axis (in reality the axis of the spiral path) is in line with the orienting force.

CHAPTER V

BEHAVIOR OF PARAMECIUM (*Continued*)

REACTIONS TO ELECTRICITY AND SPECIAL REACTIONS

1. REACTIONS TO ELECTRICITY

THE reactions of Paramecia to electricity are more complex than those to other stimuli. This is owing to certain factors peculiar to the action of the electric current, which interfere with the usual reaction method.

The gross features in the behavior under the action of electricity may be seen as follows. The Paramecia are placed in a watch-glass or other small vessel, and through the water containing them an electric current is passed (Fig. 58, *A*). Unpolarizable electrodes should be used, though the gross features in the reaction may be observed with platinum electrodes. A current such as is produced by six or eight chromic acid cells is needed. As soon as the current begins to pass, all the Paramecia swims toward the cathode or

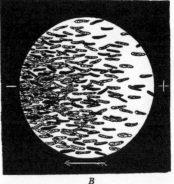

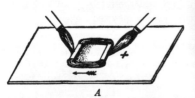

Fig. 58. — *A*, General appearance of Paramecia reacting to the electric current. After Verworn (1899). The current is passed by means of unpolarizable brush electrodes through a cell with porous walls. The infusoria have gathered at the cathodic side. *B*, Magnified view of a portion of the swarm as it moves toward the cathode. After Verworn.

negative electrode. The swarm of infusoria all moving in the same direction present a most striking appearance (Fig. 58, *B*). If while all are swimming toward the cathode the direction of the current is reversed, the Paramecia at once turn around and swim toward the new cathode. If the electrodes are small points, the Paramecia swim in curves, such as are known to be formed by the current (Fig. 59).

If while all are moving toward the cathode the current is interrupted, the group breaks up and the Paramecia scatter in all directions.

If the current is at first very weak, the Paramecia do not react sharply, only a few of them swimming toward the cathode. When the

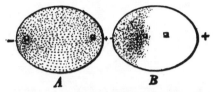

A   *B*

Fig. 59. — *A*, Curves followed by Paramecia when pointed electrodes are used. *B*, Collection of Paramecia behind the cathode, when the electrodes are placed close together. After Verworn (1899).

strength of the current is increased, more of the animals react and the movement is more rapid, till at a certain strength of current practically all are swimming rapidly to the cathode. With a further increase in the current, the rate of progression toward the cathode becomes slower. As the increase continues, the rate of swimming decreases till progress nearly or quite ceases. The animals now remain in position, with anterior ends directed toward the cathode, but not moving in either direction. Increasing the current still farther, the animals begin to swim backward toward the anode. At this time each Paramecium is seen to have become deformed, being short and thick. If the current is farther increased, the animals burst at one end and go to pieces. These remarkable phenomena were first observed by Verworn (1889 *a*).

How is this striking behavior brought about? Why do the Paramecia first all go to the cathode, then in a stronger current stop, then swim backward to the anode?

A. *Reaction to Induction Shocks*

In attempting to answer these questions, it will be best to take up first the reactions to single induction shocks. To observe the reactions accurately, the Paramecia must be placed in some viscid but not injurious substance, such as the jelly produced by allowing a few quince seeds to soak in a watch-glass of water containing the animals (Statkewitsch, 1904 *a*). This makes the movements so slow that they can be followed under the microscope. The reaction to induction shocks under these conditions has been studied especially by Statkewitsch (1903). When an induction shock is passed through a drop of such fluid containing Paramecia, the animals are found to react especially at that part of the body which is next the anode. Here the cilia are suddenly reversed, striking forward instead of backward; the ectosarc contracts sharply, and trichocysts are thrown out (Fig. 60). If the current is a very weak one, only the reversal of cilia occurs; with a stronger current the other phenomena appear. With a very powerful current, contraction and discharge of trichocysts occur also at the cathode, and with a further increase of current, over the whole body. The animal at the same time becomes deformed and usually goes to pieces.

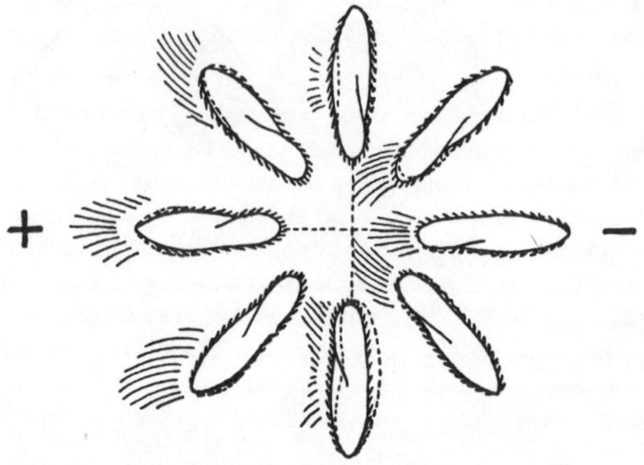

In the current of moderate strength the reversal of cilia, beginning at the anode, quickly spreads over the entire body, causing the animal to swim backward. This movement is the beginning of the avoiding reaction. After swimming backward a short distance the animal turns toward the aboral side and swims forward in a new direction. Thus the reaction to an induction shock is of essentially the same character as the reaction to other strong stimuli.

Paramecium reacts to induction shocks more readily, as might be expected, when the sensitive anterior end is directed toward the anode. When in this position, it reacts to currents that are too weak to produce reaction in specimens occupying other positions. According to Roesle (1902), Paramecium reacts more readily when the oral surface is toward the anode than when in other positions, indicating that the region about the mouth is especially sensitive. While this seems probable on general principles, it was not confirmed by the thorough work of Statkewitsch (1903). In some cases an induction shock, like a weak mechanical stimulus, causes in place of the avoiding reaction a movement forward (Roesle, 1902).

Since the animal is most stimulated when the anterior end is directed toward the anode, and this stimulation causes as a rule the avoiding reaction, one would expect that if the stimulation came repeatedly from the same direction, the animal would after a time reach a position with anterior end directed away from the anode. This is exactly what occurs. If frequent induction shocks are passed in a certain direction through

Fig. 60.—Effect of induction shocks on Paramecia in different positions. After Statkewitsch (1903). Trichocysts discharged, cilia reversed, and contraction of the ectosarc, at the anodic side or end, in a moderate current.

the water, the animals all become pointed toward the cathode and swim in that direction (Birukoff, 1899, Statkewitsch, 1903). This happens even when the current is so weak that a single induction shock causes no reaction. There is a summation of the effects of the successive shocks until a reaction is produced (Statkewitsch, 1903). As most commonly used, induction currents pass alternately in opposite directions. The induced current in one direction is due to the closing of the circuit in the primary coil, while the immediately following current induced in the opposite direction is due to the breaking of the circuit in the primary coil. The induced currents due to the breaking of the circuit are, as is well known, more powerful than those produced by the closing of the circuit. When both currents pass through the preparation alternately, Paramecia react primarily to the stronger "break" currents. They move toward the cathode of these stronger currents and are apparently not affected by the weaker "make" shocks (Birukoff, 1899, Statkewitsch, 1903 *a*).

B. Reaction to the Constant Current

If in place of induction shocks a continuous electric current is used, the result is the same as was described in the last paragraph. The Paramecia place themselves with anterior end directed toward the cathode and swim in that direction (Fig. 58).

From what we know of the behavior of Paramecium under the action of other stimuli, we might suppose that the whole secret of this behavior lies in the production of the avoiding reaction when the anterior end is directed toward the anode. This reaction, continuing until a position was reached where the anterior end was no longer stimulated, would cause it to become directed toward the cathode. If the anode stimulation still continued, now at the posterior end, the animal would continue to swim forward toward the cathode, for to stimulation at the posterior end, as we have seen, the animal responds by swimming forward. If this were the method of reaction, the behavior under the electric current would be of the same character as under the stimuli which the animal meets in its natural existence.

But a study of the exact movements of the animals shows that there is present another factor which is peculiar to the action of the electric current. To detect this the precise movements of the cilia under the action of the current must be examined. The cilia themselves may be directly observed in specimens placed in some viscous medium (see Statkewitsch, 1904 *a*). Or the effective movements of the cilia may be determined by mingling with the fluid containing them a quantity of finely ground India ink. By its aid the direction of the currents produced by the cilia becomes evident.[1] In this way we find that it is not alone at the anode that the electric current is active, but that a peculiar effect is produced also at the cathode. Here the direction of the cilia is reversed (Fig. 61) so that they point forward, and their effective stroke is forward, tending to drive the animal backward. When the electric current is weak and the animals are swimming toward the cathode, the cilia are reversed only at the anterior end (Fig. 61, 1), the reversal extending a little farther down on the oral side than elsewhere. At the anterior tip the water currents are forward instead of backward (Fig. 62, *a*), and the cilia themselves are clearly seen to be pointed forward (Fig. 61, 1). When the animal is swimming most rapidly toward the cathode, this effect is very slight; almost all the cilia of the body are beating backward in the usual way.

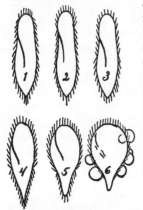

FIG. 61.—Progressive cathodic reversal of the cilia and change of form in Paramecium as the constant electric current is made stronger. The cathode is supposed to lie at the upper end. The current is weakest at 1, where only a few cilia are reversed. 2–6, Successive changes as the current is gradually increased. After Statkewitsch (1903 *a*).

If the current is made stronger, this cathodic effect increases. The cilia become reversed farther and farther back, till with a certain strength of the electric current the cilia on the anterior half of the body are striking forward, those on the posterior half backward (Fig. 61, 3).[2] The water currents produced are in opposite directions, making the animal the centre of a sort of cyclonic disturbance in the water, which gives a most extraordinary appearance (Fig. 62, *b*). The two sets of cilia oppose each other, so that the animal seems to be trying to swim in two opposite directions at once. Up to a certain strength of the electric current the posterior cilia prevail over the anterior ones, so that the animal swims forward. But the movement becomes slower and more labored as the electric current is increased, until in time the two sets of cilia balance each other. Then the animal remains in place, revolving rapidly on its long axis, or it shoots first a short distance forward, then a little backward. With a still further increase of the electric current, the cathodic effect increases to such an extent that the reversed cilia gain

---

[1] The Paramecia must be in a thin layer of fluid; this may be attained by supporting the cover-glass on thin sheets of filter paper and introducing the current through this paper.

[2] This peculiar effect was first observed by Ludloff (1895).

the upper hand, and the animal swims backward toward the anode. The cilia are now reversed even behind the middle (Fig. 61, 4, 5). The body is deformed, becoming short and thick, and pinched to a point at the anode end, while the cathode end is swollen. Finally the animal usually bursts and goes to pieces; before this happens almost all the cilia have become reversed (Fig. 61, 6).

When a Paramecium is transverse or oblique to the direction of a current at the time the circuit is closed (Fig. 63, *c*, *e*), certain striking effects are produced. If a current of medium strength is employed, such as causes reversal of about half the cilia, the following results are observed. On the anode side the cilia strike backward, as usual. On the cathode side the cilia strike forward. As a result the animal, when in a transverse position, must turn directly toward the cathode side, the cilia of both sides of the body

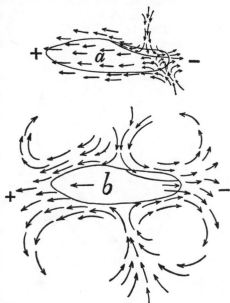

FIG. 62. — Water currents produced by the cilia in the electric current. *a*, Electric currents weak; water currents reversed only at cathodic tip. *b*, Electric currents stronger; water currents reversed over cathodic half as far back as the middle.

tending to produce this effect, as indicated by the arrows in Fig. 63, *c* and *e*. This happens even when the oral side is directed toward the cathode (Fig. 63, *e*). The animal then turns toward the oral side, — a result never produced by other stimuli, and due to the peculiar cathodic effect of the current.

This tendency to turn directly toward the cathodic side is complicated in certain positions of the animal by the usual strong tendency to turn, under the influence of stimuli, toward the aboral side, — that is, to respond by the typical avoiding reaction. If the anterior end is directed toward the anode at the time the circuit is closed, the animal invariably turns toward its aboral side, the cilia taking the position shown in Fig. 63, *b*. This method of turning is apparently due to the fact that the backward stroke of the oral cilia is more powerful than that of the opposing aboral cilia. For the same reason the animal turns toward the aboral side even when in the position shown in Fig. 63, *a*, where it would be more direct to turn toward the oral side. Between this position (*a*) and the transverse position with oral side to the cathode (*e*), there is a position in which the tendencies to turn in opposite directions are exactly balanced (*f*). The animal tries, as it were, to turn in opposite directions at the same time, so that it remains in position, though the

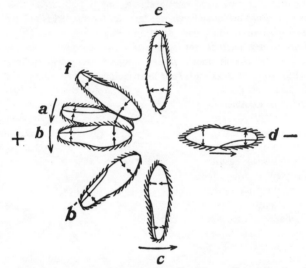

FIG. 63. — Effects of the electric current on the cilia of Paramecia, and direction of turning in different positions. The oral side is marked by an oblique line. The large arrows show the direction toward which the animal turns. The small internal arrows indicate the direction in which the cilia of the corresponding quarter of the body tend to turn the animal. In all positions save *c* and *e* the cilia of different regions oppose each other. From *a* to *d* the turning is toward the aboral side; from *d* to *f*, toward the oral side. At *f* the impulse to turn is equal in both directions, and there is no result till by revolution on the long axis the animal comes into a position with aboral side to the cathode.

cilia are beating violently, causing complicated currents in the water. This independent and opposing activity of the cilia of different parts of the body is characteristic of the effects of the electric current, and is not found in the reactions to other stimuli. In the position shown in Fig. 63, *f*, the revolution on the long axis, which is a part of the normal motion of the animal, soon interchanges the position of oral and aboral sides, whereupon the infusorian of course turns at once towards the aboral side, till its anterior end is directed toward the cathode.

Thus in a considerable preponderance of all possible cases the animal turns toward the aboral side, as it does under other stimuli. But in certain positions (from d to f, Fig. 63) it turns directly toward the oral side, a result not producible by other stimuli.

If the direction of the electric current is frequently reversed, certain peculiar effects are produced. If the reversal occurs at the moment when the anterior end has become directed toward the cathode, then the animal continues to turn toward the aboral side till the anterior end is pointed toward the new cathode. By repeated properly timed reversals, the animals can be caused to spin round and round, — always toward the aboral side.[1] If the intervals between the reversals of the current are made less, so that the animal has not yet become pointed toward the cathode, it swings back over the space through which it has turned. Thus the animals may be made to swing back and forth or turn round and round, remaining in the same spot, like animated galvanometers, — the anterior end pointing out the direction of the current.

If the rate of reversal is much increased,[2] so that the animals have scarcely time to begin swinging in a certain direction before a new reversal occurs, then certain other phenomena result; these have been described by Statkewitsch (1903, 1903 a). The Paramecia which are swimming toward one electrode when the current is closed usually continue to swim in the same direction for a time, as if reacting to only one of the current directions. Those not already pointed toward one of the electrodes usually take quickly the transverse position. Thus, soon after the beginning of the experiment, part of the animals are swimming toward the electrode at the right, part toward that at the left, while the rest are transverse. Soon those not transverse have reached the region of the electrode toward which they are swimming. Thus the Paramecia are now divided into three groups, — a group at the right swimming toward the right electrode, another at the left swimming towards the left electrode, and a central group swimming athwart the current (Fig. 64). After a time the transverse position is assumed also by those directed toward the electrodes, especially if the current is made stronger or the rate of reversal is increased. Thus at a later stage all or nearly all are transverse; they swim across the current, some toward one side of the preparation, some toward the other.

The reason for taking the transverse position when the current is rapidly reversed seems to be as follows: We have seen above that to

---

[1] As soon as a specimen has made a half revolution on its long axis, as may happen, it of course seems to spin in the opposite direction, because the aboral side has taken up a new position.

[2] The strength of the current remaining the same in both directions, not varying as in ordinary induction shocks.

single electric shocks the animals react more strongly when the anterior end is directed toward the anode. Often there is no reaction when they are in the opposite position. Consider a specimen that is oblique, as in Fig. 63, *b'*. The current comes alternately from the right and left. To the current coming from the left (anode at the left) the Paramecium reacts strongly, since its anterior end is directed toward the anode. It therefore turns its anterior end in the opposite direction, — to the right. To the opposite current, on the other hand, it reacts little or not at all, since the anterior end is not directed toward the anode. Continuing thus to react to the repeated currents from the left, it must come into the transverse position. Here the anterior end has the same relation to both currents; hence it swings as far to one side as to the other. Since it changes its position very little at any one reversal, it maintains on the whole the transverse position.

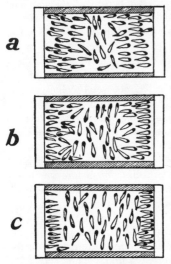

FIG. 64. — Positions taken by Paramecia in rapidly reversed currents. *a*, Positions in weak currents, or in moderate currents at the beginning of the experiment. *c* and *d*, Positions taken in stronger currents, or after the experiment has lasted for some time. After Statkewitsch (1903 *a*).

Under a constant current in one direction, the general effect of the behavior is of course to cause the animals to pass to the cathode. Here they may gather in a dense mass. But if the cathode is so placed that the Paramecia can pass behind it, they do this, thus reaching a region where the current is not acting (Fig. 59, *B*). Here they swim about in all directions. If one comes by chance again into the field of the current, it is at once returned, by the usual reaction, to the region behind the cathode. If in any other way certain areas are left free from the action of the current or with very little current, the animals gather in these free areas. Birukoff (1899) has described and figured many such cases, produced under induction shocks by the aid of electrodes of different forms; his results have been extended by Statkewitsch (1903 *a*).

It is evident that the reaction to the electric current differs fundamentally from the known behavior under other classes of stimuli. Under other stimuli the movements are coördinated, all tending toward the same end, while in the electric current different parts of the body oppose each other. The behavior thus becomes uncoördinated, lacking unity. The animal seems to strive to perform two opposite actions at once. The anterior cilia drive the animal backward, the posterior cilia forward. In certain positions (Fig. 63, *j*) part of the cilia tend to turn the animal to the right, others to the left. The action of the current is more local and direct than that of other stimuli, producing opposed reactions in different parts of the body. The whole secret of this extraordinary behavior lies in the cathodic reversal of the cilia. If this cathodic effect were non-existent, the behavior under the action of the electric current would probably be the same as under other stimuli. The reaction due to the anodic stimulation is, as we have seen, the same as that due to other strong stimuli, and in the constant current the anodic cilia strike backward in the usual way. If the anodic stimulation alone existed, the animal would doubtless become directed to the cathode by the method of trial and would swim in that direction. But as the behavior actually occurs, there is nothing like a trial of different positions. The cathodic reversal of the cilia forces the animal directly into a certain orientation. The reaction is not due, like that to chemicals, to the change in conditions as the animal passes from one region to another. It is not due to a tendency to collect about the cathode, for, as we have seen, if it is possible, the animals go beyond the cathode. Moreover, in a strong current there is no movement to the cathode, and in a still stronger current the movement is away from the cathode, though the orientation remains the same in both cases. All these peculiarities in the behavior are due to the cathodic reversal of the cilia.

What is the cause of this fundamental feature of the reaction to the electric current, — the cathodic reversal? Many theories have been proposed to account for the reaction to electricity, though often these do not touch this fundamental feature in any way. It will be better to reserve an account of these theories until we have examined the behavior of other infusoria under the action of electricity (see Chapter IX).

2. OTHER METHODS OF REACTION IN PARAMECIUM

In Paramecium there are certain methods of reacting to stimuli which we have not yet described. These are, first, local contractions of the ectosarc, and second, discharge of trichocysts. Neither of these seem to play any important part in regulating the relation of the organism to the surrounding conditions.

Slight local contractions of the ectosarc occur in response to many stimuli. Since the ectosarc of Paramecium is not known to contain contractile elements, the way in which these are brought about is unknown.

A discharge of trichocysts is produced by many different agents. The trichocysts are rodlike sacs in the ectosarc, perpendicular to the outer surface. Their contents are ejected, under certain conditions, into the water, forming long threads. According to some authors, these threads have a definite structure, and are probably preformed within the animal. Others suppose the threads to be formed by the coagulation of a fluid contained within the sacs. After discharge of the trichocysts the animal appears to be surrounded by a zone of radiating fibres (Fig. 65).

The discharge of trichocysts under the influence of stimuli has been studied especially by Massart (1901 *a*), and by Statkewitsch (1903).

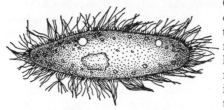

FIG. 65. — Paramecium with trichocysts discharged, as a result of the application of picric acid.

Crushing the animal causes discharge of trichocysts in the region injured. Weaker mechanical stimuli do not have this effect. If the animal is heated rapidly till it is killed, it discharges the trichocysts before dying; if heated slowly, this effect is not produced. Neither cold nor increased osmotic pressure have any effect on the trichocysts. Many chemicals produce the discharge, particularly various acids. Saturated solution of picric acid causes a sudden discharge of all the trichocysts at once. One-fourth per cent methylene blue produces a slow and irregular discharge successively from different parts of the body.[1] If any agent acts on a limited portion of the body surface, the trichocysts of only that region are discharged. Many chemicals kill the animal without discharge of the trichocysts.

A weak induction shock causes discharge of the trichocysts at the anode only (Fig. 60); a stronger shock causes discharge at both anode and cathode. A still stronger shock causes discharge of the trichocysts over the entire surface of the body (Statkewitsch, 1903).

In the discharge of the trichocysts we have a phenomenon comparable to the definite reflex actions observed in various organs of higher animals. The function of the trichocysts is uncertain. They are usually supposed to be weapons of defence. If the Paramecium is seized by an animal which is attempting to prey upon it, the trichocysts will of course be discharged from the injured region. But whether they really

---

[1] To demonstrate the discharge of the trichocysts it is convenient to use picric acid alone or picric acid to which a little aniline blue has been added. In the latter case the trichocysts become colored blue (Massart).

serve for defence seems questionable. Certainly the infusorian Didinium (Fig. 113), which is the chief enemy of Paramecium, is not hindered in the least from seizing and devouring the animal by the discharge of trichocysts. It is possible that the discharge is really an expression of injury, — a purely secondary, even pathological, phenomenon, like the formation of vesicles on the surface of an injured specimen.

LITERATURE V

A. Reaction of Paramecium to electricity : VERWORN, 1889 *a* ; LUDLOFF, 1895 ; BIRUKOFF, 1899, 1904 ; ROESLE, 1902 ; PÜTTER, 1900 ; STATKEWITSCH, 1903, 1903 *a*, 1904 ; JENNINGS, 1904 *h* ; BANCROFT, 1905 ; COEHN AND BARRATT, 1905. *B.* Discharge of trichocysts as a reaction : MASSART, 1901 *a*.

CHAPTER VI

BEHAVIOR OF PARAMECIUM (*Continued*)

BEHAVIOR UNDER TWO OR MORE STIMULI; VARIABILITY OF BE-
HAVIOR; FISSION AND CONJUGATION; DAILY LIFE; GENERAL
FEATURES OF THE BEHAVIOR

1. BEHAVIOR UNDER TWO OR MORE STIMULI

THE behavior thus far described is that which takes place under the influence of but a single kind of stimulation. But normally the conditions are as a rule more complex than this; the animal is affected by several sets of stimuli at once. What is the behavior under such conditions? If, while the Paramecium is reacting to the stimulus *a*, the stimulus *b* acts upon it, will it react in the usual way to *b*? Or will it continue to react to *a*? Or will its action form a compromise between the usual reactions to the two agents? Or will it, finally, react in a new way, different from the usual reactions to either *a* or *b*?

Let us examine first the behavior under the simultaneous action of the contact stimulus and of other usual stimuli. As we have seen, the contact stimulus often causes the animal to come to rest and behave in a characteristic manner, while other classes of stimuli usually induce the avoiding reaction or a movement forward. Thus opposite reactions are induced by the two kinds of stimuli acting separately. What will be the result when the two act together?

If the animal is at rest against a mass of vegetable matter or a bit of paper under the action of the contact stimulus, and it is then struck with the tip of a glass rod, we find that at first it may not react to the latter stimulus at all. A touch that would cause a free swimming specimen to give the avoiding reaction in a pronounced way often has no evident effect on the quiet specimen. Sometimes, however, a touch coming from behind causes the animal to move forward, still remaining in contact with the solid object; it thus creeps a short distance over the surface of the solid. Finally, a strong blow on the anterior end causes the animal to leave the solid and give the typical avoiding reaction.

Thus we find that under the simultaneous action of the two stimuli the infusorian may either react to the more effective of the two, which-ever it is, without regard to the other, or its behavior may be a sort of compromise between the usual results of both.

If specimens showing the contact reaction are heated, it is found that they do not react to the heat until a higher temperature has been reached than that necessary to cause a definite reaction in free swimming speci-mens. Thus Pütter (1900) found that at 30 degrees C. all the free specimens are strongly affected, moving about rapidly in all direc-tions, while the attached specimens remain quiet or make only slight vibratory movements. Many of them remain attached until the tem-perature has reached 37 degrees, when the free specimens are dashing about wildly. At this temperature or a somewhat lower one the at-tached specimens become free; they then dash about as furiously as the others. Thus the contact reaction interferes with the reaction to heat, preventing it until a much higher temperature has been reached than is necessary to cause reaction in free specimens.

On the other hand, both heat and cold interfere with this contact reaction. Paramecia much above or much below the usual tempera-ture do not settle against solids with which they come in contact, but respond instead by a pronounced avoiding reaction. At a still higher temperature even the avoiding reaction ceases. A Paramecium coming against a solid presses the anterior end against it and continues to try to swim forward, — succeeding only in revolving on its long axis (Massart, 1901 *a*).

Specimens in contact with a solid react less readily to chemicals than do free specimens, so that a higher concentration is required to induce the avoiding reaction. On the other hand, immersion in strong chemicals prevents the positive contact reaction; Paramecia under such conditions coming against a solid react by the avoiding reaction. In this case, then, the effect of the chemical is to change the method of reacting to another stimulus — to the solid object. Certain other chemicals have the oppo-site effect, favoring the positive contact reaction. This is notably true of carbon dioxide. In water contain-ing this substance the infusoria are strongly inclined to settle down against any object with which they may come in contact. They thus often form under these conditions dense masses attached to the glass rods used for holding up the cover-glass (Fig. 66), though usually they do not come to rest against smooth, hard objects.

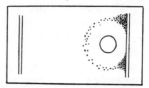

FIG. 66. — Paramecia which have formed a ring about a bubble of CO_2, and have then come to rest against the glass supporting rods, forming two dense groups.

The contact reaction may completely prevent the reaction to gravity. Paramecia placed in a tube which contains many bits of solid matter, or has its walls rough or dirty, usually do not rise to the top, but settle against the solid matter on the wall and remain. They may thus remain scattered through all parts of the tube, or may gather in any portion of it where the material inducing the reaction is found. Specimens at rest against a solid may occupy any position with reference to gravity. In similar ways the contact reaction may prevent the usual reaction to water currents.

The interference between the contact reaction and the reaction to the electric current produces a number of peculiar results. If a weak electric current is passed through a preparation containing many specimens attached to a bit of débris or to the surface of the glass, the free specimens swim at once toward the cathode, while the attached specimens do not react at all. If the current is made stronger, it produces for an instant the usual effect on the cilia of the attached specimens. The cathodic cilia strike forward, the anodic cilia backward. But this does not continue; after a moment the contact reaction resumes its sway, and the cilia have their usual positions. If the current continues, after a short time the cilia are again affected as before; then resume their original positions. This may occur many times, — the two stimuli alternating in their control of the cilia. If the current is made much stronger, the animal finally leaves the solid. It then swims directly to the cathode in the usual way. To induce this reaction in a resting specimen, it requires as a rule two or three times as intense a current as that needed for producing the same effect on free swimming animals.

If the electric stimulus is first in action and the Paramecium then comes in contact with a solid, somewhat different results are produced. If the current is weak, often the animal, swimming toward the cathode, ceases to react to the electricity on coming against the solid; it may then take up any position on the surface of the solid. If it comes against the surface film of the water, or the surface of the glass slide, it may cease its forward movement only for an instant, then, becoming free, it may swim again toward the cathode. If the current is a little stronger (such as to produce the maximum rapidity of movement toward the cathode, in free swimming specimens), a different effect is produced. The Paramecium stops against the surface of the solid, and places itself transversely or obliquely to the current, with the oral surface toward the cathode (Fig. 67). Here it remains, the current produced by the cilia being everywhere backward save in the oral groove, where it is forward. If the electric current is reversed, the oral cilia strike strongly backward, and the animal at once turns on its short axis till the oral surface faces the new cathode. It remains in this position till the current is reversed anew. Thus, when in contact with a surface, Paramecia often show a transverse orientation with reference to the electric current. At times the animal while in this position moves forward along the surface with which it is in contact, transversely to the current; on reversal of the current it turns about and moves in the opposite direction. This may often be observed if the Paramecia are placed on a slide in a thin layer of water through which the electric current is passed. Many of them in swimming come against the glass or the surface film of the water. Thereupon they begin to move transversely to the current, as just described. Meanwhile the free swimming specimens continue to pass toward the cathode.

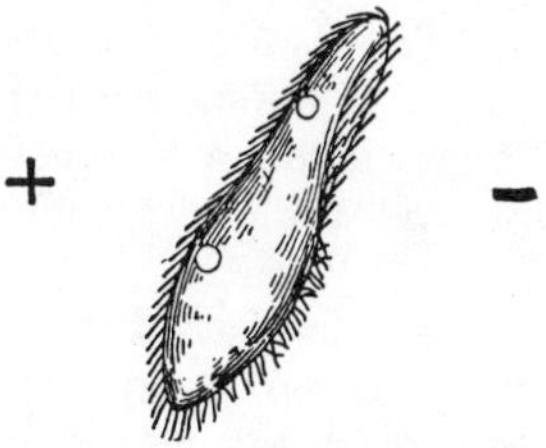

Fig. 67. — Oblique position taken by Paramecium in contact with a surface, when under the action of the electric current.

With a stronger current a still different effect is produced. The Paramecia are swimming forward in the slow, cramped manner that is characteristic for strong currents. On coming in contact with the surface film or the glass, the animals at once begin to move backward (toward the anode) instead of forward. This continues as long as the contact continues. On becoming free they swim forward again. The reason for this behavior seems to be as follows: In a strong electric current, as we know, the anterior cilia tend to drive the animal backward, the posterior cilia forward (Fig. 62, *b*); the latter prevail. The contact reaction, as we have seen, causes the cilia behind the region of contact to cease movement. When swimming forward under the conditions mentioned, the Paramecia usually come in contact with the surface at the thickest part of the body, near the middle of its length. Thereupon, owing to the contact reaction, the cilia behind this spot, driving the animal forward, cease to beat, while the cilia in front, driving it backward, continue their action. Hence, the anterior cilia gain the upper hand and force the animal backward.

Why does this contact stimulus thus interfere with the reaction to other stimuli? There are two possible factors to be considered here, one physical, the other physiological. The animal seems actually to attach itself to solids, probably by a secretion of mucus. Such a secretion is very evident in many infusoria, though it has not been demonstrated in Paramecium. This attachment would, in a purely physical way, impede the movements due to other stimuli. While it is possible that this factor may play a small part in the matter, it is clear that it is not the important or essential factor. If it were, we should see the cilia of the attached animal move in the usual manner under the influence of stimuli, though these movements would not have the usual effect. As a matter of fact, in most cases we see nothing of the kind. The cilia either do not move at all, or move in a manner different from that occurring in free specimens. The essential factor in the interference is a physiological one. When reacting to the contact stimulus, the animal is less easily affected by other stimuli, and when reacting to the other stimuli, it is less easily affected by the contact stimulus. Since the two stimuli in question require behavior of opposite character, it is indeed inevitable that one should give away to the other, or at least modify the behavior toward it; both *cannot* receive the usual reaction.

Combinations of other stimuli have been less investigated than those just considered. In any combination the reaction to gravity gives way, as we have seen, to the reaction due to other factors. Paramecia swimming upward react to other stimuli without hindrance, and Paramecia at rest against a surface often show no orientation with reference to gravity. The reactions to chemical and electrical stimuli completely supplant the reactions to gravity. In a vertical tube Paramecia may form collections in any region that becomes impregnated with carbon dioxide or may avoid any region which contains a repellent chemical. If an electric current is passed through a vertical tube, the Paramecia react to it in exactly the same manner as under other conditions, swimming toward the cathode whether this is above or below. Sosnowski (1899) and Moore (1903) have shown that many different stimuli modify the reaction to gravity, changing the direction in which the animals swim. If Paramecia in the culture fluid swim upward, mixture with tap water, or with chemicals of various sorts, often causes them to swim downward. This effect soon disappears, however, and the animals return to the top. Increase of temperature to 30 degrees (Sosnowski), or decrease to 2 degrees (Moore), often has the same temporary effect. The same result is at times produced by shaking or jarring the tube containing the animals; they go to the bottom, returning in a short time to the top. The effect of all these agents varies with different cultures of Paramecia; in some cultures the reaction to gravity is easily changed, in others with difficulty or not at all.

Reactions to chemicals often interfere with the reaction to the electric current. If through a preparation of Paramecia that are gathered in an area containing carbon dioxide, as in Fig. 68, *A*, an electric current is passed, the animals swim to the cathode side of the area, then stop. All gather in this region, seeming to make vain efforts to cross the invisible boundary (Fig. 68, *B*). Observation of individuals shows that as soon as they reach the boundary of the area of carbon dioxide, they give the avoiding reaction, in the usual way, and pass back into the area. Here they become oriented again by the electric current, and pass again to the boundary, where they react as before. Thus the reaction to the electric current prevails until a region of a sudden change in chemical character is reached; the reaction to this then supplants the reaction to the current. If the current is reversed, the animals gather in the same way at the opposite side of the area of carbon dioxide (Fig. 68, *C*). If the current is made very powerful and is long continued,

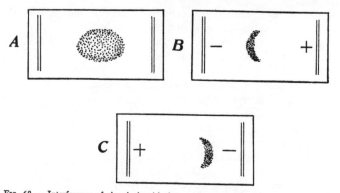

Fig. 68. — Interference of chemicals with the reaction to the electric current. At *A* Paramecia have gathered in an area containing CO_2. At *B* an electric current is passed through the preparation with cathode at the left; the animals gather at the left edge of the area of CO_2. At *C* the current has been reversed; the animals are therefore gathered at the right edge of the area.

the Paramecia are one by one caused to cross the boundary of the acid area and to swim to the cathode. If a drop of some repellent chemical — as sodium chloride or an alkali — is introduced into a preparation (Fig. 41), the Paramecia of course leave this vacant. If the electric current is passed through the preparation, the Paramecia swim toward the cathode; coming to the boundary of the drop, they swim around it, leaving it empty, and thus reach the cathode. In this case the path followed is a resultant of the operation of the two stimuli, — the orientation due to the electric current and the avoiding reaction produced by the chemical.

If the entire region next the cathode is occupied by a repellent chemical, the Paramecia may be forced by a strong and long-continued current to enter it till they are destroyed.

A very peculiar interaction of chemicals and the electric current is seen when Paramecia are placed in physiological salt solution (0.7 per cent) and the current is passed through the vessel. The strong chemical causes the animals to swim backward; the current orients them in the usual way; the result is that they swim backward to the anode. This phenomenon is to be observed in solutions of various chemicals, as acids, potassium iodide, sodium carbonate, etc. It will probably be found to occur in any solution that causes the animals to swim backward for a considerable time. It should be investigated further. As soon as the Paramecia have become accustomed to the chemical, so that they no longer swim backward within it, they react to the current in the usual way, swimming to the cathode.

Thus we find that under the action of more than one stimulus Paramecium may behave in any of the ways which we mentioned in our first paragraph as conceivable. It may react to the first stimulus without regard to the second, or to the second without regard to the first, depending on which is the more effective. Such results are often produced when both the stimuli are sufficiently strong to cause reaction if acting alone. Which stimulus shall produce its characteristic effect sometimes depends on which comes into action first. Thus, Paramecia in contact may not react to the electric current or to heat; while free Paramecia subjected to the same strength of current or degree of heat do not show the positive contact reaction. This condition of affairs seems to occur throughout the animal series; in higher animals we express the same phenomenon subjectively by saying that attention to one thing prevents attending to others.

In some cases the behavior shown is a resultant of the action of the two stimuli. Examples of this are seen in the movement along a surface under the simultaneous action of the contact reaction and a mechanical shock, or in swimming around a chemical in solution, under the influence of the electric current; or in swimming backward to the anode when in solutions of strong chemicals.

Finally, the effect of one stimulus is sometimes merely to change the method of reaction to another. Thus heat and strong chemicals cause the animal to respond to contact by the avoiding reaction in place of the positive contact reaction; carbon dioxide has the contrary effect. The modifications of the reaction to gravity above mentioned are examples of the same thing. Cases of this character have much theoretical interest. We shall return to them in considering the variability and modifiability of the reactions of Paramecium.

2. VARIABILITY AND MODIFIABILITY OF REACTIONS

We have seen in the last section that the behavior of Paramecium under a given stimulus may be determined by the simultaneous presence of other stimuli. The behavior depends not only on the stimulus to which it is primarily reacting, but also upon other external conditions. May the nature of the behavior also depend upon internal conditions? In other words, may the same animal under the same external conditions behave differently at different times? May Paramecium, like higher animals, become modified by the stimuli which it has received, or by its own reactions, so as to react for the future in a manner different from its reactions in the past?

It is difficult to obtain evidence on this question for Paramecium, because the animal moves about so rapidly that it is hardly possible to follow a given individual and determine whether its reactions do or do not change. Much more is known in regard to this matter, as we shall see later, for the fixed infusorian Stentor. But a number of significant facts have been brought out for Paramecium.

First we have the fact that the presence of a certain agent or condition may alter the method of reaction to another. Paramecia in heated water react to solids by the avoiding reaction in place of the positive contact reaction; Paramecia in a solution of carbon dioxide, on the other hand, are much more likely to respond by the positive contact reactions. Many conditions — heat, cold, chemicals, mechanical shock, etc. — alter, as we have seen, the reaction to gravity, causing the animals to swim downward instead of upward. Such phenomena indicate that the first agents alter in some way the physiological condition of the animals, so that they now react to the second agent in a changed manner. This conclusion is impressed upon the observer by the behavior of the organisms. Specimens in heated water are swimming about violently, so that we should not expect them to come to rest against solids. Those in carbon dioxide move slowly and seem in a condition predisposing to repose, so that coming to rest against solids is the reaction that might be anticipated. The interference between the two stimuli is not purely physical. There is nothing in the physical action of heat or a mechanical jar to make the animals move downward, as happens when the agents reverse the action to gravity. Indeed, in the latter case one can plainly see that the downward movement is an active one. The only explanation possible for such cases is that the animals have become changed in some way by the first stimulus, so that they now react in an altered manner to the second stimulus.

Further we find that there are great differences in the reactions of different individual Paramecia, and especially of Paramecia from different cultures. In studying the reactions to chemicals, one often finds that a few individuals swim directly into the given solution, while the majority give the avoiding reaction on coming in contact with it, and hence remain outside (Jennings, 1899 *c*, p. 373). While in a certain case individuals from one culture were repelled by $\frac{1}{40}$ per cent lithium chloride, those from another culture were found to be quite indifferent to a solution of the same chemical sixteen times as strong, swimming readily into a drop of $\frac{2}{5}$ per cent lithium chloride (Jennings, 1899 *c*, p. 374). When placed in a vertical tube, Paramecia from certain cultures gather at the top; from other cultures at the bottom; while in other cases they remain scattered throughout the tube (Sosnowski, 1899). Corresponding variations are found in the reaction to water currents. Similar differences are to be observed with regard to the positive contact reaction (Pütter, 1900, p. 253). Infusoria in certain cultures are strongly inclined to attach themselves to solids, forming dense masses on the surface; in other cultures such masses are never formed. In fresh cultures the animals are usually much inclined to attach themselves in this way; in old cultures they are not. Even in a culture where most of the animals attach themselves, there are always a number of specimens which remain persistently free. Variations are to be observed at times in the reactions to electricity (Jennings, 1904 *h*). One sometimes observes that while most of the specimens in a preparation are reacting to the electric current in a precise way, a few specimens do not react at all, swimming about at random. Sometimes single specimens will be seen swimming toward the anode, while all the rest swim toward the cathode. This is most often observed after the current has been reversed several times.

Whether the variations mentioned in the last paragraph are due to changes which have occurred during the life-time of the animals, or whether they are permanent differences between different individuals we do not know. In either case they are of importance, since they give much opportunity for the action of natural selection. This is a point to which we shall return later.

We know, however, that sometimes the behavior of the same individual varies, and in some cases we can form an idea of the nature of the change which has occurred. If a Paramecium is subjected to a strong induction shock, it fails for some time thereafter to react to weak shocks, though at the beginning it reacted to these (Statkewitsch, 1903). This result is probably due to a change in the animal such as we commonly call fatigue. To be explained possibly in a similar way is the following occasional observation. A specimen in the continuous electric current is swimming toward the cathode; on reversal of the current it retains its orientation and continues to swim forward, — now of course toward the anode. This lasts usually but a short time.

Paramecia which have been living at the usual temperatures show a temperature optimum of about 24 to 28 degrees; if they are kept for some hours at a temperature from 36 to 38 degrees, the optimum rises to 30 or 32 degrees (Mendelssohn, 1902). A change in the individuals induced in this way is commonly spoken of as acclimatization. Similar changes could doubtless be induced in the reactions to chemicals and to other stimuli; this has not yet been done.

Paramecia that have long been deprived of food behave in a somewhat different manner from normal individuals (Moore, 1903; Wallengren, 1902 *a*). But the changes in behavior are apparently due to actual structural changes in the organism, due to lack of food, and rendering it impossible for the animal to move so strongly and rapidly (Wallengren, 1902 *a*). Paramecia kept in distilled water are found to be much more sensitive to most stimuli than usual (Jennings, 1897; Wallengren, 1902 *a*); owing apparently to lack of sodium salts in the body. This condition may perhaps be called that of salt hunger. If a small quantity of some sodium salt is added to the distilled water, the Paramecia return to the usual condition (Wallengren, 1902 *a*).

Certain changes in the behavior of individuals can hardly be classified as due either to fatigue, acclimatization, or hunger. If a bit of filter paper is placed in a preparation of Paramecia, the following behavior may often be observed. An individual swims against it, gives the avoiding reaction in a slightly marked way, swimming backward a little; then it swims forward again, jerks back a shorter distance, then settles against the paper and remains. After remaining a few seconds, it may move to another position, still remaining in contact with the paper. Then it may leave the paper and go on its way. All this may happen without the slightest evident change in the outer conditions. So far as can be seen, the Paramecium first responds to the solid by the avoiding reaction, later by the positive contact reaction, and still later suspends the contact reaction, all without any change in external conditions. The changes inducing the change in reaction must then be within the animal.

Again, as we have seen, jarring Paramecia which have collected at the top of a tube often causes them to swim to the bottom of the tube (Sosnowski, Moore). The jarring itself lasts but a moment, while the Paramecia continue for some time after to swim downward. The shock must therefore have changed the physiological condition of the animals, so that they now show a change of reaction to gravity, or possibly a lack of reaction to gravity.[1]

---

[1] It is possible that the shock merely causes them to swim rapidly in any direction that is open to them. Since they are already at the top of the tube, the only direction open to them is that leading downward.

All together, it is clear that there are differences in behavior due to differences in the internal or physiological condition of the animal, — differences shown even in a single individual at different times. Some of the different physiological conditions may be characterized as fatigue, as acclimatization, as hunger, or the like. In other cases they cannot be definitely characterized. We clearly have slight beginnings of the modification of behavior through the previous experiences of the organism. The analysis of this matter will be carried farther for the behavior of unicellular organisms in the account of Stentor (Chapter X).

3. BEHAVIOR IN FISSION AND CONJUGATION

At intervals certain extraordinary episodes connected with the processes of reproduction interrupt the usual life of Paramecium. The behavior at such times seems not to differ in any notable manner from the usual behavior. We shall therefore describe it only briefly.

Fission. — At times the animal begins to divide into two by a transverse constriction at about the middle. During the early stages of the

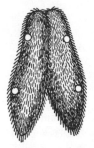

process the two halves act in unison. The currents of water are driven by the cilia in the same direction over both, and the two halves react to any stimulus as a single animal. If subjected to induction shocks the half at the anode responds by contraction of the ectosarc and discharge of trichocysts, while the cathode half does not. As the constriction separating the two halves becomes very deep, so that they are connected only by a slender strand, they begin to behave more independently. The anterior half at times changes its direction of movement, while the posterior half tries to continue straight forward. The connecting strand is strained and bent or twisted.

Fig. 69. — A pair of conjugating Paramecia.

Soon it breaks, and the two individuals are separated.

Conjugation. — In conjugation two individuals become united by their oral surfaces (Fig. 69), and a complicated process of interchange of nuclei occurs. The union of two specimens seems brought about chiefly by the usual movements and reactions of the animals, taken in connection with a physical change of the body substance in the region of the oral groove. Here the surface becomes viscid, so that if another Paramecium comes in contact with this region, the two stick together. Often two individuals may be seen at rest close together on the surface of a bit of bacterial zoöglœa. One drags its posterior end across the oral groove of the other, whereupon the two stick together (Fig. 70, *a*).

Each tries to continue its course, so that they pull in opposite directions. One may drag the other along with it, or the two may finally pull apart.

There is of course a tendency for objects to be brought against the oral groove, owing to the strong current of water that passes along this region; it is through this fact that Paramecium gets its food (compare Fig. 46). This tendency operates on other Paramecia in the neighborhood as well as on inanimate objects. If two Paramecia are close together with oral grooves facing each other (Fig. 71), this tendency is reciprocal; each tends to draw the other to its own oral surface. On

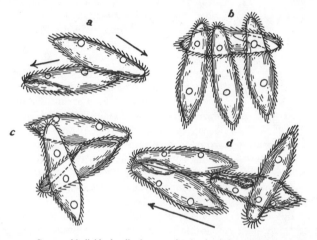

Fig. 70. — Groups of individuals adhering to each other by their oral surfaces, from cultures of Paramecia undergoing conjugation. *a*, Two attached individuals swimming in opposite directions. *b*, Three individuals attached by their oral surfaces to a fourth. *c*, Three individuals irregularly attached. *d*, A conjugating pair, swimming to the left, with a third individual attached by its oral surface to the posterior part of one of these, and a fourth individual transversely attached to the third. The third and fourth were dragged about by the first pair.

the other hand, if the aboral surfaces face each other, the currents tend to separate the two Paramecia. Hence when two Paramecia come in contact it will usually be by the oral surfaces. This often happens under usual conditions, but no conjugation results, because the oral surfaces have no tendency to adhere; the animals therefore quickly separate again. But at times when the oral surfaces are viscid, specimens which come thus in contact remain united. The succeeding internal processes fall in the field of physiology rather than that of behavior. Details concerning them will be found in text-books of zoölogy.

Thus nothing seems to be required for producing conjugation beyond the usual movements and the viscidity of the oral region. The present author has been unable, after careful study, to detect any differences in the methods of reacting during periods of conjugation. The groups formed on the surface of solids and the rapid movements of the organisms, described by Balbiani (1861, p. 441), as occurring at such

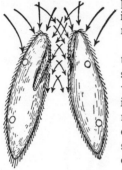 periods, are by no means peculiar to conjugating infusoria. They take place in the same manner in cultures where none are conjugating.

The significant part played in conjugation by the viscidity of the oral surfaces is demonstrated by the peculiar phenomena observed when specimens accidentally come in contact irregularly. This often happens where the animals are numerous. If any part of the body of one specimen comes by chance against the oral surface of another, the two stick together, without regard to their relative position.

FIG. 71.— Currents urging two Paramecia together when the oral sides face one another.

Often groups of three or four or more are formed in this way (Fig. 70). The individuals occupy all sorts of irregular positions, and each endeavors to swim forward in his own direction. Some are pulled backward, others sidewise, against their vigorous struggles. Often one succeeds in freeing itself, and then swims away; others remain caught in such groups indefinitely. Even moribund specimens and specimens undergoing fission sometimes thus become united irregularly with others. But the regular union of individuals by the oral surfaces is more common than the formation of irregular groups, owing to the strong tendency, produced by the usual currents, for Paramecia to come together at the oral surfaces.

During conjugation the two united individuals behave in much the same way as a single specimen. They revolve on the long axis to the left as they swim through the water, and they react to stimuli by the avoiding reaction in the usual way. The direction of turning in the avoiding reaction seems determined usually by one of the components; the pair always turn toward the aboral side of this particular individual. If subjected in the transverse position to an induction shock, only the specimen next the anode responds by ejecting trichocysts (Statkewitsch, 1903).

4. THE DAILY LIFE OF PARAMECIUM

Let us now try to form a picture of the behavior of Paramecium in its daily life under natural conditions. An individual is swimming freely in a pool, parallel with the surface and some distance below it.

No other stimulus acting, it begins to respond to the changes in distribution of its internal contents due to the fact that it is not in line with gravity. It tries various new positions until its anterior end is directed upward, and continues in that direction. It thus reaches the surface film. To this it responds by the avoiding reaction, finding a new position and swimming along near the surface of the water. Now there is a strong mechanical jar, — some one throws a stone into the water, perhaps. The Paramecium starts back, tries certain new directions, and finishes by reacting to gravity in the reverse way from its former reaction; it now swims downward. But this soon brings it into water that is notably lacking in oxygen. To this change it responds as before, trying new directions till it has come near the surface again. Swimming forward here, it approaches a region where the sun has been shining strongly into the pool, heating the water. The Paramecium receives some of this heated water in the current passing from the anterior end down the oral groove. Thereupon it pauses, swings its anterior end about in a circle, and finding that the water coming from one of the directions thus tried is not heated, it proceeds forward in that direction. This course leads it perhaps into the region of a fresh plant stem which has lately been crushed and has fallen into the water. The plant juice, oozing out, alters markedly the chemical constitution of the water. The Paramecium soon receives some of this altered water in its ciliary current. Again it pauses, or if the chemical was strong, swims backward a distance. Then it again swings the anterior end around in a circle (Fig. 38) till it finds a direction from which it receives no more of this chemical; in this direction it swims forward.

Thus the animal swims about, continually hesitating as it reaches regions where the conditions differ, trying new directions, and changing its course frequently. Every faint influence in the water affects it, for the animal is very sensitive. Other Paramecia swim about in the same way. They do not avoid each other, but often strike together; then one or both draw back and turn in another direction. The animal may strike in the same way against stones or the sides of a glass vessel. In such cases it may be compelled to try successively many different directions before it succeeds in avoiding the obstacle, — acting like a blind·man who finds a stone wall in his course.

After a time our animal comes against a decayed, softened leaf. At first it draws back slightly, then starts forward again, and places itself against the leaf. The body cilia cease their action, while the oral cilia carry a strong stream of water to the mouth. It so happens that this leaf has lately fallen into the water and has no bacteria upon it, so that the Paramecium receives no food. Nevertheless the animal "tries" it for a while. Other Paramecia may gather in the same way, but after a considerable time they one by one leave the dead leaf. Our Paramecium swims about again, being directed hither and thither by the various changes in the chemical constitution or temperature of the water, till it comes to a region containing more carbon dioxide in solution than usual. It gives no sign of perceiving this, save perhaps by swimming a little less energetically than before. The area containing carbon dioxide is small, and soon the animal comes to its outer boundary, where the water drawn to its oral groove contains no carbon dioxide. It stops, and tries different regions, by swinging its anterior end around in a circle, till it again finds a direction from which it receives carbonic acid; in that direction it swims forward. Since it behaves in the same way whenever it comes to the outer boundary of the carbonic acid, it remains swimming back and forth within this region, and thus in time explores it very thoroughly. Finally it comes upon the source of the carbon dioxide, — a large mass of bacteria, embedded in zoöglœa, that are giving off this substance. The infusorian places itself against the mass of zoöglœa, suspends the activity of the body cilia, and brings a strong current of water along the oral groove to the mouth. This current removes some of the bacteria from the zoöglœa and carries them to the mouth, where they are swallowed. While the animal is thus occupied, other Paramecia in their headlong course may strike against it. But now it does not react to such a shock at all; it remains in place, engaged with its food taking. After the animal has been in this position for some time, the sun begins to shine strongly on this part of the pool, heating the water. All the free-swimming Paramecia in this region thereupon begin to swim rapidly about, repeatedly backing and trying new directions, till a direction is found that leads to a cooler region. But our Paramecium, busy with its food-getting, does not react to the heat at all. The water becomes hotter and hotter, and after a time our infusorian moves about a little, turning over or shifting its position, but still remaining against the zoöglœa. All the free swimming specimens have left this region long ago. As the water becomes still hotter, our Paramecium suddenly leaves the mass of zoöglœa and now dashes about frantically under the influence of the great heat. It first swims backward, then forward, and tries one direction after another. Fortunately one of these directions soon lead it toward a cooler region. In this direction it continues and its behavior becomes more composed. It now swims about quietly, as it did at first, till it finds another mass of bacteria and resumes the process of obtaining food.

In this way the daily life of the animal continues. It constantly feels its way about, trying in a systematic way all sorts of conditions, and retiring from those that are harmful. Its behavior is in principle much like that of a blind and deaf person, or one that feels his way about in the dark. It is a continual process of proving all things and holding to that which is good.

5. FEATURES OF GENERAL SIGNIFICANCE IN THE BEHAVIOR OF PARAMECIUM

A. The Action System

Passing in review the behavior of Paramecium, we find that the animal has a certain set of actions, by some combination of which its behavior under all sorts of conditions is made up. The number of different factors in this set of actions is small, and they are combined into a coördinated system, so that we may call the whole set taken together the *action system.* The action system of Paramecium is based chiefly on the spiral course, with its three factors of forward movement, revolution on the long axis, and swerving toward the aboral side. The behavior under most conditions is determined by variations in these three factors. Such variations, combined in a typical manner, produce what we have called the avoiding reaction. Other elements in the action system are the resumption of forward movement, in response to stimulation, and the coming to rest against solid objects in what we have called the positive contact reaction. Subordinate activities, playing little part in the behavior, are the contractions of the ectosarc and the discharge of trichocysts.

The action system thus includes only a small number of definite movements. By one or another of these, or by some combination of them, we may expect the organism to respond to any stimulus which acts upon it. We cannot expect each kind of stimulation to have a specific effect, different from that produced by other stimuli, for all any stimulus can do is to set in operation certain features of the action system. Many different stimuli acting on this one organism therefore necessarily produce the same effect. Different organisms have different action systems, so that the same agent acting on different organisms may produce entirely different effects. The nature of the behavior under given conditions depends as much (or more) on the action system of the animal as on the nature of the conditions. In studying the behavior of any organism the most important step is therefore to work out its action system, — the characteristic set of movements by which its behavior under all sorts of conditions is brought about.

The most important features of the action system of Paramecium are those shown in what we have called the avoiding reaction. This, as we have seen, consists essentially in reversing, stopping, or slowing up the forward motion, then swerving more than usual toward the aboral side, while at the same time the rate of revolution on the long axis is decreased. By this combination of movements Paramecium responds to most effective stimuli that act upon it. By it are produced both negative and positive reactions.

B. *Causes of the Reactions, and Effects produced by them*

Examination has shown us that the cause for this reaction is some change in the conditions; usually some change in the relation of the animal to the environmental conditions. Such changes are brought about chiefly by the movements of the animal. In certain cases they are due to the direction of movement, carrying the animal into environmental conditions which stimulate it; in other cases they are due to the axial position taken by the animal, this resulting in internal or external disturbances which act as stimuli.

These stimuli produce, as we have seen, not a single, simple, definitely directed movement, comparable to the typical reflex act. On the contrary, stimulation is followed by varied movements, made up of several simultaneous or successive factors, each of which may vary, as we have seen in detail, more or less independently of the others. These movements produce varied effects, as follows: (1) They place the animal successively and in a systematic way in many different axial positions (see Fig. 38); (2) they cause it to move successively and systematically in many different directions; (3) they subject it successively to many different environmental conditions, — of temperature, light, chemicals, mechanical stimuli, etc. Now, it is evident that in this way the animal is practically certain to reach finally a position, direction of movement, or environmental condition, that removes the cause of stimulation, since the latter was due to something wrong in one of these respects. The reaction then ceases, since its cause has ceased; the animal therefore retains the axial position, direction of motion, or environmental condition thus reached. The method of reaction is then of such a character as to bring about whatever is required for putting an end to the stimulation, — whether this requirement is one of orientation, of general direction of locomotion, or of the retention of certain environmental conditions.

Thus the behavior and reactions of Paramecium consist on the whole in performing movements which subject the organism to varied conditions (using this word in the widest sense), with rejection of certain of these conditions, and retention of others. It may be characterized briefly as a selection from among the varied conditions brought about by varied movements.

The fundamental question for this method of behavior is, Why does the organism reject certain conditions and retain others? We find that the animal rejects, on the whole, such things as are injurious to it, and accepts those that are beneficial. There are perhaps some exceptions to this, but these are rare and only noticeable because exceptional; in a general view the relation of rejection and acceptance to injury and benefit is evident. It results in keeping the animals from entering temperatures that are above or below those favorable for the life processes, in causing them to avoid injurious chemicals of all sorts, in saving them from mechanical injuries, and in keeping them in regions containing food and oxygen. Clearly, the animal rejects injurious things, and accepts those that are beneficial.

How does this happen? We meet here the same question that we find in higher organisms and man. How does it happen that in man the response to heat and cold beyond the optimum is by drawing back, just as it is in Paramecium? How does it happen that in both cases there is a tendency to reject things injurious and retain things beneficial? We shall attempt in a later chapter to bring out the relations involved in this problem, in such a way as to make it possibly a little more intelligible; here we shall content ourselves with pointing out the identity of the problem in the infusorian and in man.

LITERATURE VI

A. Interference between contact and other stimuli: PÜTTER, 1900; JENNINGS, 1897, 1904 *h.*
B. Heat and other stimuli: MENDELSSOHN, 1902 *a;* MASSART, 1901 *a.*
C. Gravity and other stimuli: SOSNOWSKI, 1899; MOORE, 1903.
D. Behavior in conjugation: JENNINGS, 1904 *h;* BALBIANI, 1861.

CHAPTER VII

THE BEHAVIOR OF OTHER INFUSORIA

ACTION SYSTEMS. REACTIONS TO CONTACT, TO CHEMICALS, TO HEAT AND COLD

THE infusoria form a large and varied group of organisms. In the present chapter we shall try to show how far the behavior of Paramecium is typical for the group, and to bring out important differences found in the behavior of other species. Certain features of behavior are better illustrated in other infusoria than in Paramecium; these we shall treat in detail. This is notably true of the reactions to light, and to a less degree of the reactions to certain other stimuli. Certain infusoria are much more favorable for a study of the modifiability of reactions than Paramecium, so that we shall examine these relations with care.

I. THE ACTION SYSTEM

We found that Paramecium has a certain set of ways of acting, — of "habits," one might call them, — of which its behavior under most conditions is made up. These are few in number and combined into a connected system, which we have called the "action system." The action system of Paramecium is typical of what we find throughout the infusoria, including both the flagellates and the ciliates. But it becomes modified among different species, in accordance with their varying structure and the conditions under which they live. Practically all the infusoria agree with Paramecium in swimming in a spiral when passing freely through the water, and in the fact that when stimulated they turn toward a certain side, defined by the structure of the organism. But some species instead of swimming freely usually creep along surfaces, while others are attached by one end to solid objects, remaining in the same spot indefinitely. These different methods of life necessitate changes in the action system. We shall take up briefly a number of species, bringing out the essential features of the action system.

A. Flagellata

The free swimming flagellates move in a spiral, keeping a certain side of the body always toward the outside of the spiral,[1] just as Paramecium does. By means of the flagella they draw a cone of water from in front to the anterior end of the body, as happens in Paramecium. Among the flagellates the behavior has been most precisely studied in Chilomonas and Euglena (Jennings, 1900, 1900 a and b).

Chilomonas. — Chilomonas is an unsymmetrical organism, of an irregularly oblong form. The body is compressed sideways and bears an oblique notch at the broader anterior end (Fig. 72). Of the two anterior angles which lie on either side of the notch, one (x) is larger and lies more to the right than the other (y). From the notch arise two long flagella, by the aid of which the animal swims. Chilomonas often occurs in uncounted millions in water containing decaying vegetation.

In swimming, Chilomonas revolves on its long axis, at the same time swerving toward the smaller of the two angles at the anterior end (Fig. 72, y). The path followed thus becomes a spiral (Fig. 73).

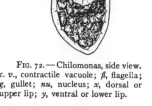

FIG. 72.— Chilomonas, side view. *c. v.*, contractile vacuole; *fl*, flagella; *g*, gullet; *nu*, nucleus; *x*, dorsal or upper lip; *y*, ventral or lower lip.

The animal often comes to rest against solid objects; it is then attached by one of the two flagella, while the other is free.

To most effective stimuli Chilomonas responds by an avoiding reaction similar to that of Paramecium. Its forward movement becomes slower, ceases, or is transformed into a movement backward. Then the animal turns more strongly toward the side which bears the smaller angle, and finally starts forward again. Thus the path is altered. The reaction consists essentially in pointing the anterior end successively in many directions, toward one of which the animal finally swims. The different factors in the reaction vary with the intensity of the stimulation, just as they do in Paramecium. The reaction may be repeated, as in the animal last named, until it finally carries the organism away from the stimulating region. Thus it is clear that in Chilomonas, as in

[1] This was first observed by Naegeli (1860).

Paramecium, the method running through the behavior is that of the selection of certain conditions through the production of varied movements. When stimulated the animal "tries" many different directions till one is found in which stimulation ceases. This reaction is known to be produced in Chilomonas by heat, by the drying up of the water containing the animals, by mechanical stimulation, by various chemicals, by passage from water containing certain chemicals (acid) to water containing none, and by the electric current. We shall take up certain details of the reactions of Chilomonas in the sections which deal with the different classes of stimuli.

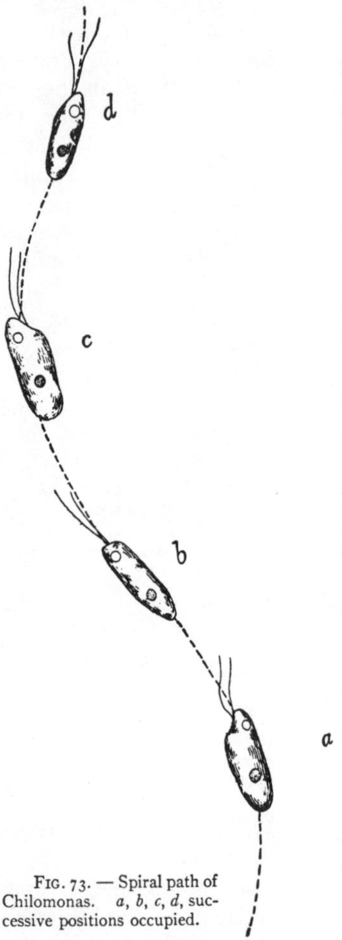

Euglena. — *Euglena viridis* (Fig. 74), like Chilomonas, swims in a spiral. The larger lip (Fig. 74, *x*) is always toward the outer side of the spiral (Fig. 94). When stimulated by coming in contact with a weak chemical, by a mechanical shock, or by a change in the intensity of light, Euglena responds by an avoiding reaction similar to that of Paramecium and Chilomonas. The forward motion becomes slower, ceases, or (more rarely) is transformed into a backward motion. Then the organism

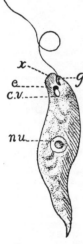

FIG. 74. — *Euglena viridis,* after Kent. *c. v.,* reservoir of the contractile vacuole; *e,* eye spot; *g,* gullet; *nu,* nucleus; *x,* larger or upper lip.

FIG. 73. — Spiral path of Chilomonas. *a, b, c, d,* successive positions occupied.

swerves more strongly than usual toward the larger lip. Thus the spiral becomes wider and the organism becomes pointed successively in many directions (see Fig. 91). In one of these directions it finally swims forward, repeating the reaction if again stimulated. We shall have occasion to describe in detail the reactions of Euglena to light (Chapter VIII).

To most very intense stimuli Euglena responds by contracting into a sphere and beginning to encyst.

The behavior of most other flagellates is not known in detail, since the organisms are usually very minute and their precise movements can be followed only with much difficulty. *Cryptomonas ovata* is known to respond to stimuli in essentially the same way as Euglena (Jennings, 1904 *a*), — the swerving being toward the more convex surface. The flagellate swarm spores of various algæ react in much the same way, as is shown by the descriptions of Naegeli (1860) and Strasburger (1878), though the precise details have not been worked out as they have for Chilomonas, Euglena, and Cryptomonas. Naegeli (*l. c.*, p. 101) describes the behavior of the flagellate swarm spores on coming against a mechanical obstacle, as follows: They swim backward, turn to one side, then swim forward in the changed direction. This is exactly what Chilomonas does, as we have seen. Similar observations have been made on flagellates by various investigators, but only in the species we have named has the side toward which the organism turns been determined.

B. Ciliata

In many free swimming ciliates the action system is known to be essentially similar to that of Paramecium. All swim in spirals, swerving toward a certain side, and react to stimuli by backing and swerving more than usual toward a structurally defined side. *Loxodes rostrum*

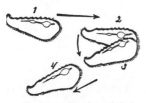

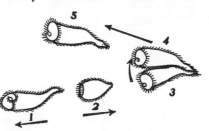

FIG. 75. — Reaction of *Loxo-phyllum meleagris.* 1–4, successive positions.

FIG. 76. — Methods of reaction to strong stimuli in Stentor. The individual at 1 is stimulated; it thereupon swims backward (2, 3), turns toward the right aboral side (3, 4), and swims forward (5).

in reacting turns toward the aboral side. *Loxophyllum meleagris* reacts as a rule by turning toward the oral side (Fig. 75). *Stentor polymorphus, Stentor cæruleus,* and *Stentor ræselii* (Fig. 31, *b*), when free swimming, react by turning toward the right aboral side (Fig. 76). *Bursaria truncatella* reacts to most stimuli by swimming backward and turning toward the right side (Fig. 77). *Spirostomum ambiguum* and *Spirostomum tenue* swim backward and turn toward the aboral side. *Opalina ranarum* turns toward the more convex (right) side, Nycto-

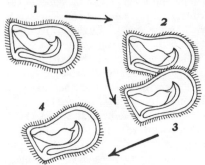

therus toward the aboral side (Fig. 78). Many of these organisms show an additional reaction to strong stimuli, consisting in a marked contraction of the body. This is particularly noticeable in Spirostomum and Stentor.

Many of the Ciliata do not as a rule swim freely through the water, but creep along surfaces, keeping one side against the surface. This is true at times of most of the organisms

FIG. 77. — Reaction of Bursaria, ventral view. 1-5, successive positions occupied.

mentioned in the foregoing paragraph. It is much more usual in certain other ciliates, belonging to the group of Hypotricha (Fig. 31, *f*; Fig. 81). In these animals the cilia of one side of the body are specially modified for creeping, while the opposite side bears either few and weak cilia or none at all. The Hypotricha are usually found running about on the bottom, or on the surface of objects in the water. In addition to their creeping movements, they produce by means of strong peristomal cilia a vortex leading back to the mouth. These animals of course do not revolve on the long axis as they progress, and the corresponding feature is likewise lacking in the reactions to stimuli. On coming in contact with an obstacle, or when otherwise stimulated, they stop or move backward a distance, then turn toward a certain structurally marked side, keeping in contact with the substratum and not revolving on the long axis. This renders it much easier to observe the precise method of reacting than in Paramecium, where the rapid revolution on the long axis is very confusing.

FIG. 78. — Nyctotherus. The arrow to the right shows the direction of turning in response to stimulation, while the three interior arrows indicate the direction of beat of the cilia. After Dale (1901).

As examples of the creeping infusoria, the following may be mentioned: —

Stylonychia (Fig. 31, *f*), Oxytricha, and other Hypotricha react to most stimuli by moving backward and turning to the right (Fig. 79). These organisms are particularly favorable for the study of the reaction method. The body is flat, and the right and left sides are very easily distinguished, so that the direction of turning after stimulation can be determined with the greatest ease. In many respects the Hypotricha are among the most favorable objects to be found among unicellular animals for studying behavior.

Microthorax sulcatus usually creeps along the bottom, and reacts to most stimuli by turning suddenly toward the convex ("dorsal") edge. The turning may or may not be preceded by a start backward.

Colpidium colpoda (Fig. 31, *d*) usually moves forward with one side against the substratum, following a curve with its oral edge on the concave side of the curve. When stimulated mechanically or chemically, it turns toward the aboral side and continues its course (Fig. 80).

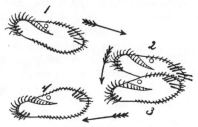

Fig. 79. — Reaction of Oxytricha, ventral view. 1–4, successive positions.

In some cases the reaction to strong stimulation takes on special features. For example, in *Pleuronema chrysalis*, in *Halteria grandinella*, and in various Hypotricha, there are powerful bristle-like cirri,

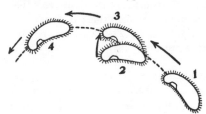

Fig. 80. — Path of Colpidium. At 2 it is slightly stimulated; it thereupon turns toward the aboral side (3–4) and continues its curved course.

by means of which the animal may leap suddenly backward or to one side. These are probably to be considered strongly marked avoiding reactions, not differing in principle from what we find in Paramecium or Oxytricha.

All the species which usually move along on a surface may at times swim freely through the water. They then as a rule revolve on the long axis, both when progressing and in the avoiding reaction. On the other hand, almost all the species which characteristically swim freely through the water do at times move along surfaces. They may then react to stimuli in the same way as do the Hypotricha. Such forms as Bursaria and Loxophyllum are transitional between the free swimming species and those that creep along surfaces; they are found about as often in one situation as in the other.

In the Ciliata thus far considered the reaction method is evidently that of the selection of certain environmental conditions through the productions of varied movements. When its movement leads to stimulation, the animal responds by trying many new directions, till one is found which does not lead to stimulation. The reaction is less flexible in the ciliates which creep along surfaces than in the free swimming ones. In the former, owing to the lack of revolution on the long axis, all the directions tried lie in a single plane. But under many powerful stimuli even these species usually leàve the surface on which they are moving; they then react in the freer way characteristic of unattached organisms, trying directions lying in many different planes.

There exists also a large number of ciliates which become more or less permanently attached by the part of the body opposite the mouth. This attached portion is usually drawn out to form a slender stalk or foot. Examples of such infusoria are Stentor (Fig. 31, *b*) and Vorticella (Fig. 31, *c*). Some of these species are found attached under all usual conditions; such are Vorticella and Carchesium. Others are frequently found swimming freely; this is the case, for example, with *Stentor cæruleus*. Some infusoria become fixed in only a temporary way, by a mucous secretion. Such are Spirostomum and Urocentrum, which are often found suspended from solid objects by a thread of mucus (Fig. 82). Even the species which are most firmly fixed may under powerful stimuli detach themselves and swim away. The heads of Vorticella and Carchesium thus at times detach themselves from their stalks and swim about like Paramecium. At such times they may also creep over surfaces, just as do the Hypotricha. The behavior when free is essentially similar in its main features to that of Paramecium or Oxytricha.

In the attached condition the mouth and peristome are usually above, surrounded by a wreath of large cilia. These cilia are in continual movement, in such a way as to bring a current of water from above to the mouth. Some fixed infusoria contract at intervals with marked regularity, even when there is no external stimulation. Such is the case in Vorticella. The reactions to stimuli are much modified as compared with those of the free swimming species. The avoiding reaction becomes broken up into a number of factors, any one of which may take place more or less independently of the others. Thus, *Stentor roeselii* may respond to stimulation either by a reversal of the cilia, driving away the water currents, by bending over toward the right aboral side, or by withdrawing into its tube. Each of these reactions corresponds to a certain definite feature in the avoiding reaction of free infusoria. Owing to the disintegration of the avoiding reaction into independent parts, the behavior of these fixed infusoria become more varied and more highly developed than that of the unattached species. We shall have occasion to treat of this in detail later, in our account of the modifiability of reactions in Protozoa.

2. REACTION TO MECHANICAL STIMULI

In the responses of infusoria to contact with solid objects we may distinguish the same two reaction types that we found in Paramecium. The animal may react in what might be called a "negative" way, avoiding the object, or it may react "positively," placing itself against the solid body.

The negative response to contact with solid bodies is the typical "avoiding reaction." The animal moves backward, turns toward a certain definite side, then swims forward again. In other words, it tries a new direction. If this leads again against the obstacle, the animal again reacts in the same way, and this is repeated, till through frequent trials the obstacle is avoided.

There are certain important points regarding the relation of the direction of movement in this avoiding reaction to the part of the body that is stimulated. Since the animal usually swims forward under natural conditions, it will as a rule come in contact with large solid objects at its anterior end. Further, small objects may be carried by the ciliary currents to the oral side. Thus the movement backward and the turning toward the aboral side in the avoiding reaction remove the animal from the source of stimulation. But experimentally other parts of the body can be stimulated. Thus in Oxytricha (Fig. 79), we may with the tip of a fine glass rod stimulate either the left (oral), or the right (aboral), side. In either case the animal backs and turns to the right. If the right side is repeatedly stimulated, the animal continually wheels toward the stimulated side; if the left side is touched, it wheels continually away from the stimulated side. Thus the direction of movement in the reaction is not determined by the side stimulated, but by the structural relations of the organism. On the other hand, if we stimulate the posterior end sharply, the animal does not respond by the typical avoiding reaction, but simply runs forward. The direction of movement is in this case determined by the part stimulated. These results have been found to hold also in many other infusoria.

Experiments of the kind just described have shown that the anterior end is as a rule much more sensitive than the remainder of the body surface. A light touch, having no effect at the posterior end, produces a strong reaction when applied to the anterior end.

It is a general rule that unlocalized mechanical stimuli, such as are produced by jarring the vessel containing the animals, have the same effect as stimuli applied to the anterior end; they induce the avoiding reaction.

In the positive contact reaction, the animal places itself in contact with the solid object and remains against it.   It may now continue quiet, while the oral cilia bring a current of water containing food to the mouth.

FIG. 81. — Side view of Stylonychia creeping along a surface.   After Pütter (1900).

But sometimes the animal runs over the surface of the solid, using its cilia as if they were legs.   This, as we have seen, is the common method of locomotion in the Hypotricha.   A side view of one of the Hypotricha while creeping along a surface is shown in Fig. 81.   In other cases the animal secretes a layer or thread of mucus and thereby attaches itself to the solid.   Attached in this way by a long thread (Fig. 82), Spirostomum and Urocentrum often remain in a certain position, revolving on the long axis.   The thread is usually quite invisible, but by passing a needle between the solid object and the animal, the latter may often be pulled backward by the thread of mucus. In still other cases the infusorian reacts to solid objects by fixing its posterior end firmly, remaining in this place for long periods, like a plant.   How this occurs in Stentor is described in Chapter X.

The contact reaction is often directed toward very minute objects, as we have set forth in detail in the case of Paramecium.   It then serves the purpose of helping to obtain food.   In some of the fixed infusoria such behavior is especially striking.   Thus, if a small object touches gently one side of the disk of Stentor, the animal may bend over toward it.   This reaction may be seen when a small organism in swimming about comes against the disk of the animal, then attempts to swim away.   The Stentor bends in that direction, so as to keep in contact with the organism as long as possible.   At the same time, of course, the ciliary vor-

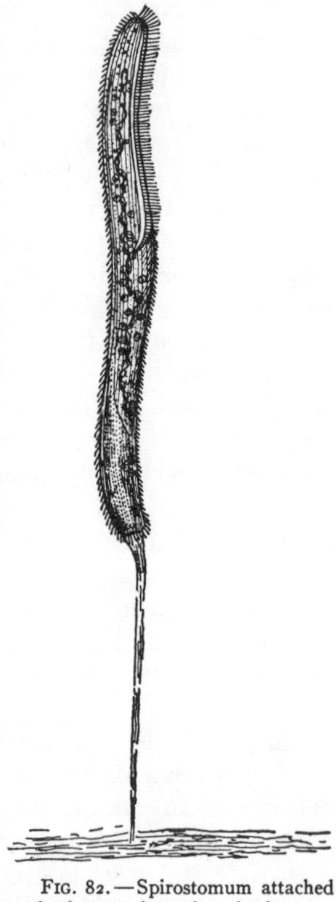

FIG. 82. — Spirostomum attached to the bottom by a thread of mucus and remaining stationary with anterior end upward.

tex tends to draw the prey to the Stentor's mouth.   This reaction may be produced experimentally by attaching a bit of soft, flocculent débris to the tip of a fine glass rod, and allowing this to touch the disk of Stentor, then drawing it gently to one side. The Stentor follows it, often bending far over (Fig. 83). The animal may thus bend in any direction — to the right, to the left, or toward oral or aboral side.

When infusoria are in contact with solids, their behavior always becomes much modified. The spiral movement of course ceases, and the reaction to many stimuli — especially such reactions as depend largely on the spiral movement — either cease or become changed. Animals that when free place the axis of swimming in line with gravity, usually take up, when in contact with solids, any position without reference to gravity. To high temperatures attached specimens respond much less readily than do free swimming ones. *Stentor cæruleus* responds readily to light when free swimming, directing its anterior end away from the source of light; when attached, it does not react in this way. Many infusoria show a modified reaction to the electric current when in contact with solids. The flagellates Chilomonas, Trachelomonas, Polytoma, and Peridinium react readily to the electric current when free swimming; not at all when in contact (Pütter, 1900, p. 246). Most ciliates when in contact with solids react

FIG. 83. — *Stentor ræselii* bending over to remain in contact with a shred of débris which is pulled by the experimenter to the right.

less readily to the electric current, and frequently when the reaction does occur, it is of a different character from usual. While free specimens place themselves in line with the current, attached infusoria often take up a transverse or oblique position with the peristome or oral side directed toward the cathode, — just as happens in Paramecium. This is true in general for the Hypotricha.

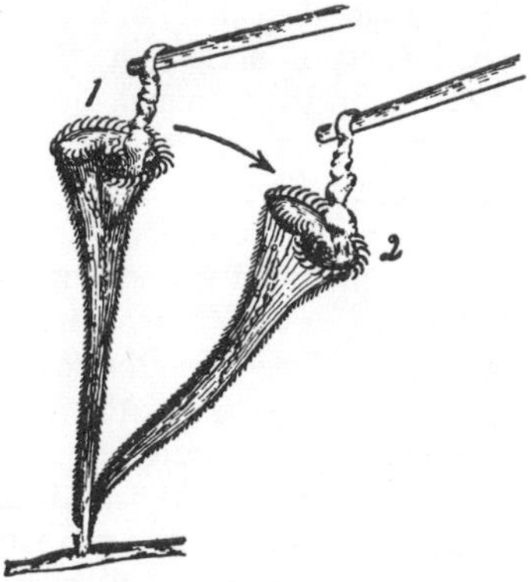

What is the cause of the interference of the positive contact reaction with the reaction to other stimuli? It is necessary, as we have seen in our discussion of this reaction in Paramecium, to distinguish two factors in the contact reaction; one physical, the other physiological. The physical factor is found in the fact that the organism actually adheres to the surface of the solid, — in many cases, at least, by means of a mucous secretion. This physical adhesion would of course tend to prevent that rapid movement under the influence of a stimulus which is shown by free individuals. Thus, the animal might attempt to react in the usual way, — showing the same ciliary movements as free individuals, — but might find itself stuck, and unable to escape. Doubtless sometimes this condition of affairs is realized; it is described, for example, by Pütter as present in the reaction of attached specimens of Colpidium and some other infusoria. But in many cases this physical factor will not account for the observed behavior. Infusoria in contact may take different positions without difficulty, and could easily place themselves in line with gravity, yet as a rule they do not do so. Attached Stentors could easily bend into a position with anterior end away from the light, yet their position shows no relation to the direction of the light rays. There is nothing in the physical adherence to a surface that should compel the animal to take a transverse position in the electric current, rather than a position parallel to the current, yet this is what occurs in attached specimens. It is clear that there is a physiological factor involved. Contact with solids tends to make the animal act in one way, the other stimulus in another; hence the two *must* interfere. If we object, as some authors have done, to the admission that the contact reaction interferes with the reaction to other stimuli, we are compelled to admit in any case that the reactions to other stimuli do interfere with the contact reaction, and one admission has as much theoretical significance as the other. It is evident that when two agents influencing the organism in opposite ways act simultaneously, the effect of one must give way to that of the other, or the two must combine to produce a resultant. It is impossible that each should produce its characteristic effect. The interference of the contact reaction with the reactions to other stimuli is one of the most striking phenomena to be observed in the behavior of these lower organisms. It is always necessary to distinguish carefully the behavior of free swimming specimens from those that are in contact with surfaces, for the two differ radically.

3. REACTION TO CHEMICALS

The reactions to chemical stimuli take place in all accurately known cases through the typical avoiding reaction. As a rule the motor organs of the infusoria, both flagellates and ciliates, act in such a way that a current of water passes from in front of the animal to the anterior end and mouth, as illustrated for Paramecium in Fig. 35. Thus when a chemical is dissolved in the water, a "sample" of it is brought to the most sensitive part of the body. If the chemical is of such a nature as to act as a stimulus, the animal swims more slowly, stops, or moves backward, turns toward the customary side (usually the aboral side), until it no longer receives the chemical, then moves forward in the new direction. Thus the region containing the chemical is avoided. In many cases this reaction takes place in a very pronounced manner; the animal shoots far backward, whirls rapidly toward the one side, and repeats the reaction many times. In other cases the reaction is less pronounced, and motion merely becomes a little slower as long as the chemical is received in the ciliary current, while at the same time the animal quietly swings its anterior end about in a circle (as in Fig. 37 or 38). This continues until it finds a direction from which no more of the chemical is received; in that direction it swims forward. If the movements of the animal are not precisely observed, the method by which the reaction occurs may in such cases be easily misunderstood.

There are various chemicals in which certain infusoria gather, producing collections like those formed by Paramecium in acids (Fig. 43). In all cases in which the facts are accurately known, these collections are formed in the same way as are those of Paramecia. The animals enter without reaction into the region where the substance is present, then respond by the avoiding reaction whenever they come to the outer boundary of the area containing the substance. Thus every individual that enters the area of the chemical remains, and in the course of a longer or shorter period a collection is formed here. In many cases this indirect method of gathering together is strikingly evident, and the individuals may be clearly seen to move about within the area containing the chemical, in the manner represented in Fig. 44. If the infusoria observed are very minute, so that differentiations of the body are to be seen only with great difficulty, if their movements are rapid, and if in the avoiding reaction they do not swim backward, but merely stop and turn toward one (structurally defined) side, at the same time revolving on the long axis, then the reaction method is not so evident on a cursory examination. In such cases, if the relation of the direction of turning to the structural differentiations of the body and to the revolution on the long axis are not carefully determined, the animal will be supposed to turn *directly*, without variations of any sort, into the chemical. This was formerly supposed to be the universal method of reaction to chemicals. The cause for the turning was supposed to be found in the difference in the concentration of the chemical on the two sides of the organism. The animal turned directly toward the side of greater concentration ("positive chemotaxis") or of less concentration ("negative chemotaxis"). This method of reacting to chemicals is no longer supposed to exist for infusoria by any one familiar with the reaction method described in the foregoing pages, so far as I am aware, save in the case of certain very minute organisms, — fern spermatozoids, Saprolegnia swarm spores, and the flagellate *Trepomonas agilis* (Rothert, 1901,

p. 388). But it is notable that in none of these cases has the relation of the direction of turning to the differentiations of the body been observed, and this is the crucial point for determining the nature of the reactions. The fact that it is only for these very difficult objects that the direct turning is maintained must make us cautious in accepting this exceptional result.[1]

Let us now leave the method of reacting, and turn to certain more general phenomena. In what chemicals do infusoria gather? What chemicals do they avoid?

In no other infusoria is the behavior toward different chemicals so well known as in Paramecium. Chilomonas collects in acids in general, and especially in solutions of carbon dioxide, just as Paramecium does. Spontaneous gatherings are often formed by Chilomonas, and it seems probable that these are due, as in Paramecium, to the carbon dioxide produced by the animals themselves (Jennings and Moore, 1902). *Cyclidium glaucoma* and *Colpidium colpoda* likewise collect in carbonic and other acids. Opalina, Nyctotherus, and *Balantidium entozoön*, living in an alkaline medium, gather in acids, but if transferred to an acid medium, they gather in alkali (Dale, 1901). Many other infusoria show no tendency to gather in acids. *Loxocephalus granulosus* and *Oxytricha æruginosa* form spontaneous collections resembling precisely those of Paramecium, but they are not due to the same cause. These species do not collect in solutions of carbon dioxide, nor in other acids. When they are mingled with Paramecia in the same preparation, they collect in one region, while the Paramecia collect in another. It is apparent that Loxocephalus and Oxytricha produce some substance to which the collections are due, and that this substance is not carbon dioxide. A number of other infusoria form spontaneous collections, the cause of which has not been investigated. Many of the commonest species do not form such collections.

There are many chemicals in which one or another species of infusoria have been found to collect. Most of the details are of comparatively little general interest from the standpoint of animal behavior, so that we shall not take them up here. An excellent summary of these results will be found in Davenport's "Experimental Morphology" (Vol. I, pp. 32–45). Certain general features are important for our purposes; these we may bring out briefly.

First, from the way the collections are brought about, it is evident that whether given infusoria tend to collect in a certain solution or not depends on the nature of the solution in which they are already found. This has been illustrated in detail for Paramecium. Paramecia in

---

[1] For a discussion of related points, see Chapter XIV.

strong salt solution collect in weak salt solutions or in tap water; Paramecia in tap water collect in distilled water; Paramecia in distilled water collect in weak acids. In the same way, if two solutions are open to any given infusorian, they tend to collect in that one by which they are least repelled. Thus "attraction," as determined by the formation of collections, is a relative matter; the infusoria, like higher organisms, often have to put up with merely that by which they are least repelled. To say that a certain infusorian gathers in a given substance *A*, therefore, signifies little more than that it is less repelled by this substance *A*, than by the substance in which it was found at the time the experiment was tried.

Most flagellates and ciliates are repelled by strong solutions of chemicals of almost all sorts. This is true even for strong solutions of the same substances in which they collect when the solutions are weak. In such substances we can therefore distinguish an optimum concentration. Below the optimum the organisms are indifferent, while above the optimum they are repelled. Expressing the facts more concretely, at the indifferent concentration no reaction is caused when the organism passes into the solution or out of it; at the optimum concentration no reaction is caused when the organism passes into the solution, but the avoiding reaction is induced on passing out, while at concentrations above the optimum the organisms react at passing inward. The result is then in every case that they tend to gather in the optimum.

The reaction is in each case caused by a change from one concentration to another. The amount of change necessary to cause the reaction has been shown, in the case of fern spermatozoids (Pfeffer, 1884), to bear a definite relation to the concentration of the solution in which the organisms are immersed. In other words, the amount of change necessary to cause the reaction varies according to Weber's law. Thus in the fern spermatozoids the concentration of malic acid necessary to produce a collection of the organisms must be about thirty times that in which the organisms are already immersed.

Massart (1891) found that specimens of *Polytoma uvella* in his cultures were not repelled by chemicals even in the strongest solutions. Such cases are very exceptional; other investigators have found that even this same organism (from other cultures) is repelled by various chemicals (Pfeffer, 1904, p. 808, note).

The variability and inconstancy of the reactions of infusoria to chemicals deserves emphasis. Whether infusoria of a given species react to a certain chemical or not, and how they react, depends upon the past and present conditions of existence of the individuals. The general outlines of the reactions can be determined for any species, but the details, especially from a quantitative standpoint, vary in accordance with the environmental influences acting upon the individuals in question. As a rule, infusoria collect in solutions of substances which may serve them as food. This is almost invariably true for substances which form the usual food of the organism under natural conditions. When the amount of oxygen present in the water is low, most infusoria collect about bubbles of air or other sources of oxygen.

Infusoria sometimes gather in substances which do not serve for food or respiration, but which serve other important purposes in the physiology of the species concerned. Thus, the flagellate spermatozoids of ferns were found by Pfeffer to gather in solutions of malic acid. This substance is found in the fern prothalli, and probably occurs in the mouth of the archegonium, into which the spermatozoids must enter in order that fertilization may take place. The tendency to collect in malic acid then doubtless plays a part in bringing about fertilization in ferns. The collection of Paramecia in carbon dioxide seems to be another case of a reaction which is useful to the organisms, though the substance causing it does not itself serve as food.

Many infusoria collect, under certain circumstances, in substances which do not serve as food and are not known to play any useful part in the biology of the animal. Thus, Pfeffer found that the flagellate *Bodo saltans* gathers in most of the salts of potassium, as well as in various salts of lithium, sodium, rubidium, cæsium, ammonium, calcium, strontium, barium, and magnesium. This signifies only, as we have already seen, that they are less repelled by solutions of these substances than by the fluid in which they are situated. In most cases, as soon as a substance is sufficiently concentrated to be injurious it becomes repellent.

Whether the repellent effect of chemicals is due to the chemical properties of the solution, or to its osmotic pressure, has been rigidly determined only for Paramecium. In this animal, as we have seen, the osmotic pressure is usually not the cause of the reaction. There is much evidence that this is true for most species, but accurate quantitative evidence is needed on this point.

4.   REACTION TO HEAT AND COLD

Infusoria in general react to heat and cold in much the same way as does Paramecium, — through the avoiding reaction. The way the reaction occurs is most easily seen in the Hypotricha. The phenomena to be observed are of special interest, because they show clearly how a movement of a large number of individuals in a certain uniform direction ("orientation") may be brought about by the selection of varied movements.

The common hypotrichan *Oxytricha fallax*, abundant in vegetable infusions, is well fitted for the study of this reaction. A large number of specimens are placed on a slide or trough. When one end of the trough is gradually heated by passing water at a temperature of 40 degrees beneath it, the Oxytrichas at this end are seen to become very active, darting about in all directions (Fig. 84). As the temperature rises, they give the avoiding reaction, — darting backward, and turning to the right. This is alternated with rapid dashes forward. Whenever a specimen passes toward the warmer end of the trough, or when it comes in contact with the sides or end, it responds with the avoiding reaction. But a specimen passing away from the heated region, in the direction of the arrow at 14 (Fig. 84), does not give the reaction, because it is passing from a hot to a cool region. The

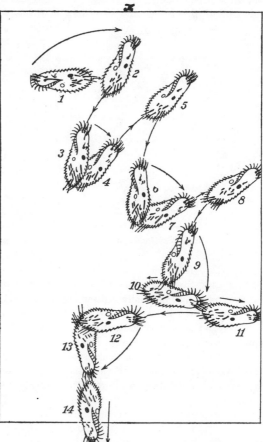

FIG. 84. — Reaction of Oxytricha to heat. The slide is heated at the end *x*. An Oxytricha in position 1 reacts as indicated by the arrows, repeatedly moving backward, turning to the right, and moving forward, thus occupying successively the positions 1–14. When it finally becomes directed away from the heat, as at 13–14, it ceases to change its direction of movement, but continues to move straight ahead, thus reaching a cooler region.

result is that all the specimens which swim in any direction but that toward the cooler water are quickly stopped and turned, while all that pass toward the cooler water continue in that direction.   Since all the specimens in the heated region are moving very rapidly and turning at very brief intervals, in a short time all will have become directed toward the cool water.   Hence soon after the water has been heated at one end of the trough, a stream of Oxytrichas will be seen passing toward the cool water.   The animals are all "oriented" in a common direction, but the orientation has taken place by *exclusion* — through the fact that movement in any other direction is at once stopped.

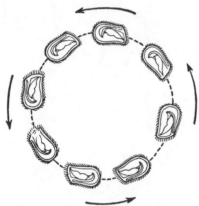

FIG. 85. — Bursaria swimming backward in a circle when heated.   Ventral view.

If one end is cooled to 10 degrees C. or below, while the other is left at the usual temperature, the Oxytrichas react in this same way in the cold region; hence they leave it, as they before left the heated region.   The reaction in the case of cold is much less striking and less complete than that produced by heat.   This is because the cold has the effect not only of producing the avoiding reaction, but also that of making the movements slower, and of finally benumbing the animals, so that they cease to move.   Thus it takes much longer for the animals to pass out of a cold region than out of a warm region, and many of them do not succeed in escaping before the cold has stopped their movements.

The reaction of Oxytricha is essentially similar to that of Paramecium.   But in Oxytricha the method of reaction is much more evident, because the movements are slower, and there is usually no revolution on the long axis.

In many other infusoria the reaction to heat and cold has been shown to take place in the same manner as in Oxytricha.   In some species the individuals show this type of behavior, yet with slight modifications that are such as to make the reaction quite ineffective, so that the animals do not escape from the heated region, and are finally killed.   This may be observed in *Bursaria truncatella.*   If one end of a trough containing specimens of Bursaria is heated, the animals respond with the avoiding reaction, as Oxytricha does.   They begin to swim backward, and at the same time to circle to the right (Fig. 85). But they do not alternate this with movement forward, as Paramecium and Oxytricha do, and they do not revolve on the long axis. Bursaria simply continues the reaction once begun, and this of course has little tendency to remove the organisms from the heated region. They circle about till they die. Among different infusoria all gradations may be found, from the ineffective reaction of Bursaria through the moderately rapid but effective behavior of Oxytricha to the quick movements of Paramecium, which can be followed only with much difficulty.

Mendelssohn (1902) has determined the optimum temperature for a considerable number of infusoria. He finds the following values: *Paramecium aurelia*, 24–28 degrees; *P. bursaria*, 23–25; *Pleuronema*, 25–27; *Colpoda*, 25–31; *Spirostomum teres*, 24–33; *Coleps*, 28–31; *Stentor*, 25–28; *Chlorogonium*, 23–30. As a rule the organism is stimulated by temperatures both above and below the optimum, so that it seeks the optimum region. But in rare cases a higher temperature acts as a stimulus, while a lower temperature does not. This is true, according to Mendelssohn, in Pleuronema.

If the entire vessel containing the infusoria is heated, or if the animals are dropped into heated water, the avoiding reaction is produced, just as when the heat is applied from one side. The animals swim backward and turn to one side. It is thus evident that there need not be differences of temperature in different parts of the body in order to produce the avoiding reaction. In the experiment just mentioned the animal "tries" swimming in many different directions, but of course does not find a direction that takes it away from the heated region.

LITERATURE VII

BEHAVIOR OF INFUSORIA IN GENERAL

A. Action systems, methods of movement and reaction: JENNINGS, 1900, 1899 *b*, 1902; PÜTTER, 1904; NAEGELI, 1860; ROTHERT, 1901.
B. Reactions to contact with solids: PÜTTER, 1900.
C. Reactions to chemicals: PFEFFER, 1884, 1888; MASSART, 1889, 1891; ROTHERT, 1901, 1903; GARREY, 1900; DALE, 1901; GREELEY, 1904; JENNINGS, 1900 *a*, 1900 *b*; JENNINGS AND MOORE, 1902.
D. Reactions to heat and cold: JENNINGS, 1904; MENDELSSOHN, 1902, 1902 *a*, 1902 *b*.

CHAPTER VIII

REACTIONS OF INFUSORIA TO LIGHT AND TO GRAVITY

1. REACTIONS TO LIGHT

LIKE Paramecium, most colorless infusoria do not react at all to light of ordinary intensity. But many species of infusoria are colored, and these commonly react in a decided manner even to the light supplied by the natural conditions of existence. Some react positively; they gather in lighted regions or swim toward the source of light. Others are negative, avoiding light regions and swimming away from the source of light. We shall take up as examples the behavior of a negative organism, *Stentor cæruleus*, and of a positive organism, *Euglena viridis*.

A. Negative Reaction to Light: Stentor cæruleus

The blue Stentor is a trumpet-shaped organism, with a circle of large adoral cilia or membranellæ surrounding the large end or peristome. This circle leads to the mouth, lying at one side of the disklike peristome. The remainder of the body is covered with finer cilia.[1] The animal is colored a deep blue. Stentor is often attached to solid objects by its pointed end or foot, but it is likewise found at times swimming freely.

We shall have occasion to study the general features of the behavior of Stentor, particularly when attached, in a later section (Chapter X). Here we need to recall only the facts that in response to strong stimulation it may contract, becoming shorter and thicker, and that when free swimming it has an avoiding reaction similar to that of Paramecium. When stimulated, it stops or swims backward, turns toward the right aboral side, and continues forward in the new direction (Fig. 76). This is the reaction produced by mechanical stimulation, by heat, and by chemical stimulation acting either on the anterior end or on the body as a whole. The results of localized stimulation have shown clearly that the anterior end or peristome is more sensitive than the remainder of the body surface.

---

[1] For a figure of another species of Stentor, resembling in essentials the present one, see Fig. 31, *b*.

The blue Stentor tends to gather in shaded regions, and when subjected to light coming from one side it moves away from the source of light. Thus, if a glass vessel containing Stentors is placed near a window, the animals swim away from the source of light, and are soon found to be collected on the side opposite the window.

How is this result brought about? Just what is the cause of the reaction to light, and what is the behavior of the Stentors in reaching the shaded regions?

In arranging experiments which shall answer these questions, let us first try the effects of sudden strong changes in the intensity of the light affecting the animals. This may be done by placing a flat-bottomed glass vessel containing many Stentors in a shallow layer of water on the stage of the microscope in a dark room. From beneath, strong light is sent directly upward through the opening of the diaphragm by means of the substage mirror, while all other light is completely excluded. In this way a circular area in the middle of the field is strongly illuminated, while the remainder of the vessel containing the Stentors is in darkness.[1]

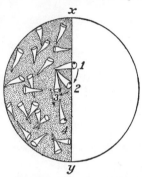

FIG. 86. — Reaction of Stentor at passing from a dark to a light region (1–4).

The Stentors in the darkness swim about in all directions, but as soon as one comes to the lighted area it at once responds by the avoiding reaction — it swims backward and turns toward the right aboral side (Fig. 86, 1–4). Thus its course is changed and it does not enter the lighted area. Since every Stentor reacts in this way, the lighted area[2] remains empty. Usually the avoiding reaction occurs as soon as the anterior end of the Stentor has reached the lighted region. In other cases the entire Stentor passes completely into the lighted area, then reacts in the usual manner, thus passing back into the dark.

---

[1] By using a projection lantern as the source of light the field of the microscope is projected on the ceiling, or, by the use of a mirror to reflect the light at right angles, on the ordinary projection screen. When thus projected, the behavior of the Stentors is observable with the greatest ease.

[2] The light is passed first through a thick layer of ice water, in order to remove the heat as far as possible. The fact that the reactions are not due to heat is shown in the following manner. Specimens of Paramecium, an organism which is more sensitive to heat than Stentor, but is nòt sensitive to light, are mingled with the Stentors. The Paramecia pass into the lighted region without hesitation, showing that this region is not heated sufficiently to affect them; the heat then cannot affect the Stentors.

Thus an area lighted from below acts in the same manner as a region containing a strong chemical. The animals keep out of both by the avoiding reaction.

We may now arrange the conditions so that the light shall come from one side, while at the same time differences in illumination shall exist in different regions. This may be done by placing the glass vessel containing the Stentors near a source of light which falls obliquely from one side, then shading a portion of the vessel with a screen. We may first so place the screen that the vessel is divided into right and left halves, at equal distances from the source of light, but one shaded, the other illuminated (Fig. 87). The Stentors are at the beginning scattered throughout the dish and are moving in all directions. Stentors in the illuminated half whose path lies in the proper direction pass into the shaded region without reaction. Since nearly all keep in motion for a long time, after an interval nearly all will have passed into the shaded half.

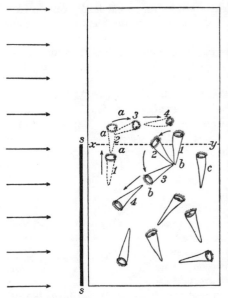

FIG. 87. — Reaction to light in Stentor. The light comes from the left, as indicated by the arrows. *S–S* is a screen shading one half the vessel, so that the line *x–y* is the boundary of the shadow. At *b*, 1–4, is shown the reaction of a Stentor on reaching this boundary line. (The dotted outline *a*, 1–4, shows the reaction that would occur if the light caused increased activity in the cilia of the side which it strikes.)

Stentors in the shaded half respond by the avoiding reaction as soon as they come to the boundary of the lighted area. That is, they swim backward and turn toward the right aboral side (Fig. 87, *b*). Thus they remain within the shaded area, and after a short time most of the Stentors in the vessel are to be found in the shaded half.

It is evident that the Stentors do not simply turn and swim parallel with the light rays from the source of light. If this were the method of reaction, a Stentor coming to the boundary *x–y*, Fig. 87, would turn and swim directly toward the side *y*. This it does not do. The direction of turning depends upon the position of the right aboral side; the animal may even turn toward the source of light. The essential point is the swimming back into the shaded region, without reference to the direction from which the light comes.

Similar phenomena are observed if the side of the vessel next to the source of light is shaded, the shadow of the screen reaching to the middle

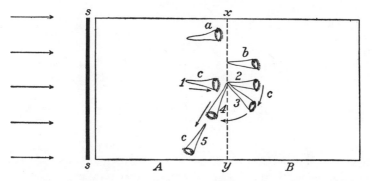

FIG. 88. — Reaction of Stentor to light when one half the vessel next the source of light is shaded by a screen *S–S* (as indicated in Fig. 89). On reaching the line *x–y*, where it would pass into the light, the animal responds as shown at *c*, 1–5.

of the vessel, so that the side farthest from the source of light is illuminated (Figs. 88 and 89). Under such circumstances the Stentors gather in the shaded area, next to the window. A specimen in the shaded area which swims toward the lighted side is of course moving when it comes

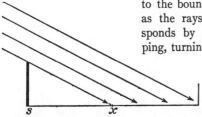

FIG. 89. — Side view of the conditions in the experiment shown in Fig. 88. The arrows show the direction of the rays of light.

to the boundary line in the same direction as the rays of light. It nevertheless responds by the avoiding reaction, — stopping, turning toward the right aboral side, and swimming back to the shadow. This often happens when the animal has completely passed the boundary and is entirely within the lighted area (Fig. 88, *b*). In passing back into the darkened area it now swims of course directly toward the source of light.

All together, then, our experiments thus far have shown that the cause of the avoiding reaction is the change from darkness to light. At every such change, Stentor responds by the avoiding reaction; that is, it tries swimming in other directions until it is no longer subjected to the light.

Let us now arrange the conditions in such a way that all parts of the vessel are equally illuminated and the light comes from one side. This may be done by placing the Stentors in a glass vessel with plane sides, at one side of the source of light, as a window or an electric lamp. Movement from one part of the vessel to another cannot cause a change from darkness to light, for all parts are equally lighted.[1] Yet the Stentors

Fig. 90. — Method of observing the reaction of Stentor to light. *A* and *B* are two electric lights, which can be extinguished or illuminated separately.

usually, after a short interval, turn and swim away from the source of light, after a time reaching that side of the vessel farthest from the lamp or window. If the animals are observed as they turn, it is found that the turning is brought about through the avoiding reaction. A short time after the light is directed upon them, they swim more slowly or cease the forward movement, and begin to swerve more strongly toward the right aboral side, thus swinging the anterior end about in a circle. The direction of movement thus becomes changed; in the new direction the animal swims forward. If its anterior end is still not directed away from the source of light, the avoiding reaction is repeated; the animal continues to try new directions till the anterior end is directed away from the lighted side. In that direction it continues to move, so that it finally comes to the side opposite the window or lamp.[2]

[1] There is of course an infinitesimal difference in the illumination of different parts of the vessel, due to the fact that one part is nearer the source of light than another. The experiment succeeds equally well when the sun is employed as the source of light, in which case the difference of illumination in different regions is practically infinitely minute. The reaction cannot be therefore conceived as due to these differences. Experiments show that the differences in illumination necessary to produce reaction are much greater than those obtaining in different parts of a vessel thus lighted from one side.

[2] The reaction may be obtained by focussing the Braus-Drüner binocular microscope on a shallow vessel of Stentors swimming about at random in a diffuse light, then allowing a strong light from an electric lamp or a brightly lighted window to fall upon them from one side. In order to have the reaction repeated many times, so as to give opportunity for careful study, the vessel containing the Stentors may be placed between two electric lights, as in Fig. 90. One of these lights can be extinguished at the same instant that the other is brought into action ; by repeating this process the direction of the light rays is repeatedly reversed. At each reversal the Stentors react in the way described in the text.

Why does the animal react in this way, even when the vessel is not divided into regions of light and darkness, but is lighted from one side? The essential problem is, Why does a specimen swimming transversely or obli ely to the direction of the light rays give the avoiding reaction and continue this until the anterior end is directed away from the source of light?

To understand this, certain facts need to be recalled. We know that the anterior end is much more sensitive than the remainder of the body. We know that an increase in illumination causes the avoiding reaction. We know that this is true even when the anterior end alone is subjected to such a change. Now, Stentor swims in a spiral of some width, so that its anterior end swings always in a circle, and is pointed successively in many different directions. If the animal is swimming transversely or obliquely to the direction of the light rays, the anterior end in one phase of the spiral path is directed more nearly toward the source of light, in another phase more nearly away from it, so as to be partly shaded, — as is illustrated for Euglena in Fig. 94. The result is, of course, that the sensitive anterior end is subjected to repeated changes in intensity of illumination; at one instant it is shaded, at the next the light shines directly upon it. As we know from other experiments, the change from light to darkness produces no reaction, while the changes from darkness to light produce the avoiding reaction. Every time, therefore, that the anterior end swings into the light, the avoiding reaction is caused; the animal therefore swings its anterior end in a large circle, trying many directions. Every time it swings its anterior end away from the source of light into the shadow of its body, on the other hand, no reaction is produced; the position thus reached is therefore retained. This process continues, the animal trying new directions every time its anterior end swings toward the light, until in a short time the anterior end must inevitably become directed away from the light. In this position the anterior end is no longer subjected to changes in illumination, for the axis of the course coincides with the axis of the light rays, and the body maintains a constant angle with the axis of the course. The amount of light received by the anterior end therefore remains constant. Hence there is no further cause for reaction, and the organism retains the position with anterior end directed away from the source of light.

Attached specimens of Stentor do not become oriented with reference to the light. They may occupy any position with reference to the direction from which the light comes, even though the light shines directly on the anterior end. We have seen previously that contact interferes with many of the reactions of organisms. But if the animals are subjected to a sudden, powerful increase in the intensity of the light falling upon them, they often contract (Mast, 1906), and later bend in various directions, till they have become accustomed to the light.

To sum up, the orientation of the free Stentor in line with the light rays, with its anterior end directed away from the source of light, is due to the fact that an increase of illumination at the sensitive anterior end induces the avoiding reaction. As a necessary result the oriented Stentor swimming in a spiral path tries new directions of movement until it finds one where such changes of illumination no longer occur. Such a direction is found only in orientation with the anterior end directed away from the source of light. From a knowledge of the spiral course and the fact that increase of illumination at the anterior end causes the avoiding reaction, this result could be predicted. The reaction to light, like that to most other stimuli, is based on the method of trial of differently directed movements, till one puts an end to the stimulation.

B. Positive Reaction to Light: Euglena viridis

Euglena is not closely related to Stentor; it is a flagellate, while Stentor is a ciliate. If we find similar principles governing the reaction to light in these widely separated organisms, it is probable that these principles are valid for the infusoria in general.

Euglena viridis (Fig. 74) is a fish-shaped green organism, often found abundantly in the water of stagnant roadside pools, giving them a green color. At the anterior end is a notch from which there extends a single long flagellum, by the lashing of which Euglena swims. Within the body are chlorophyll masses, giving the organism its green color. Near the anterior end, close to the side bearing the larger lip of the notch, — the "dorsal" side, — is a red pigment spot, usually known as the eye spot. As we have seen previously, the "action system" of Euglena resembles in essentials that of Paramecium. It swims in a spiral (Fig. 94), and to most stimuli it responds by an avoiding reaction which consists in stopping or backing, then turning more strongly than usual toward the "dorsal" side.

If the light is not too strong, Euglenæ gather in lighted areas, and when the light comes from one side, they swim toward the source of light. Thus in the culture jar the organisms are usually found on the side next the window or other source of light. In very powerful light, such as the direct rays of the sun, however, Euglena swims away from the source of light. How is this behavior brought about?

Let us first study the effect of changes in the intensity of the light. The Euglenæ are placed on a slide in a thin layer of water, and are ex-

amined with the microscope in the neighborhood of a window. Soon
all the Euglenæ are seen swimming toward the window. Now the

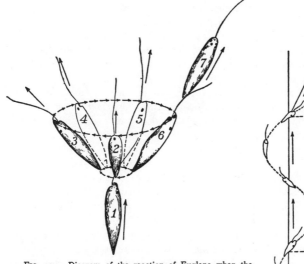

Fig. 91. — Diagram of the reaction of Euglena when the
light is decreased. The organism is swimming forward at 1;
when it reaches 2 it is shaded. It thereupon swerves toward
the dorsal side, at the same time continuing to revolve on the
long axis, so that its anterior end describes a circle, the Euglena occupying successively the positions 2–6. From any of
these it may start forward in the directions indicated by the
arrows.

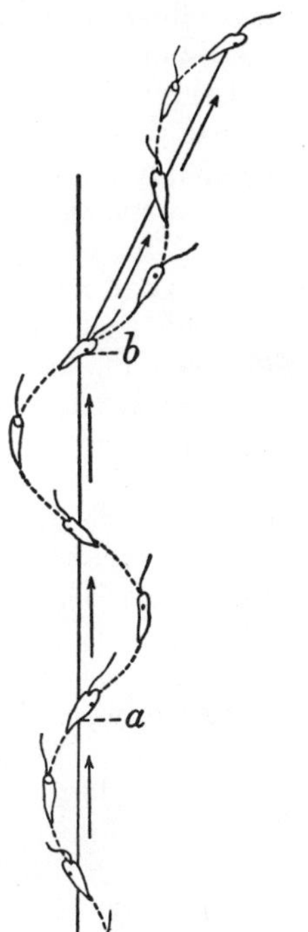

light is decreased by placing the hand or a
screen between them and the window. At once
all give the avoiding reaction; that is, they stop
or swim backward an instant, then swerve
strongly toward the dorsal side, so that the anterior end swings about a circle (Fig. 91). If the
light is decreased strongly, the anterior end describes a wide circle or may even turn through
an angle of 180 degrees, so that the direction
of movement is reversed. If only a little of the
light is cut off, the anterior end describes only a
narrow circle. The organisms soon resume the
forward movement, but now the axis of the
spiral path coincides with one of the directions
indicated by the anterior end in swinging about

Fig. 92. — Change of
direction in the spiral path
of the Euglena, as a result
of a slightly marked reaction. At *a* the illumination
is decreased, causing the
organism to swerve toward
the dorsal side; thus widening the spiral path. At *b*
the ordinary swimming in a
narrow spiral is resumed;
since at this point the organism was necessarily more
inclined to the axis of the
spiral than before the reaction, the new course lies at
an angle to the previous one.

a circle. In other words, the direction of the path has been changed (Fig. 92). The whole action may be expressed as follows: when the light is suddenly decreased, the organism tries successively many different directions, finally following one of these.

The reaction is a very sharp and striking one, and produces a most peculiar impression. At first all the Euglenæ are swimming in parallel lines toward the window. As soon as the shadow of the hand falls upon the preparation, the regularity is destroyed; every Euglena turns strongly and may appear to oscillate from side to side. This apparent oscillation is due to the swerving toward the dorsal side, combined with the revolution on the long axis. The organism swings thus first to the right, then upward, then to the left, then down, etc. (see Fig. 91).

This reaction occurs whenever the light is decreased in any way. Thus, in place of cutting off the light coming from the window, that coming from the mirror of the microscope may be decreased by closing the iris diaphragm. The Euglenæ react in the manner above described, though they soon resume their movements toward the window. Again, if the light from the window is decreased only slightly, the Euglenæ react in the manner described, thus changing their direction of movement; very soon, however, they swim again toward the window. The same reaction occurs in Euglenæ that are for any reason not swimming toward the source of light. Even if a specimen is swimming away from the window, it gives the avoiding reaction in the usual way when the light from the window is decreased.

It is clear that the reaction is due to the decrease in the intensity of light, not to a change in the direction of the light rays. In the first and second experiments mentioned in the preceding paragraph, the Euglenæ are, some time after the light is decreased, swimming in the same direction as they were before, though at the moment of decrease there is a reaction.

Engelmann (1882) tried shading parts of the body of Euglena. He found that a shadow which is cast on the body of the organism without affecting the anterior one-third produces no effect whatever. On the other hand, a shadow affecting only the anterior tip — if even only the part in front of the eye spot — causes the same reaction as shading the entire body. Thus it is clear that the anterior end is more sensitive to light than the remainder of the body. These results of Engelmann are of much importance for understanding the remainder of the reaction to light.

If Euglenæ are placed on a slide and a certain spot is lighted from below by the mirror of the microscope, a dense collection is in the course of time formed in the lighted region. Observations show that the Euglenæ in the darker portion swim about at random; many of them thus pass into the lighted region. There is no reaction at passing from the dark to the light. In the lighted region they likewise swim about in all directions. But as soon as an individual reaches the outer boundary of

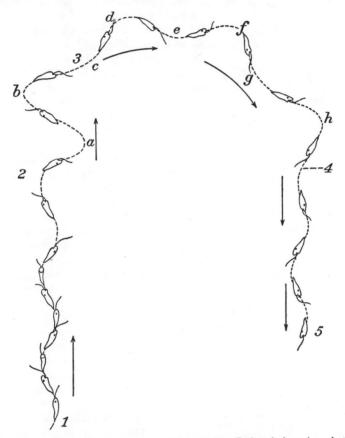

Fig. 93. — Illustration of the devious path followed by Euglena in becoming oriented when the direction of the light is reversed. From 1 to 2 the light comes from above; at 2 it is reversed. The amount of wandering (*a–h*) varies in different cases.

the lighted area, it gives the typical avoiding reaction; it backs, turns toward the dorsal side, and thus reënters the lighted area. This reaction frequently occurs as soon as the anterior tip is pushed into the shade. In other cases the reaction does not occur till the Euglena has passed completely into the dark; it then turns and passes back into the light. At the boundary of the lighted area the organism is, of course, subjected to a sudden decrease in illumination, and this, our previous experiments have shown us, is the cause of the avoiding reaction. Whenever lighted or shaded areas are open to Euglenæ, the organisms gather in the lighted areas in the way just described.

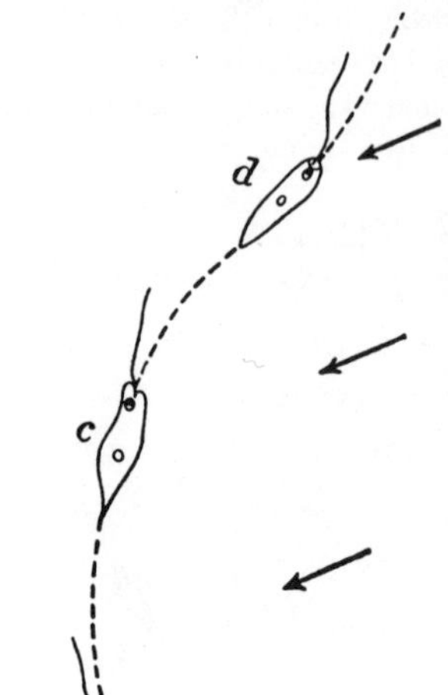

If the entire area containing the Euglenæ is illuminated from one side, the organisms swim toward the side from which the light comes. That is, they become oriented with anterior end toward the source of light. If we watch them as they become oriented, we find that the orientation takes place, as in Stentor, through the avoiding reaction. The course of events is about as follows : The Euglenæ are swimming about at random in a diffuse light, when a stronger light is allowed to fall upon them from one side. Thereupon the forward movement becomes slower and the Euglenæ begin to swerve farther than usual toward the dorsal side. Thus the spiral path becomes wider and the anterior end swings about in a larger circle and is pointed successively in many different directions. In some part of its swinging in a circle the anterior end of course becomes directed more nearly toward the light; thereupon the amount of swinging decreases, so that the Euglena tends to retain a certain position so reached. In other parts of the swinging in a circle the anterior end becomes less exposed to the light; thereupon the swaying increases, so that the organism does not retain this position, but swings to another. The result is that in its spiral course it successively swerves strongly toward the source of light, then slightly

Fig. 94. — Spiral path of Euglena. *a, b, c, d,* successive positions taken. The arrows at the right indicate the direction of an incoming force, as light, showing how the relation of the body axis and the anterior end to such a force changes continually. At *d* the body axis is nearly parallel to the lines of force, and the anterior end is directly illuminated. At *b* the axis is nearly transverse, and the sensitive anterior end is largely shaded, so as to receive but little light.

away from it, until by a continuation of this process the anterior end is directed toward the light. In this position it swims forward. The course of Euglena in becoming oriented is shown in Fig. 93.

This behavior is intelligible when we recall the effect of the spiral course in causing changes in the intensity of the light affecting the anterior end. The anterior end is, as we have seen, the part most sensitive to light; it may be compared with the eye of a higher animal. In a

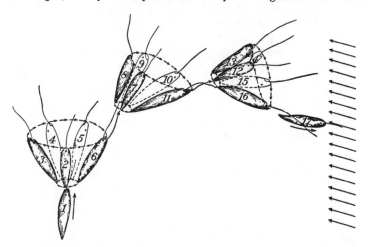

FIG. 95. — Diagram of the method by which Euglena becomes oriented with anterior end toward the source of light. At 1 the organism is swimming toward the source of light. When it reaches the position 2, the light is changed, so as to come from the direction indicated by the arrows at the right. As a consequence of the decrease of illumination thus caused, the organism swerves strongly toward the dorsal side, at the same time continuing to revolve on the long axis. It thus occupies successively the positions 2–6. In passing from 3 to 6 the illumination of the anterior end is increased, hence the swerving nearly ceases. In the next phase of the spiral therefore the organism swerves but a little, — from 7 to 8. But this movement causes the anterior end to become partly shaded, and this decrease of illumination again induces a strong swerving toward the dorsal side. Hence, in the next phase of the spiral the organism swings far, through 9 and 10, to 11. Thus it continually swerves much toward the source of light and a little away from it, till it reaches the position 16. Now it is directed toward the source of light, and such swerving as occurs in the spiral course neither increases nor decreases the illumination of the anterior end. Hence there is no further cause for reaction; the Euglena continues its usual forward movement, which now takes it toward the source of light.

Euglena swimming obliquely or transversely to the rays of light, as in Fig. 94, the illumination of the anterior end changes greatly with each turn in the spiral. At *d* the light is shining almost directly upon the anterior end, while at *b* the organism is nearly tranverse, so that the anterior end is partly shaded. The effect is like that of turning an eye first toward the sun, then away from it; though the movement is slight, the change in illumination produced is great. The variations in illumination due to the spiral course are doubtless much accentuated by the fact that one side of the anterior end bears a pigment spot, which in certain positions of the unoriented Euglena cuts off the light. A decrease of illumination causes, as we know, the avoiding reaction; the anterior end swings in a wider circle (Fig. 91). This still further increases the variations in the illumination of the anterior end. Every time the illumination is decreased, this causes the animal to swerve still more; so that its anterior end becomes pointed in many different directions, till it comes into one where such changes in illumination no longer occur. Such a position is found when the animal is swimming toward the source of light. Now the axis of the body retains always the same relation to the direction to the rays of light, so that the anterior end is not subjected to variations in intensity of illumination. There is then no further cause for reaction. Orientation is thus reached by trying various directions. This will be best understood by an examination of Fig. 95, together with its explanation.

Euglena responds most readily to light of a blue color (Engelmann, 1882). Passage from blue light to light of other colors has essentially the same effect as passage from stronger to weaker light. If the difference between the two is sufficiently decided, Euglena responds by the avoiding reaction in passing to the other color; it therefore remains in the blue. If a small spectrum is thrown on a slide containing many Euglenæ, they gather in larger numbers in the blue, — especially in the near vicinity of the Frauenhofer's line *F*.

Very strong light, such as direct sunlight, has an effect on Euglena precisely the opposite of that produced by weaker light. If the organisms are subjected suddenly to sunlight, they give the avoiding reaction. They tend therefore to gather in less lighted regions. If the sunlight falls upon them from one side, they become oriented with anterior ends away from the source of light, and swim in that direction. The orientation takes place in exactly the way described above, save that now it is the increase of light at the anterior end that causes the avoiding reaction. If a vessel is placed in such a position that the sun shines on it from one side, while the half of the vessel away from the sun is shaded with a board, the following result is produced: The Euglenæ gather in a band at the edge of the shadow (Fig. 96). They do not pass into the dark area beneath the shadow, nor do they remain in the region affected by direct sunlight, but in an area of intermediate illumination.[1]

---

[1] This experiment is due to Famintzin (1867, p. 21).

We can thus distinguish an optimum intensity of light, in which Euglena tends to remain. Movement toward either a greater or a less intensity of light causes the avoiding reaction, with its trial of different positions and directions of movements, till a position or direction is found which leads toward the optimum, or retains the optimum intensity undiminished. Or, in other words, after Euglena receives an amount of light which we might call "enough," it avoids more light, and also less light. That degree of light in which it tends to remain seems to be about the amount which is most favorable to its life activities. Euglena requires light for assimilating carbon dioxide by the

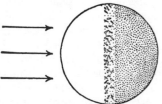

FIG. 96. — Diagram to illustrate the results of Famintzin's experiment. The light comes from the direction indicated by the arrows, while the opposite side of the vessel is shaded, as indicated by the dots. The Euglenæ gather in the intermediate region, across the middle.

aid of its chlorophyll, just as do higher plants. If confined to darkness, it soon ceases activity, contracts into a sphere, and becomes encysted. On the other hand, direct sunlight is very injurious to it; if long continued it causes the organism to fall to the bottom and die. Euglena avoids both the higher and the lower intensities that are injurious to it.

C. *Negative and Positive Reactions compared*

Thus in both negative organisms (Stentor) and positive organisms (Euglena), the determining cause of the reaction is a change in the intensity of light, and the reaction takes place by the usual method of the performance of varied movements, subjecting the animal successively to different conditions. When the sensitive anterior end is subjected alternately to light and shade, the organism "tries" other directions of movement till it finds one where such changes are not produced. In Stentor it is an increase in light that causes this reaction; in Euglena is it usually a decrease that causes the reaction, though when the light is very strong an increase may have the same effect.

D. *Reactions to Light in Other Infusoria*

The reactions of other infusoria to light are similar in character, so far as known, to those of Stentor and Euglena. In only a few other cases have details of the avoiding reaction been worked out as thoroughly as for the two species mentioned. But all that we know of the reactions of infusoria to light is consistent with the method of reaction known to exist in Stentor and Euglena; indeed, the evidence seems clear that these reactions take place in essentially the same way throughout the group. In *Cryptomonas ovata,* and less completely in the swarm spores of Chlamydomonas and Cutleria, the present writer has observed that the reaction to light is of the same character as in Euglena. We shall pass in review certain general features of the reaction in other infusoria, as described by various authors.

As we have before noted, most colorless infusoria give no indication of sensitiveness to light. But color is not absolutely necessary in order that reaction to light may occur, as is shown by the fact that Amœba reacts to light. Even in the infusoria, colorless species may react to light when such behavior is distinctly beneficial to the organism. A species of Chytridium, a colorless flagellate that is parasitic on the green organism Hæmatococcus, reacts to light in the same manner as Hæmatococcus, collecting as a rule in lighted regions, or at the side of the vessel next the source of light (Strasburger, 1878). This, of course, aids it in finding its prey, which collects in the same regions. Several other colorless infusoria that are parasitic on green flagellates have been found to react to light in the same manner as their prey. Verworn (1889, Nachschrift) found that the colorless ciliate *Pleuronema chrysalis* reacts to a sudden increase in the intensity of light by a rapid leaping movement, — evidently a strongly marked avoiding reaction. Certain colorless infusoria react, as we shall see later, to ultra-violet light.

In the green ciliate *Paramecium bursaria* the reaction to light depends, according to Engelmann (1882), on the amount of oxygen in the water. This animal contains chlorophyll, which produces oxygen in the light. When there is little oxygen in the water, the organism gathers in lighted regions, thus of course increasing its store of oxygen. When the individuals in the light come to the boundary of a dark region, "they turn around at once into the light, as if the darkness was unpleasant to them" (*l.c.,* p. 393). The response is thus clearly an avoiding reaction, like that of Stentor. When the water contains much oxygen, on the other hand, *Paramecium bursaria* avoids the light. On reaching a lighted area the animals react in the way above characterized, and return into the darkness. When they gather in light, it is especially in the red rays of the spectrum that they collect; these are the rays in which the chlorophyll is most active. When they avoid light, it is again the red rays that are most effective in producing the avoiding reaction.

Hertel (1904) found that *Paramecium bursaria, Epistylis plicatilis, Stentor polymorphus,* and Carchesium react to ultra-violet light, of

280 $\mu\mu$ wave length.   In the two species last named the chief reaction observed was a sudden contraction.   Epistylis bends to one side under the action of the light, while *Paramecium bursaria* reacts in essentially the same manner as to ordinary light, as described above.   All died quickly under the action of powerful ultra-violet light.

The flagellate swarm spores of many algæ react to light.   Their behavior in this reaction has been studied especially by Strasburger (1878).   These swarm spores (Fig. 97) usually resemble Euglena in essential features, though they may differ in form, in the number of flagella, and in other details.   They contain chlorophyll or other coloring matter, and usually a red eye spot.   The action system of the spores is similar to that of Euglena. They swim in a spiral path, keeping a certain side always toward the axis of the spiral (Naegeli, 1860, p. 96). On coming to an obstacle, they react by turning to one side (Naegeli, *l.c.*), with or without a previous start backward.   It is probable that the turning in response to a stimulus is always toward the side directed outward in the spiral path, as it is in Euglena, Chilomonas, and Cryptomonas.   The movements of the swarm spores, so far as known, exactly resemble those of the organisms just named.   It is further without doubt true that the anterior end is in the swarm spores, as in other infusoria, the most sensitive part of the body.   The swarm spores are much smaller than Euglena, so that the details of the behavior are less easy to determine.

Fig. 97.— Examples of swarm spores, after Schenck.   *a*, Hæmatococcus pluvialis; *b*, Ulothrix zonata; *c*, Botrydium granulatum, gamete; *d*, Cladophora glomerata; *e*, Œdogonium.

Strasburger found that when the light is weak, all the colored swarm spores[1] swim toward the lighted side of a drop (positive reaction). When the light is strong, some swim away from the lighted side (negative reaction).   If different parts of a drop or a vessel are unequally illuminated, the swarm spores gather in the lighted region.   The phenomena are thus in general similar to those found in Euglena.   There are certain variations among the different swarm spores.   Thus, Strasburger found that Botrydium and Cryptomonas are positive even in the strongest light, while in a weak light Cryptomonas is indifferent.   But in most species there is, as in Euglena, an optimum.   In light below

<hr />

[1] Strasburger studied the swarm spores of Hæmatococcus lacustris, Ulothrix, Chætomorpha, Ulva, Botrydium, Bryopsis, Œdogonium, Vaucheria, and Scytosiphon, as well as the flagellate Cryptomonas (called Chilomonas by Strasburger), and the colorless swarm spores of Chytridium and Saprolegnia.

the optimum they are positive; in light above the optimum they are negative.

Strasburger did not determine the precise movements of the organisms in the reaction to light. That is, he did not determine toward which side they turn in becoming oriented. But in other respects his account is so excellent that, with the fuller results on Euglena as a key, it is not difficult to analyze out the precise factors in the behavior.

If the light affecting the organisms is suddenly decreased in intensity, Strasburger found that the swarm spores (Botrydium and Ulva) suddenly turn toward one side (*l.c.*, p. 25). In Bryopsis this reaction was produced also when the light was suddenly increased. In all the swarm spores it was evident that as soon as the light was decreased by the interposition of a screen the path became more crooked (*l.c.*, p. 27). In other words, the spiral became wider, owing to the increased swerving toward a certain side. In these respects the swarm spores precisely resemble Euglena. It is clear that they react to a sudden decrease in illumination by an avoiding reaction, which consists in turning more or less strongly toward a certain side, with or without a cessation of the revolution on the long axis; in this way the direction of progress is changed.

As would be expected from this method of response, the organisms react at passing from a light to a dark region. If a ring is placed over the drop containing the organisms, so that only a central circle is illuminated, the positive organisms gather in the illuminated circle (Fig. 98, *A*). Here they swim toward the window from which the light comes, but on reaching the edge of the shadow, they turn back into the lighted region (*l.c.*, p. 28). Often the organism passes completely into the shadow before reacting, then it turns and swims back into the light. Thus it does not react till a short time after the moment of change. If a narrow band of shadow passes across the middle of the drop, transversely to the direction from which the light is coming, this usually does not stop the organisms, because of this interval of time which elapses before their reaction; before they begin to react they have passed completely across the band into the lighted region beyond. But if a larger vessel is used and a broader transverse band of shadow passes across it (Fig. 98, *B*), this does stop the organisms. They gather on the edge of the shadow without passing across it. In many other ways Strasburger shows that when the area containing the swarm spores is unequally illuminated, the positive organisms collect in the more illumined region. In this they precisely resemble Euglena, as Strasburger himself noted. The behavior is of course a direct result of the production of the avoiding reaction by a decrease in light.

If the experiments were made with swarm spores that were negative to the intensity of light used, they gathered of course in the shadow instead of in the light. If a board was placed across the middle of the vessel from right to left, such swarm spores formed a collection in the partly shaded region at the edge of the board (as in Fig. 96), where they found the optimum degree of illumination. They were repelled both by the strong light and by the deep shadow.

Thus it is clear that in the swarm spores, as in Euglena and Stentor, a change in the intensity of illumination produces reaction. But a certain amount of change is required before any effect is produced. If the intensity of illumination changes only very gradually from one

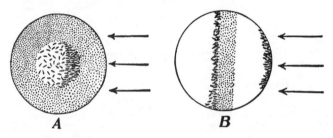

Fig. 98. — Diagrams to illustrate the results of some of Strasburger's experiments with positive swarm spores (original). *A*, the margins of the drop are shaded (as indicated by the dots); the organisms gather in the lighted centre. *B*, a broad band of shadow lies transversely across the drop; the organisms swim toward the light, but are stopped by the shadow. Thus two groups are formed, one at the side of the drop next the light, the other in a corresponding position at the edge of the shadow.

region to another, the difference in intensity between succeeding points is insufficient to cause reaction. Hence under these circumstances the organisms remain scattered and move about without reaction. Strasburger showed this in the following way. He used a hollow wedge-shaped prism, 20 cm. long, filled with a partly opaque solution of humic acid in ammonia. Through this the light was passed. At the thin end of the wedge nearly all the light was transmitted; at the thick end little or none, and there was a gradual transition from light to dark between the two ends. This prism was placed over the drop containing the swarm spores, and the light was allowed to fall directly from above (Fig. 99, *X*). The drop being very small in comparison to the length of the wedge-shaped prism, there was of course but little difference in the illumination of its two sides, and the transition from one to the other was very gradual. Under these conditions the swarm spores remained scattered throughout the drop. The change in passing from one region to another was not sufficiently marked to cause reaction.[1]

When the entire area is equally lighted and the light comes from one side, the positive swarm spores swim toward the source of light. If the light is made strong, most species swim away from its source. In this behavior the agreement with Euglena is complete. The orien-

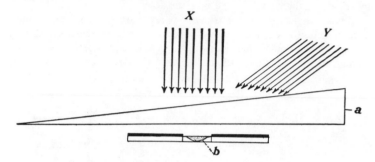

FIG. 99. — Diagram of the conditions in Strasburger's experiments with a wedge-shaped prism, constructed from the data furnished by Strasburger. *a*, prism 20 cm. in length, filled with a translucent fluid. *b*, hanging drop containing the swarm spores. *X*, rays of light coming from above, as in the first experiments. *Y*, rays coming obliquely from the thicker end of the wedge, as in the second set of experiments. The figure is one half natural size.

tation takes place gradually, by a series of trials, as in Euglena. Strasburger paid no special attention to this point, but the present writer has observed that this is true in Cryptomonas, Chlamydomonas, and the swarm spores of the marine alga Cutleria, as well as in Euglena, and Strasburger (1878, p. 24) notes incidentally that it is true in Hæmatococcus.[2]

It seems clear, then, that the reaction takes place in the same manner

---

[1] It is curious that Strasburger drew from this experiment the erroneous conclusion that variations in the intensity to light play no part in the reaction. The only essential difference between this experiment and the previous ones (Fig. 98) is that in the previous experiments the change of illumination in passing from one region to another is sudden and pronounced, while in the present experiments it is slow and gradual. The logical conclusion is that the lack of reaction in the present experiment is due to the slightness of the change in passing from one part of the preparation to another. When we consider that the prism was 20 cm. in length, and was placed over a mere drop, it is evident that the difference in illumination in different parts of the drop was excessively small. We know that for the effective action of all stimuli a certain threshold amount of change is necessary, so that the results are exactly what might be anticipated. Our account of Euglena shows beyond doubt that a change in intensity of illumination does cause reaction. Strasburger himself (*l.c.*, p. 25) observed the same fact in swarm spores, though he paid little heed to this observation in the remainder of the work.

[2] He says that when the direction of the light is changed, the swarm spores become oriented " Nach verschiedenen Schwankungen."

in the swarm spores as in Euglena. As set forth on page 139, the movement toward or from the source of light, in a field of which all parts are equally lighted, is due to the fact that in the unoriented individuals the sensitive anterior end is subjected to frequent changes in the intensity of illumination. It is first directly lighted, then shaded. These changes induce reaction. By the method of trial the organism then comes into a position such that these changes cease. Such a position is found only in orientation. All these relations evidently hold equally well for the swarm spores; for details the reader may refer to the account of the behavior of Euglena.

What happens if the field containing the organism is lighted from one side, and there are at the same time variations in the intensity of light in different parts of the field? Strasburger devised certain experiments to answer this question. These experiments have become celebrated, and an immense amount of ingenuity has been expended in endeavoring to interpret them in one way or another. Strasburger's experiments involved the use of the wedge-shaped prism shown in Fig. 99. This prism was placed over the drop containing the swarm spores, in such a way that the light came obliquely from the direction of the thick end of the wedge, as in Fig. 99, Y. Now the intensity of illumination is greater on the side farthest away from the source of light, and decreases as we pass toward the source of light. Will the positive swarm spores move toward the source of light, and thus into a region of less illumination, or will they rather move into the region of greater illumination, and thus away from the source of light?

Strasburger found that the positive swarm spores move toward the source of light, and hence into the region of less illumination. It is extraordinary that this result should have occasioned the surprise and comment which have been bestowed upon it. Strasburger's previous experiment with perpendicular light (Fig. 99, X) had shown that the variations in intensity of illumination in different parts of a drop under this prism were too slight to cause reaction, the organisms remaining scattered throughout the drop. Evidently so far as the organisms were concerned these slight variations did not exist; they were not perceived. Therefore, when the light comes from one side, the organisms react exactly as they do when such variations do not exist. They swim toward the source of light for the same reason that they do when the prism is not present. The experiment consists essentially in making the differences in the intensity in neighboring regions so slight that they are unperceived. We need not, therefore, be surprised that the organisms fail to react to them.

The experiments show, what they were designed to show, that the reason for swimming toward the source of light is not the progression into a lighter region. But they do not indicate in the least that the reactions are not due to changes in intensity of illumination. So long as turning the sensitive anterior end away from the source of light causes a greater decrease in its illumination than does movement into the slightly less illuminated region, the organism will move toward the source of light. If the difference in intensity of light in different parts of the drop were increased till the change in illumination due to progression is greater than the change due to swinging the anterior end away from the source of light, then the positive organisms would gather in the more illuminated regions. This is the condition of affairs in the experiment shown in Fig. 98.

In the swarm spores, as in Euglena, the positive reaction usually changes to a negative one when the light is much increased. We can thus distinguish an optimum intensity of light, to which the organisms may be said to be attuned. Either increase or decrease from the optimum causes the avoiding reaction. Often the organisms are positive when placed at some distance from a window, but become negative when brought nearer. There is much variation among different species, and even among different individuals of the same species, as to the amount of light that causes this change from positive to negative. Sometimes, with a given intensity of light, half the individuals of Ulothrix are found to be positive, the other half negative (Strasburger, *l.c.* p. 17). The same individual is seen at times to be at first positive, later negative. Some of the influences which modify the reaction to light are known. Certain swarm spores are attuned to a stronger light in the early stages of development than in the later stages. Specimens grown in shaded regions seem attuned to less intense light than those living in well-lighted cultures. That is, the organisms are attuned more nearly to the light to which they are accustomed. But subjection to darkness sometimes causes negative organisms to become for a short time positive. Hæmatococcus is negative in a certain intensity of light, gathering at the negative side of the drop. Now the preparation is covered and left in the dark for a few minutes, then the cover is removed. At once the Hæmatococci leave the negative side and swim toward the light for a short distance. But this lasts only a moment. After reaching the middle of the drop, they swim back again to the negative side. An increase of temperature increases the tendency to a positive reaction to strong light; a decrease of temperature has the opposite effect. Lack of oxygen increases the tendency to a positive reaction. This is accounted for by the fact that the green organisms produce oxygen in the light.

A change in the intensity of light does not as a rule produce its characteristic effect immediately, but requires a definite interval of time. When the light is faint and the organisms are swimming toward it, if the light is suddenly increased to an intensity to which they are negative, the swarm spores continue to swim toward it for some time. The interval may amount to as much as half a minute. At the end of this period they turn and swim away from the light. Again, when the organisms are swimming away from a strong light, a sudden decrease in illumination causes them to become positive only after some seconds. But in some species there is no such delay in the effects of a change of illumination.

To sum up, we find that the reactions to light occur in the infusoria in essentially the same way as do the reactions to most other stimuli, through the avoiding reaction; that is, by the method of trying movements in different directions. The cause of reaction is a change in the intensity of light, — primarily that affecting the sensitive anterior end. Changes in intensity may be produced either (1) by the progression of the organism into a region of greater or less illumination, or (2) by the swinging of the sensitive anterior end toward or away from the source of light, so that it is shaded at one moment and strongly lighted the next. Usually these two classes of changes work in unison; when they are opposed, the organism reacts in accordance with that which is stronger. When the second class of changes above mentioned is the determining factor, the organism continues to react by trial till these changes cease. This results in producing orien ation with anterior end directed toward or away from the source of light. In strong light the effect of an increase or decrease of intensity is often the reverse of that observed in weak light.

2. REACTION TO GRAVITY AND TO CENTRIFUGAL FORCE

A considerable number of infusoria have been found to react to gravity in much the same way as does Paramecium (Jensen, 1893). As a rule, when placed in vertical tubes, they rise to the upper end. The following infusoria have been found to behave in this way: Among the flagellates: Euglena, Chlamydomonas, Hæmatococcus, Polytoma, Chromulina; among the ciliates: *Paramecium bursaria*, Urostyla. *Spirostomum ambiguum* takes at times a vertical position in the water a short distance above the bottom, with anterior end upward. Under these circumstances it is anchored by an invisible thread of mucus, as may be observed by passing a glass rod between it and the bottom (Fig. 82). The stationary position oriented with reference to gravity seems to be the result of a slight activity of the cilia, tending to cause movement upward, combined with the downward pull of the thread at the posterior end. Jensen found that *Colpoda cucullus*, *Colpidium colpoda*, *Ophryoglena flava*, and *Coleps hirtus* showed no clear reaction to gravity.

There is reason to suppose that reaction to gravity, where it occurs, is brought about in the same manner as in Paramecium. The details given in the account of Paramecium therefore need not be repeated here.

As a general rule the reaction to gravity is easily masked by reactions to other stimuli. It is shown in a marked way only when other effective stimuli are largely absent, and in cases of conflict with other reactions, it is usually the reaction to gravity that gives way. In some cases the action of other agents causes the reaction to gravity to become reversed, just as in Paramecium. Massart (1891 *a*) finds that this effect is produced in Chromulina by lowering the temperature to 5–7 degrees C.

A number of infusoria are known to react to centrifugal force in the same way as to gravity. They swim in the opposite direction from that in which the centrifugal force tends to carry them, just as Paramecium does. It is probable that in all cases centrifugal force could be substituted for gravity without essential alteration of the reactions. Schwarz (1884) found that Euglena and Chlamydomonas react to centrifugal force when it is equal to about $\frac{1}{2}$ the force of gravity, and continues the reaction till the centrifugal force is about $8\frac{1}{2}$ times gravity. Above this they are passively carried in the direction of action of the centrifugal force.

LITERATURE VIII

Behavior of Infusoria in General

A. Reactions to light: JENNINGS, 1904 *a*; STRASBURGER, 1878; ENGELMANN, 1882; MAST, 1906; FAMINTZIN, 1867; HERTEL, 1904; HOLT AND LEE, 1901; HOLMES, 1903; OLTMANNS, 1892.

B. Reactions to gravity: JENSEN, 1893; MASSART, 1891 *a*; SCHWARZ, 1884.

CHAPTER IX

REACTIONS OF INFUSORIA TO THE ELECTRIC CURRENT

1. DIVERSE REACTIONS OF DIFFERENT SPECIES OF INFUSORIA

THERE is great diversity in the gross features of the behavior of different infusoria under the action of the continuous electric current. Some swim, like Paramecium, to the cathode; some to the anode; some take a transverse position; some swim to one electrode in a weak current, to the other in a strong current; some, finally, do not react at all. Yet, in spite of this great diversity, we find the fundamental effect of the current on the motor organs to be almost identically the same throughout the series. In all infusoria having cilia in different regions of the body, the cilia of the cathode region strike forward, those of the anode region backward, just as we have seen to be the case in Paramecium. How the organisms move under these conditions depends on the peculiarities of structure and of the action system of the infusorian in question. We shall review here the different types of behavior under the action of electricity, endeavoring to show how each is brought about.

A. Reaction to Induction Shocks

We may again take up, first, the reactions to single induction shocks, studied by Roesle (1902) and Statkewitsch (1903). In all infusoria investigated the reaction to moderately strong induction shocks is essentially similar to the reaction to other stimuli. The animal usually responds to the shock by the avoiding reaction, which begins with a reversal of the cilia in that part of the body directed toward the anode. In some cases, however, the induction shock causes, like a weak mechanical stimulus, a mere movement forward (Roesle, 1902). If the shock is a powerful one, the body may contract in the anode region, or, in the case of very contractile species, such as Lacrymaria and Spirostomum, the entire body may contract. Reaction takes place most readily as a rule when the sensitive anterior end is directed toward the anode, or especially, according to Roesle, when the mouth opening is precisely directed toward the anode. When the animal is in the transverse position, it is least affected by the induction shock, and in many cases it is less affected when the aboral side is directed toward the anode, than in the opposite position.

B. *Reaction to the Constant Current*

Under the action of the constant current there are a few infusoria which do not react at all, so far as known. This is the case, for example, with *Euglena viridis*. Even with powerful currents it shows no reaction.

The larger number of free ciliate infusoria swim under the influence of the constant current to the cathode, while a few swim to the anode or take a transverse position. A considerable number of flagellates swim to the anode, though some swim to the cathode.

The reaction of the flagellates has been little studied in any precise way. Owing to their minuteness it is usually very difficult to determine their exact movements. According to Verworn (1889 *b*), Trachelomonas and Peridinium swim to the cathode; *Polytomella uvella*, *Cryptomonas ovata*, and *Chilomonas paramecium* to the anode. In stronger currents some of the individuals of Chilomonas swim to the cathode. The reason for the diversity in the reactions of different flagellates has not been determined. In the case of Trachelomonas, according to Verworn, the flagellum is strongly stimulated when directed toward the anode. The result is that it strikes strongly in such a way as to turn the organism around, — doubtless by a typical avoiding reaction similar to that described on page 111 for Chilomonas. On reaching a position with anterior end directed to the cathode, it is no longer effectively stimulated; it therefore continues to move toward the cathode. In Chilomonas the orientation to the electric current is known to be brought about through the typical avoiding reaction. That is, the animal turns toward the smaller lip (Fig. 72, *y*), till orientation is attained (Pearl, 1900). Since in the flagellates the motor organs are all at one end, all bear the same relation to cathode or anode, so that we cannot expect any opposition in the action of the different flagella, such as we find in the cilia of different regions in Paramecium. There is thus no sign in the flagellates of that lack of coördination or of an apparent attempt to move in two directions at once, which we find in Paramecium.

Among the Ciliata, most species, under usual conditions, turn the anterior end to the cathode and move toward that electrode. But Opalina moves, usually, to the anode, and Spirostomum as a rule takes a transverse position. Certain variations in the reactions under different conditions will be brought out later.

Among the organisms which pass to the cathode, the manner in which orientation takes place varies in different species. The direct effect of the current is, as in Paramecium, to cause the cilia on the cathode side to strike forward, while those on the anode side strike backward. This would result, taken by itself, in turning the animals directly, by the shortest path, toward the cathode. But in many species, as our study of the reactions to other stimuli has shown us, there is a strong tendency to turn toward one side rather than the other, usually toward the aboral side, — that opposite the peristome. The cilia of the peristome are usually more powerful than those of the remainder of the body, so that the direction in which the animal turns depends largely upon the way these cilia strike. When the peristomal cilia strike strongly backward, the organism turns toward the opposite or aboral side, with little regard to the beat of the remainder of the cilia. These peristomal cilia are as a rule limited to one of the four quarters into which the surface of the body can be divided, as illustrated in Fig. 100. They, of course, beat backward when either the end bearing them, or the side bearing them, is directed toward the anode (1–3, Fig. 100), so that in these positions the animal turns toward the aboral side in order to reach the position of orientation, just as it does in response to other stimuli. It is only when the side bearing the peristome is directed toward the cathode that these cilia beat forward, and hence tend

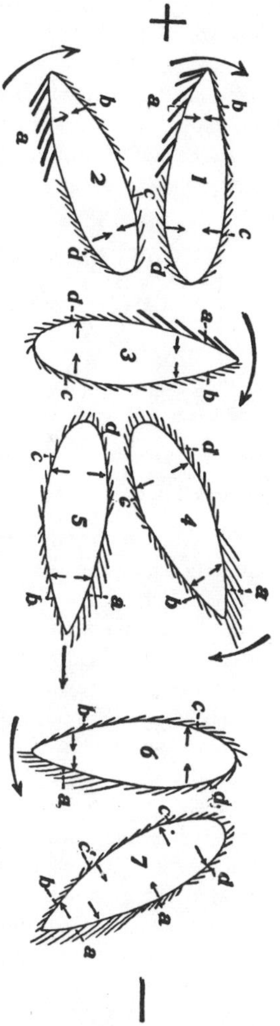

FIG. 100. — Diagrams to illustrate the movements of infusoria under the action of the electric current when the peristomal cilia (*a*) are strongly developed. The small arrows within the outlines show the directions in which the adjacent cilia tend to turn the organism; the large external arrows show the actual direction of turning. In positions 1, 2, and 3, the organism turns toward the side opposite the peristome (the aboral side *b*), urged thereto by the powerful backward beat of the peristomal cilia *a*. In position 4 the turning is still toward the aboral side, because *b* and *d* are more numerous than *a* and *c*. In positions 5 and 7 there is a condition approaching equilibrium. In position 6 the turning is toward the oral or peristomal side *a*, — all the cilia concurring to produce this result. (The peristomal cilia when beating backward are more powerful than the others, and are therefore represented by heavy lines.)

to turn the organism toward the oral or peristomal side (Fig. 100, 6). Under these circumstances, another principle requires consideration. Normally the peristomal cilia strike backward. When they strike forward, they develop much less energy, — less turning power, — than when they strike backward. Therefore, when in the position shown at 6, Fig. 100, the turning is much less rapid than in other positions, and may easily be prevented by a slight resistance. These relations will be understood by an examination of the diagram (Fig. 100).

In Paramecium, as we have seen, the same condition of affairs is exemplified to a certain degree, so that the organism turns toward the oral side in all positions save from d to f, Fig. 63. In the Hypotricha (Oxytricha and Stylonychia) this condition is most typically exemplified. A large share of the body cilia are absent or have taken the function of legs, while the peristomal cilia are very powerful. In almost all cases these organisms become oriented to the electric current by turning toward the aboral (right) side. It is only when the peristomal cilia are squarely facing the cathode (Fig. 100, 6) that the animal may turn toward the oral (left) side. In this position the peristomal cilia beat forward, and all the cilia of the body aid in turning the organism toward the oral side. On reaching a position with anterior end directed to the cathode the peristomal cilia are directed forward, but their beating has become so weak as to be almost without effect. The animal, therefore, retains this position.

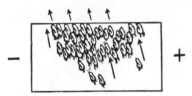

FIG. 101. — Transverse (or oblique) position and movement of Oxytricha under the action of the electric current, when the animals are in contact with the substratum. The peristome is directed toward the cathode.

When specimens of the Hypotricha are in contact with a surface, as is usually the case, the forward beat of the peristomal cilia is often so weak and ineffective in the transverse or oblique position (Fig. 100, 6) that it does not turn the animal against the resistance offered by the attachment of the ventral cilia. Such specimens, therefore, remain in the transverse or oblique position, the anterior end usually slightly inclined toward the cathode, as in Fig. 101. In this position they run forward. When the current is reversed, so that the anode lies next the peristome, the powerful peristomal cilia strike backward. The animals, therefore, turn toward the aboral (right) side till they have again become nearly transverse to the current. They then move forward in the direction so indicated. Similar phenomena are at times to be observed in other ciliates, not belonging to the Hypotricha. This is true, as we have seen, even for Paramecium.

Thus we can distinguish two factors in the turning produced by the electric current. The first is a tendency to turn directly toward the cathode, the second a tendency to turn toward a structurally defined side, — usually the aboral side. The conflict of these tendencies when the animal is in certain positions, and their mutual reënforcement in other positions, often give rise to peculiar and complicated phenomena.

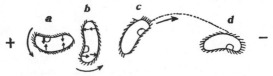

FIG. 102. — Diagrams of the reaction of Colpidium to the electric current when in various positions. Based on the descriptions and figures given by Pearl (1900).

Thus, in Colpidium, as described by Pearl (1900), we have the following different methods of reacting to the electric current. (It should be premised that Colpidium tends under ordinary conditions to turn toward the aboral side.)

(1) When the anterior end is directed approximately toward the anode, or in any position in which the aboral side is nearest the cathode,

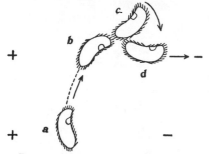

FIG. 103. — Diagram of one method by which Colpidium reacts to the electric current when transverse with the oral side to the cathode. Constructed from data given by Pearl (1900).

Colpidium turns toward the aboral side (Fig. 102, *a*, *b*), till the anterior end is directed toward the cathode. Both the factors mentioned above coöperate to produce this result.

(2) When the animal is nearly transverse, or is oblique, with the oral side next to the cathode, it usually swims slowly forward, and at the same time gradually turns toward the *oral* side till it becomes oriented (Fig. 102, *c–d*). The two tendencies mentioned above oppose each other in this case, and the first one overcomes the second.

(3) But in other cases when the animal is in the position described in the last paragraph (Fig. 103, *a*) it reacts in another way. It moves forward, slowly turning toward the oral side (Fig. 103, *a–b*), then turns on its long axis (*b-c*) (as happens in ordinary locomotion). This brings the aboral side next to the cathode (*c*). Now the animal turns suddenly toward the aboral side till the anterior end is directed toward the cathode (Fig. 103, *d*). In this case, then, the two tendencies mentioned above oppose each other till the revolution on the long axis occurs, then they reënforce each other.

(4) If Colpidium is squarely transverse, with oral side to the cathode (Fig. 104, 1), or especially if the anterior end is a little inclined toward

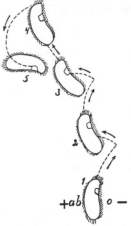

the anode, the organism often starts transversely to the current. Suddenly it jerks its body a little toward the aboral side (Fig. 104, 1–2), then moves forward again. Again it jerks toward the aboral side (3), again moves forward, and repeats this behavior until the anterior end is directed toward the anode. Then it turns steadily toward the aboral side till the anterior end is directed toward the cathode (Fig. 104, 4–5). In this behavior the two tendencies mentioned oppose each other, as in case 2, but the second one prevails over the first.

Various combinations of these different reaction types may occur, making the behavior of Colpidium under the electric current very complicated. Similarly varied behavior is often observed in other infusoria, through the action of similar causes.

FIG. 104. — Another method of reaction to the electric current in Colpidium. After Pearl (1900).

In such infusoria as Stentor, where the peristomal cilia form a circle surrounding the anterior end, there is no reason for such a conflict of tendencies. The peristomal cilia are divided by an electric current coming from one side, so that the animal turns directly away from the side on which these cilia strike backward (Fig. 105). If the anterior end is directed toward the anode at the beginning, the animal doubtless turns as usual toward the right aboral side. In other positions the usual method of turning seems to have no effect on the reactions. In Vorticella and other infusorians resembling Stentor in the distribution of the cilia, the orientation to the current would doubtless take place in the same direct manner, though this has never been determined.

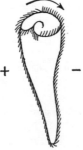

FIG. 105. — Reaction of Stentor when transverse to the current. It turns directly toward the cathode, all the cilia concurring to produce this effect.

In Spirostomum and Opalina, the conflict of the two tendencies mentioned above leads to certain very remarkable and complex results. Under usual conditions Spirostomum takes a transverse position in the electric current, while Opalina swims to the anode. The gross features of the behavior thus differ markedly from those shown by most other infusoria.

But Wallengren has shown that the effect of the current is in these infusoria of essentially the same character as in others. Let us examine briefly the facts as set forth by Wallengren (1902 and 1903).

Spirostomum (Fig. 106) is a very long, slender infusorian, easily bent in any direction, and very contractile. The peristomal cilia are very large and numerous, extending from the anterior end along one side to a point behind the middle. Whether striking forward or backward, the beating of these cilia is decidedly more effective than that of the cilia on the opposite side of the body. It is to this fact, taken in connection with the slenderness and suppleness of the body, that most of the peculiarities in the reaction of Spirostomum to the electric current are due.

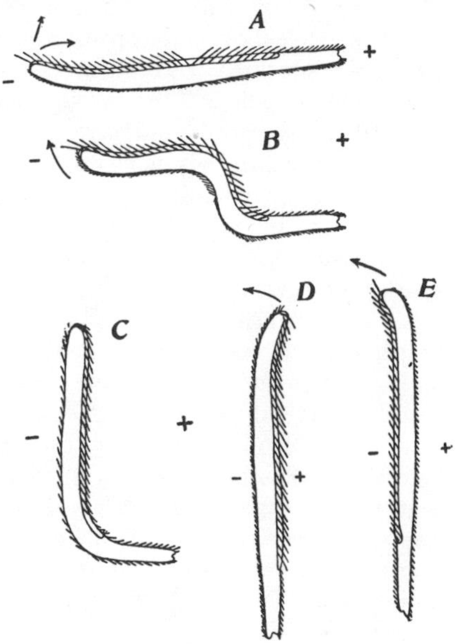

In a very weak current, such as does not cause contraction of the body, Spirostomum swims to the cathode. The cilia on the anodic part of the body strike backward, those in the cathodic region forward, just as happens in Paramecium. As a result, the animal takes a position with anterior end directed to the cathode, in essentially the

FIG. 106. — Diagrams illustrating reaction of Spirostomum to the electric current. *A*, *B*, *D*, and *E* after Wallengren (1903).

same manner as does Paramecium, — usually turning to the aboral side, but in certain cases toward the oral side. When the anterior end is directed toward the cathode, the cilia on the cathodic half of the body are partly directed forward, but with the weak current most of them still strike most strongly backward. Those of the anode half of course strike backward, so that the general result is to drive the animal forward to the cathode. Sometimes Spirostomum under these conditions comes against the bottom or other solid object; it may then nearly or quite cease to move forward. The facts thus far are quite parallel to those observed in Paramecium.

As the electric current is made stronger, the cilia on the cathodic half of the body strike more powerfully forward, and at a certain strength their effect, tending to drive the animal backward, becomes about equal to that of the anodic cilia, tending to drive it forward. The result is that the animals move neither forward nor backward, or only very slowly in one direction or the other. They thus sink to the bottom before much progress has been made. Now, if in this position the anterior end is directed toward the cathode (Fig. 106, A), of course the cilia of the anterior (cathodic) half of the body tend to push the animal backward, while those of the opposite half tend to push it forward. This push in opposite directions bends the supple body near its middle. Moreover, in the cathodic half the peristomal cilia have a more powerful forward stroke than do the ordinary cilia on the opposite side, hence the anterior half of the body tends to bend toward the peristomal or oral side. The general result is that the animal is bent into the position shown in Fig. 106, B. The bending of the anterior part of the body toward the oral side continues, until this part of the body becomes transverse to the current (Fig. 106, C). The body may now become completely straightened (Fig. 106, D), or it may not. But in either case the peristome is now turned toward the anode. The powerful peristomal cilia therefore strike backward, causing the anterior end to swing toward the aboral side, directing it again toward the cathode, as indicated by the arrow in D. On becoming directed toward the cathode, the original condition (Fig. 106, A) is restored. The animal therefore again takes the positions B, C, and D. It thus continues to squirm from side to side. But during its movements Spirostomum, like Paramecium, frequently revolves on its long axis. This often happens when in the position shown in Fig. 106, C, so that the animal becomes placed transversely to the current, with peristome to the cathode (Fig. 106, E). In this position the peristomal cilia are directed forward and have therefore comparatively little motor effect. If at the same time the animal comes in contact with the bottom, the contact reaction may overcome for a time this slight motor effect, so that the animal lies nearly quiet, in the transverse position. If now the current is reversed, so that the peristome is at the anode (Fig. 106, D), the animal at once swings again toward the aboral side. Even if the current is not reversed, the animal usually does not remain long in the position shown at E. The peristomal cilia being more effective than the opposing ones, gradually swing the anterior half toward the oral side. Soon a bending takes place again, as in B, and the organism is forced to squirm about from side to side, as before.

Thus Spirostomum finds in a strong current no position of equilibrium, because the peristomal cilia have always a more powerful effect than the opposing ones, and because the opposed action of the cilia on the anodic and cathodic halves of the body soon bends the slender body. It thus squirms about from one side of the transverse position to the other, taking many shapes besides those figured. It remains quiet only for certain periods in the transverse position with peristome to the cathode, when it is in contact with a surface: this is a result of the interference of the contact reaction with the reaction to the electric current. Under the action of the current alone, the reaction of Spirostomum does not tend to bring it to a position where it is not effectively stimulated, for no such position exists. In this respect the electric stimulus shows again a marked contrast with other stimuli.

In *Opalina ranarum* the first marked effect of the electric current is to cause the animals to swim to the anode instead of to the cathode. Its reaction seems thus in striking contrast with that of other ciliate infusoria. We must examine the reaction in Opalina, following Wallengren (1902), to see how this result is brought about.

Opalina is a large, flat, disk-shaped, parasitic infusorian, living in the large intestine of the frog. For experimental work it is examined in physiological salt solution, as it soon dies in water. There is no mouth, since food is obtained by absorption over the entire body surface. The body is closely set with fine cilia. The anterior end of the body is more pointed than the posterior. From the anterior portion there extends backward at one edge a convex region, ending at a sort of notch in the middle of the body (Fig. 107, *x*). This convex region is set with cilia having, as we shall see, a somewhat different function from those of the remainder of the body. The side bearing this convexity is usually known as the right side.

Opalina swims with anterior end in front, at the same time usually revolving on its long axis. When stimulated by contact with a solid, or in other ways, it turns toward the side bearing the convexity — the right side. Observation shows that this movement is due to the fact that the cilia on the convexity of the right side now strike forward instead of backward, thus necessarily turning the animal toward the side bearing them. In this way the typical avoiding reaction of Opalina is produced.

If a preparation of Opalina in physiological salt solution is subjected to the action of a weak electric current, the animals swim to the anode. Examining the individuals, it is found that the cilia on the anode half of the body strike backward, those on the cathode half forward, exactly as in Paramecium. Why then does Opalina swim to the anode instead of to the cathode?

The secret of this difference lies in the following facts. The cilia of the convexity of the right side (Fig. 107, *x*) are very easily reversed by a weak current. The cilia of the opposite side, on the other hand, are little affected by a weak current. Their usual backward stroke is decreased in power, and doubtless some of the cilia are reversed, but the general effect of their action is still to drive the animal forward. Let us suppose that the Opalina is at first transverse to the electric current, with right side to the cathode, as in Fig. 107, 1. As soon as the current

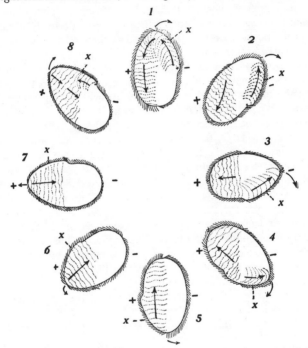

Fig. 107. — Diagrams of the movements of the cilia, and of the direction of turning, in the reaction of Opalina to the electric current. After Wallengren (1902).

begins to act, the cilia of the right (cathodic) side become directed forward, while those of the left (anodic) side remain directed backward. The result is of course to turn the animal to the right, toward the cathode. Thus the specimen passes through the position shown in Fig. 107, 2, and comes into a position with the anterior end directed toward the cathode (3). The cilia of the anterior part of the body are now directed partly forward, those of the posterior half backward. In this position, as we know, Paramecium remains; indeed, the whole reaction thus far is essentially like that of Paramecium. But in Opalina, so long as the current is weak, only the cilia on the convexity of the right side strike powerfully with their reversed stroke, — these being the cilia that are reversed in the usual avoiding reaction. The other reversed cilia strike only weakly. In consequence the animal must turn toward the right side, reaching the position shown in Fig. 107, 4. Here most of the strong cilia x of the convexity are still striking forward, hence the animal still turns toward the right. A little beyond 4, — between this and 5, — the animal reaches a position where the tendencies to turn in opposite directions are equal.[1] But the turning which has been initiated in positions 1–4, as a rule has given the animal sufficient momentum to carry it past this dead point, so that it reaches the anode pointing position (Fig. 107, 7). Here the cilia of both sides of the anterior end are directed backward. When striking backward the cilia x of the convexity are no more powerful than those of the opposite side. Hence there is now no tendency to turn farther, and the anode-pointing position is retained. Since the backward stroke of the anterior cilia is more powerful than the forward stroke of the reversed posterior cilia, the animal is carried forward to the anode. Thus in a weak current the position with anterior end directed to the anode is the stable one, so that in the course of time, after some oscillation, the animals reach this position and swim toward the anode.

Now if the current is considerably increased in strength, the cathodic cilia are caused to strike more strongly forward than before. Their motor effect therefore nearly equals that of the anodic cilia, so that the forward movement toward the anode is made much slower. If at the time the current is made the Opalina is in an oblique position, as will usually be the case, or if as a reaction to other stimuli during the passage of the current it passes out of the position with anterior end to the anode, then another effect is produced. Suppose it comes thus into the position shown in Fig. 107, 8. Then the larger number of cilia tend to turn it to the right, as is shown by the arrows at 8. It thus comes into position 1, where all the cilia assist in turning it to the right; it continues in the same way through position 2 to position 3, with anterior end pointing to the cathode. With a weak current, as we have seen, this position is not a stable one; the stronger forward beating of the cilia on the convexity of the right side cause the animal to continue to turn to the right. But with a stronger current this becomes changed. Since even in a weak current the cilia of this convexity strike as strongly forward as they can, their forward stroke is not increased when the current is

[1] If the animal at this point or earlier turns on its long axis, as it frequently does in its usual locomotion, it must now swing back through the cathode-pointing position, till it again reaches a position corresponding to 4 or 5.

made stronger. But as the current is increased, the forward stroke of the cilia on the left side of the anterior half of the body becomes more powerful, — just as happens with all the anterior cilia in Paramecium. Hence, when the current reaches a certain strength, the cilia of the left side, in an Opalina pointing toward the cathode, beat as strongly forward as do those of the right side. There is then no cause for turning toward either the right or the left. The position with anterior end directed toward the cathode has become a stable one. Thus, when a strong current is passed through a preparation of Opalinæ, most of them become directed after a time toward the cathode, and swim slowly in that direction. A number may be at first directed toward the anode, but as soon as these by any chance get out of the anode-pointing position, they also become directed toward the cathode.

With a still more powerful current the Opalinæ retain nearly or quite the position with anterior end to the cathode, but move backward (or sometimes sideways) toward the anode. Wallengren believes that this is a passive movement due to the cataphoric action of the electric current. In Paramecium, as we have seen, there is a similar movement under these conditions, but due to the fact that the cathodic cilia beat more effectively forward than do the anodic cilia backward.

Thus altogether we find that in Opalina the electric current acts on the motor organs in fundamentally the same way as in Paramecium. But owing to peculiarities of the action system of Opalina, this results, with a weak current, in movement forward toward the anode; with a stronger current in movement forward toward the cathode; with a still stronger current in movement backward or sideways toward the anode.

2. SUMMARY

Reviewing our results as to the effect of the continuous electric current on the ciliate infusoria, we find a complete agreement throughout in the action of the current on the motor organs, with the greatest possible diversity in the resulting movements of the animals. In all cases the cilia in the anode region strike backward, as in the normal forward movement, while the cilia of the cathode region are reversed, striking forward. With different strengths of current, and with infusoria of different action systems, this results sometimes in movement forward to the cathode; sometimes in movement forward to the anode; sometimes in a cessation of movement, the anterior end continuing to point to the cathode; sometimes in a backward movement to the anode; sometimes in a position transverse to the current, the animal either remaining at rest or moving across the current. These variations depend upon the differences in the strength of beat of the cilia of different regions of the body under currents of different strength. The different effects produced may be classified, as to their causes, in the following way: —

1. The orientation with anterior end to the cathode is due to the fact that the cilia of the cathodic side strike forward; of the anodic side backward. This may be assisted or hindered by the usual tendency of the organisms to turn when stimulated toward a certain structurally defined side.

2. The movement toward the cathode in weak or moderate currents is due to the fact that under these conditions the backward stroke of the anodic cilia is more powerful than the forward stroke of the cathodic cilia.

3. The cessation of progression in a stronger current, with retention of the cathode-pointing orientation, is due to the fact that as the current is increased the forward stroke of the cathodic cilia becomes more powerful, till it equals the backward stroke of the anodic cilia.

4. The swimming backward toward the anode in a still stronger current is due to a continued increase in the power of the forward stroke of the cathodic cilia, so that they overcome the tendency of the anodic cilia to drive the animal forward. (In Opalina, Wallengren believes that this backward movement is due, at least partly, to the cataphoric effect of the current.)

5. The unstable transverse position seen in some cases (Spirostomum) is due primarily to the fact that the cilia of one side of the elongated body are more powerful, when striking either backward or forward, than are the corresponding cilia of the opposite side. As a result, neither the position with anterior end to the cathode nor that with anterior end to the anode is a stable one, and the animal is compelled to oscillate about a transverse position. This result is accentuated by the slenderness and suppleness of the body in these species.

6. The orientation with anterior end to the anode seen in certain cases (Opalina in a weak current) is due to the fact that the cilia of one side of the anterior half of the body are more readily reversed than the opposing cilia, and their reversed stroke is more powerful, though their usual backward stroke is not. The result is that the position with anterior end to the cathode becomes unstable, while the position with anterior end to the anode is stable so long as accidental causes do not produce slight deviations from it.

7. The transverse or oblique position, at rest or with movement athwart the current, is due to interference between the contact reaction and the effect of the current. This position is maintained only when the more powerful cilia of the peristome are striking forward; that is,

when the peristome is directed toward the cathode. When the peristomal cilia are thus striking forward, their action is comparatively ineffective, so that it does not overcome the attachment to the substratum, in the contact reaction.

3. THEORIES OF THE REACTION TO ELECTRICITY

What is the cause of the reaction to the electric current? The most striking phenomenon in a general view is usually a movement of the organisms *en masse* toward the cathode or anode. It is well known that the electric current has the property of carrying small bodies suspended in a fluid toward the cathode or anode, depending on the conditions. This phenomenon is commonly known as cataphoric action, or as electrical convection. When the movement of small organisms toward one of the electrodes is mentioned, the first thought that comes to mind is of course the possibility that they are thus passively carried by the cataphoric action of the current. But this view can be maintained only on the basis of an extraordinarily superficial acquaintance with the facts. Careful study shows, as we have seen, that the current has definite and striking effects on the cilia, and that it is to these effects that the peculiarities of movement under the action of the current are due. Nevertheless, the theory that the phenomena are passive movements due to the cataphoric action of the current continues to be brought gravely forward at intervals, and doubtless this will continue. The fundamental fallacy of this theory is the idea that we must account in some way by the action of the current for the fact that the organisms move. This is quite unnecessary, for they move equally without the action of the current. The movement is spontaneous, so far as the electric current is concerned. It takes place by the agency of the motor organs of the animal, driven by internal energy, and acting upon the resistance furnished by the water. It is only the changed direction of the movement that the electric current must account for. There is no place for the agency of the cataphoric action in transporting the animals, for they are visibly transporting themselves, just as they were before the cataphoric action began. It is absolutely clear that the movements of the cilia, described in the preceding pages, are at the bottom of the observed behavior, and any explanation of the reaction to electricity must account for the influence of this agent on the cilia. This the theories of passive movement by cataphoric action make no attempt to do.

The clearest disproof of the theory that the movement is a passive one due to cataphoresis is of course the well-established positive proof that the movement is an active reaction of the organism. But the theory can be disproved on other grounds. Statkewitsch (1903 a) shows that dead or stupefied Paramecia that are suspended in viscous fluids are not moved by cataphoric action, while living Paramecia in the same fluids swim to the cathode. Dead or stupefied Paramecia placed in water in a perpendicular tube through which an electric current is passed sink slowly and steadily to the bottom, whatever the direction of the current, while living specimens pass upward when the cathode is above. If the anode is above and a very strong current is used, the living animals swim backward to the anode, as described on page 98. They therefore move upward against gravity, while dead or stupefied specimens with the same current sink slowly to the bottom of the tube. It is thus clear that neither the forward movement to the cathode nor the backward movement toward the anode is directly due to the cataphoric action of the current, for this action is not capable of producing the observed movements.

The cataphoresis might of course act in some way as a stimulus to induce the observed active movements of the cilia. This is apparently the view toward which Carlgren (1899, 1905 a) and Pearl (1900) are inclined. This is of course a theory of a radically different character from that which we have been considering. Just how this effect would be produced through the known physical action of the current has not been shown.

Coehn and Barratt (1905) hold that Paramecia in ordinary water become positively charged, through the escape into the water of the negative ions of the electrolytes which the body holds, while the positive ions are retained. As a result of this positive charge, the electric current tends to carry the animals to the cathode; the infusoria are held to follow this tendency and swim with the pull of the current toward the cathode. In a solution containing more electrolytes, it is held that the positive ions escape from the protoplasm; hence the animals become negatively charged. They therefore pass to the anode when placed in a solution of sodium chloride or sodium carbonate. This theory leaves unaccounted for precisely the essential feature of the reactions, — the cathodic reversal of the cilia. It likewise fails to account for the fact that as the current becomes stronger the passage to the cathode ceases and the animals begin to swim backward to the anode, and for the further fact that individuals which have become accustomed to a solution of sodium chloride or carbonate no longer swim to the anode, but pass to the cathode as usual. These facts appear to be absolutely fatal to the view under consideration. Little is to be hoped of any theory that neglects what is clearly the fundamental phenomenon in these reactions, — the cathodic reversal of the cilia.

Another theory has held that the reaction to the electric current is due to the electrolytic effect of the current on the fluid containing the animals (Loeb and Budgett, 1897). The water of course contains electrolytes. These are separated by the current into their component ions, and the products of this electrolysis may be deposited on opposite poles of a body immersed in the fluid. There is some reason to suppose that an alkali may be deposited on that portion of the surface of the infusorian where the current is entering its protoplasm (the anodic surface), an acid where it is leaving the protoplasm (the cathodic surface). The relative amount of such action is unknown, but the suggestion is made that the observed effects of the current are due to these chemicals. This very interesting and suggestive theory seems, however, not to be supported by other known facts. The effects of different chemicals on the ciliary action are known, and it is not true that acids produce continued reversal of the cilia, alkalies the opposite effect, as would be necessary in order to make this explanation satisfactory. Any effective chemical, either acid or alkali, produces, as we know, the avoiding reaction, with its succession of coördinated changes in the ciliary movements. Again, as Ludloff (1895) and Statkewitsch (1903) show, the characteristic anodic and cathodic effects do not correspond throughout to the regions where the current is entering or leaving the protoplasm. If a Paramecium has an oblique position, as in Fig. 108, the current enters the body on the entire left side, and leaves the body on the entire right side. Hence, on the theory we are considering, all the cilia of the left side ought to act alike, and in the opposite manner from the cilia of the right side. But this is not true. On the left side the cilia of the region b beat forward, those of c backward; on the right side the cilia a strike forward, d backward. A similar distribution of the discharge of trichocysts under the influence of the induction shock is shown to exist by Statkewitsch. The distribution of the effects of the current on the cilia and on the trichocysts therefore does not correspond to the distribution of the regions where the current is entering and leaving the protoplasm; hence the latter cannot explain the former.

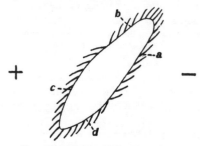

FIG. 108.— Diagram of the effects of the electric current on the cilia, showing that the regions where the cilia are directed forward and backward, respectively, do not correspond to the regions where the current is leaving and entering the body.

Another theory, somewhat less definite than the one last mentioned, but widely accepted, is the following. The electric current is conceived to have a polarizing effect on the organism, resulting in the different action of the cilia on the two halves. At the anodic half the current is considered to cause a backward movement of the cilia, or "contractile stroke"; at the cathodic half, a forward movement or "expansive stroke" (Verworn, 1899; Ludloff, 1895). The precise cause of this action is not given, but as supporting the possibility of this view, the experiments of Kühne (1864, page 99) and Roux (1891) on the polarizing effects of the current may be cited. Kühne showed that the violet-colored cells of Tradescantia become under the influence of the electric current red at the anodic end, green at the cathodic end, — indicating that the anodic end becomes acid, the cathodic end alkaline. Roux showed that under the electric current the frog's egg becomes divided into two halves of different color. Furthermore, the two halves of a cell in the electric current become physically somewhat different, owing to the cataphoric action. There is a tendency for the fluids of the body to be carried to one end, — the cathodic, — while the solids are carried to the other, — the anodic. As a result of such chemical or physical polarization, or of both, it is then conceivable that the body of the infusorian may become divided into two halves, differing in such a way that the cilia act in opposite directions. On this view the backward stroke of the cilia on the anodic half of the body is as much a specific effect of the current as is the forward stroke of the cathodic cilia. Opposed to this view is the consideration that the action of the anodic cilia is as a matter of fact not different from that in the unaffected animal, and the further fact that the cathodic effect is limited, in a weak current, to only the cathodic tip of the animal. If both the backward and the forward positions of the cilia are specific effects of the current, it is difficult to see why the former should prevail so strongly over the latter in a weak current. On the other hand, if we consider the cathodic action alone as a specific effect of the current, interfering with the normal backward stroke of the cilia, then it becomes at once intelligible that this interference should be least in a weak current, and should increase as the current becomes more powerful. In producing its characteristic effect chiefly at the cathode, the action of the electric current on infusoria agrees with its action on muscle, as Bancroft (1905) has recently pointed out.

The most thorough study of the fundamental changes produced by the electric current is that made by Statkewitsch (1903 a), and his conclusions are entitled to high consideration. Statkewitsch subjected Paramecia that had been stained in the living condition with certain chemical indicators, — neutral red and phenol-phtalein, — to the influence of the electric current. He found that the current caused chemical changes within the protoplasm, the endoplasmic granules and vacuoles becoming more alkaline in reaction. Statkewitsch therefore concludes that the peculiar effect of the electric current on the cilia is due to a disturbance in the usual equilibrium of the chemical processes taking place in the protoplasm. The results of this disturbance are first shown, so far as the ciliary action is concerned, in the cathodic region, spreading thence over the remainder of the body, as illustrated in Fig. 61.

For any satisfactory theory of the reaction to the electric current, one thing is essential; it must account for the cathodic reversal of the cilia. It is perfectly clear that this is the characteristic feature of this reaction, and a theory that will account for this reversal will at once clear up the curious and apparently contradictory effects produced under various conditions. Theories which do not take this into account are at the present time anachronisms; they fail to touch the real problem.

Whatever be the cause, it is clear that the behavior of infusoria under the action of the electric current differs radically from the behavior under other conditions. The position taken by the organism is not attained by trial of varied directions of movement, as in the reactions to most other stimuli, but in a more direct way. Different parts of the body are differently affected by the current, so that the behavior is not co-ordinated and directed toward a unified end, as in the reactions to other stimuli. The motor organs of the different parts of the body tend to drive the animal in different directions. The movement actually occurring is a resultant of these differently directed factors. It is therefore sometimes in one direction, sometimes in another, depending on the relative strength of the opposing factors. The animal thus does not approach an optimum nor cease to be stimulated, whatever the direction taken. Sometimes indeed no position of even comparatively stable equilibrium is possible (Spirostomum).

These peculiarities of the reaction to the electric current are due to the forced reversal of the cilia in the cathodic region of the body, — an effect not produced by any other agent. If the current produced only its anodic effect, the reaction to electricity would be, so far as the evidence indicates, precisely like that to other agents. The cathodic reversal of the cilia interferes with the normal behavior of the organism. Thus the action of the infusoria under the electric current is not typical of the behavior under other stimuli. It may be compared to the behavior of an organism that is mechanically held by clamps and thus prevented from showing its natural behavior. It is interesting to note that this cramped and incoherent behavior is found only under the influence of an agent that never acts on the animals in their natural existence. The reaction to electricity is purely a laboratory product.

LITERATURE IX

REACTIONS OF INFUSORIA TO ELECTRICITY

A. Reactions to induction shocks: ROESLE, 1902; STATKEWITSCH, 1903; BIRU-KOFF, 1899.

B. Reactions to the constant current: STATKEWITSCH, 1903 *a*, 1904; WALLEN-GREN, 1902, 1903; PEARL, 1900; VERWORN, 1889 *a*, 1889 *b*, 1896; LOEB AND BUD-GETT, 1897; DALE, 1901; CARLGREN, 1899, 1905; BANCROFT, 1905; COEHN AND BARRATT, 1905.

CHAPTER X

MODIFIABILITY OF BEHAVIOR IN INFUSORIA, AND BEHAVIOR UNDER NATURAL CONDITIONS. FOOD HABITS

1. MODIFIABILITY OF BEHAVIOR

WE have seen that in Paramecium the behavior varies to a certain extent in different individuals or under different conditions. Similar variations might be described for other free swimming infusoria. But these observations do not tell us whether the behavior may change in the same individual or not. Does a given individual always react in the same way to the same stimulus under the same conditions? Or may the individual itself change, so that it behaves differently even when the external conditions remain the same, — as we know to be the case in higher animals? To answer these questions it is necessary to follow continuously the behavior of a single individual, and this can be done most satisfactorily in attached organisms, such as Stentor and Vorticella. We shall base our account on the usual behavior of *Stentor ræselii*, which illustrates well the points in which we are at present interested.

Stentor ræselii Ehr. (Fig. 109) is a colorless or whitish, trumpet-shaped animal, consisting of a slender, stalklike body, bearing at its end a broadly expanded disk, the peristome. The surface of the body is covered with longitudinal rows of fine cilia, while the edge of the disk is surrounded by a circlet of large compound peristomal cilia or membranellæ. These make a spiral turn, passing on the left side into the large buccal pouch, which leads to the mouth. The mouth thus lies on the edge of the disk, nearly in the middle of what may be called the oral or ventral surface of the body. The smaller end of the body is known as the foot; here the internal protoplasm is exposed, sending out fine pseudopodia, by which the animal attaches itself.

Stentor ræselii is usually attached to a water plant or a bit of débris by the foot, and the lower half of the body is surrounded by the so-called tube. This is a very irregular sheath formed by a mucus-like secretion from the surface of the body, in which are embedded flocculent materials of all sorts. It is frequently nearly transparent, so as to be almost invisible. *Stentor ræselii* is found in marshy pools, where much dead vegetation is present, but where decay is taking place only slowly.

In the extended animal the peristomal cilia are in continual motion. When finely ground India ink or carmine is added to the water, the currents caused by the cilia are seen to be as follows: The mouth of the animal forms the bottom of a vortex, toward which the water above the disk descends from all sides (Fig. 109). Only the particles near

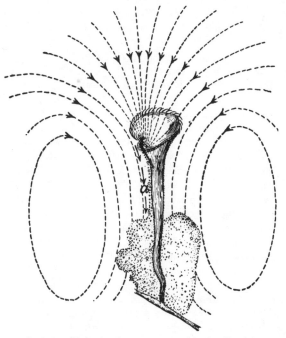

FIG. 109. — *Stentor ræselii*, showing the currents caused by the cilia of the peristome.

the axis of the vortex really strike the disk; those a little to one side shoot by the edges without touching. Particles which reach the disk pass to the left, toward the buccal pouch, following thus a spiral course. Reaching the buccal pouch, they are whirled about within it a few times; then they either pass into the mouth, at the bottom of the pouch, or they are whirled out over the edge of the pouch, at the mid-ventral notch. In the latter case they usually pass backward along the mid-ventral line of the body (Fig. 109, *a*), till they reach the edge of the tube. To this they may cling, thus aiding to build up the tube.

When stimulated, *Stentor ræselii* may contract into its tube, taking then a short oblong or conical form (Fig. 110). Such contractions do not as a rule take place save in response to well-marked stimuli. When not disturbed in any way, the animal remains extended, with cilia in active operation.

Let us try the effect of disturbing the animal very slightly. While the disk is widely spread and the cilia are actively at work, we cause a

fine current of water to act upon the disk, in the following way. A long tube is drawn to a very fine capillary point and filled with water. The capillary tip is brought near the Stentor, while the long tube is held nearly perpendicular. The pressure causes a jet of water from the tip to strike the disk of the animal. Like a flash it contracts into its tube. In about half a minute it extends again, and the cilia

FIG. 110.—*Stentor ræselii* contracted into its tube.

resume their activity. Now we cause the current to act again upon the disk. This time the animal does not contract, but continues its normal activities without regard to the current of water. This experiment may be repeated on other individuals; invariably they react to the current the first time, then no longer react. The same results are obtained with other fixed infusoria: Epistylis and Carchesium. By using other very faint stimuli, such as that produced by touching the surface film of the water close to the organism, or by slightly jarring the object to which it is attached, the same results are obtained. To the first stimulus they respond sharply; to the second and following ones they do not respond at all, even if long continued.

Thus the organism becomes changed in some way after its first reaction, for to the same stimulus, under the same external conditions, it no longer reacts. What is the nature of this internal change? The first suggestion that rises to the mind in explanation of such a cessation of reaction is that it may be due to fatigue. The distinction between fatigue and other changes of condition is an important one, for the following reason. Fatigue is due to what may be called a failure. It is an imperfection inherent perhaps in the nature of the material of which organisms are composed, preventing them from doing what might be to their advantage. Changes of reaction due to other causes might on the other hand be regulatory, tending to the advantage of the organism. Higher animals often react strongly by a "start," to the first incidence of sudden harmless stimuli, then no longer react, and this cessation is evidently a regulation of behavior that is to the interest of the organism. We must then determine whether the failure of the infusorian to react to the second stimulation is due to fatigue or to some other cause.

It seems improbable that the change of behavior is due to fatigue, since the change occurs after but a single stimulation and a single reaction. It could hardly be supposed that these would fatigue the animal to such an extent as to prevent further contractions. And if we use stronger stimuli, we find that the animal continues to contract successively every time the stimulus is applied, for an hour or more. It is evident that the failure to contract after the first stimulation cannot be due to fatigue of the contractile apparatus.

If we make the stimulation somewhat stronger than in our first experiments, as may be done by touching the animal lightly with a capillary glass rod, the behavior is a little different. The animal may react the first and second times, then cease to react, or it may react half a dozen times, or more, then cease. If we continue the stimuli, we find a change in the behavior. The animal instead of contracting bends into a new position, and it may do this repeatedly. This shows that the failure to contract is not due to a failure to perceive the stimulus, — in other words, to a fatigue of the perceptive power, — for the bending into a new position shows that the stimulus is perceived, though the reaction differs from the first one.

Our results thus far show that after responding once or a few times to very weak stimulation, the organism becomes changed, so that it no longer reacts as before, and that this change is not due to fatigue, either of the contractile apparatus or of the perceptive power. The behavior may then be of the same regulatory character as is the similar behavior in higher animals. Indeed, so far as the objective evidence goes, this behavior in Stentor precisely resembles that of higher animals, and is to the same degree in the interest of the organism.

With still stronger stimulation, produced by touching the animal with the capillary glass rod, another curious phenomenon often shows itself. The animal may react to each of the first half dozen strokes, then cease to react; then after a few more strokes react again, then cease to react till a large number have been given, and so continue. A typical series, giving the number of strokes before contraction is produced, is the following, obtained from experiments with an individual of Epistylis: —

1 — 22 — 10 — 3 — 3 — 1 — 1 — 22 — 59 — 125 (continuous blows for one minute) — ($\frac{3}{4}$ minutes) — ($1\frac{1}{2}$ minutes) — ($4\frac{1}{2}$ minutes).

During such experiments the organism, when it does not contract, continually changes its position, as if trying to escape the blows. The reason for the contraction at irregular intervals which become longer as the experiment continues, is not clear. Possibly fatigue may have something to do with this matter.

The stimuli with which we have thus far dealt are not directly injurious, and do not interfere in the long run with the normal functions of the organism, so that the power of becoming accustomed to them and ceasing to react is useful. Let us now examine the behavior under conditions which are harmless when acting for a short time, but which, when continued, do interfere with the normal functions. Such conditions may be produced by bringing a large quantity of fine particles, such as India ink or carmine, by means of a capillary pipette, into the water currents which are carried to the disk of Stentor (Fig. 111).

Under these circumstances the normal movements are at first not changed. The particles of carmine are taken into the pouch and into

the mouth, whence they pass into the internal protoplasm. If the cloud of particles is very dense, or if it is accompanied by a slight chemical stimulus, as is usually the case with the carmine grains, this behavior lasts but a short time; then a definite reaction supervenes. The animal bends to one side — always, in the case of Stentor, toward the aboral side. It thus as a rule avoids the cloud of particles, unless the latter is very large. This simple method of reaction turns out to be more effective in getting rid of stimuli of all sorts than might be expected. If the first reaction is not successful, it is usually repeated one or more times. This reaction corresponds closely with the "avoiding reaction" of free-swimming infusoria, and like the latter, is usually accompanied by revolution on the long axis, — the animal twisting on its stalk two or three times as it bends toward the aboral side.

FIG. 111. — A cloud of carmine is introduced into the water currents passing to the mouth of Stentor.

If the repeated turning toward one side does not relieve the animal, so that the particles of carmine continue to come in a dense cloud, another reaction is tried. The ciliary movement is suddenly reversed in direction, so that the particles against the disk and in the pouch are thrown off. The water current is driven away from the disk instead of toward it. This lasts but an instant, then the current is continued in the usual way. If the particles continue to come, the reversal is repeated two or three times in rapid succession. If this fails to relieve the organism, the next reaction — contraction — usually supervenes.

Sometimes the reversal of the current takes place before the turning away described first; it may then be followed by the turning away. But usually the two reactions are tried in the order we have given.

If the Stentor does not get rid of the stimulation in either of the ways just described, it contracts into its tube. In this way it of course escapes the stimulation completely, but at the expense of suspending its activity and losing all opportunity to obtain food. The animal usually remains in the tube about half a minute, then extends. When its body has reached about two-thirds its original length, the ciliary disk begins to unfold and the cilia to act, causing currents of water to reach the disk, as before.

We have now reached a specially interesting point in the experiment. Suppose that the water currents again bring the carmine grains. The stimulus and all the external conditions are the same as they were at the beginning? Will the Stentor behave as it did at the beginning? Will it at first not react, then bend to one side, then reverse the current, then contract, passing anew through the whole series of reactions? Or shall we find that it has become changed by the experiences it has passed through, so that it will now contract again into its tube as soon as stimulated?

We find the latter to be the case. As soon as the carmine again reaches its disk, it at once contracts again. This may be repeated many times, as often as the particles come to the disk, for ten or fifteen minutes. Now the animal after each contraction stays a little longer in the tube than it did at first. Finally it ceases to extend, but contracts repeatedly and violently while still enclosed in its tube. In this way the attachment of its foot to the object on which it is situated is broken, and the animal is free. Now it leaves its tube and swims away. In leaving the tube it may swim forward out of the anterior end of the tube; but if this brings it into the region of the cloud of carmine, it often forces its way backward through the substance of the tube, and thus gains the outside. Here it swims away, to form a new tube elsewhere.

While swimming freely after leaving its tube, Stentor shows the characteristic behavior of the free-swimming infusoria, such as Paramecium. Upon this, therefore, we need not dwell, passing at once to the behavior in becoming reattached and forming a new tube.

On coming to the surface film of the water, or the surface of solid objects, the free-swimming Stentor behaves in a peculiar way. It applies its partially unfolded disk to the surface and creeps rapidly over it, the ventral side of the body being bent over close to the surface. It may thus creep over a heap of débris, following all the irregularities of the surface rapidly and neatly, seeming to explore it thoroughly. This may last for some time, then the animal may leave the débris and swim about again. Other heaps of débris or the surfaces of solids are explored in the same way. Finally, after ten or twenty minutes or more, one of these is selected for the formation of a new tube. It may be seen that as the Stentor moves about a viscid mucus is secreted over the surface of the body. To this mucus particles of débris stick and are trailed behind the swimming animal. In a certain region, perhaps between two masses of débris, the animal stops and begins to move backward and forward with an oscillatory motion, through a distance about two-thirds its contracted length. This movement, in precisely the same place, is kept up for about two minutes, while the mucus

from the surface is rapidly secreted. The movement compacts this mucus into a short tube or sheath, — the tube in which the Stentor is to live. The process is represented in Fig. 112. Next the tip of the foot is pressed against the débris at the bottom of the tube. There it adheres by means of fine pseudopodia sent out from the internal protoplasm. Now the Stentor extends to full length, and we find it

FIG. 112. — Oscillating movement of Stentor, by which it forms a new tube. 1–2, alternating positions. *a*, the secreted mucus; *b*, masses of débris.

in the usual attached condition, with the lower half of the body surrounded by a transparent tube of mucus. The Stentor has thus moved away from the place where it was subjected to the mass of carmine particles, and has established itself in another situation.

The behavior just described shows clearly that the same individual does not react always in the same way to the same stimulus. The stimulus and the other external conditions remaining the same, the organism responds by a series of reactions becoming of more and more pronounced character, until by one of them it rids itself of the stimulation. Under the conditions described — when a dense cloud of carmine is added to the water — the changes in the behavior may be summed up as follows: —

(1) No reaction at first: the organism continues its normal activities for a short time.

(2) Then a slight reaction by turning into a new position, — a seeming attempt to keep up the normal activities and yet get rid of the stimulation.

(3) If this is unsuccessful, we have next a slight interruption of the normal activities, in a momentary reversal of the ciliary current, tending to get rid of the source of stimulation.

(4) If the stimulus still persists, the animal breaks off its normal activity completely by contracting strongly — devoting itself entirely, as it were, to getting rid of the stimulation, though retaining the possibility of resuming its normal activity in the same place at any moment.

(5) Finally, if all these reactions remain ineffective, the animal not only gives up completely its usual activities, but puts in operation another set, having a much more radical effect in separating the animal from the stimulating agent. It abandons its tube, swims away, and forms another one in a situation where the stimulus does not act upon it.

The behavior of Stentor under the conditions given is evidently a special form of the method of the selection of certain conditions through varied activities, — a form which we have not met before. The organism "tries" one method of action; if this fails, it tries another, till one succeeds. Like other behavior based on this method, it is not a specific reaction to any one stimulus, but is seen whenever analogous conditions are produced in any way. Thus we may use in place of carmine other substances. Chemicals of different kinds produce a similar series of reactions. A decided change in osmotic pressure has a somewhat similar effect. There are variations in the details of the reaction series under different conditions. Sometimes one step or another is omitted, or the order of the different steps is varied. But it remains true that under conditions which gradually interfere with the normal activities of the organism, the behavior consists in "trying" successively different reactions, till one is found that affords relief. The production of any given step in the behavior cannot be explained as a necessary consequence of the preceding step. On the contrary, the bringing into operation of any given step depends upon the ineffectiveness of the preceding ones in getting rid of the stimulating condition. The series may cease at any point, as soon as the stimulus disappears. Moreover, it is evident that the succeeding steps are not mere accentuations of the preceding ones, but differ completely in character from them, being based upon different methods of getting rid of the stimulation.

All our results on Stentor then show clearly that the same organism may react to the same stimulus in various different ways. It may react at first, then cease to react if the stimulus does not interfere with its normal activities; it may react at first by a very pronounced reaction (contraction), then later by a very slight reaction (bending over to one side); or it may respond, if the stimulus does interfere with its normal functions, by a whole series of different reactions, becoming of a more and more pronounced character. Since in each of these cases the external conditions remain throughout the same, *the change in reaction must be due to a change in the organism.* The organism which reacts to the carmine grains by contracting or by leaving its tube must be different in some way from the organism which reacted to the same stimulus by bending to one side. No structural change is evident, so that all we can say is that *the physiological state of the organism has changed.* The same organism in different physiological states reacts differently to the same stimuli. It is evident that the anatomical structure of the organism and the different physical or chemical action of the stimulating agents are not sufficient to account for the reactions. The varying physiological states of the animal are equally important factors. In Stentor we are compelled to assume at least five different physiological states to account for the five different reactions given under the same conditions. We shall later find much occasion to realize the importance of physiological states in determining behavior.

These relations may be stated from another point of view, which leads to interesting questions. The present physiological state of an organism depends upon its past history, so that we can say directly that the behavior of such an organism as Stentor under given conditions depends on its past history. This statement we know is markedly true for higher organisms. What a higher animal does under certain conditions depends upon its experience: — that is, upon its past history. In the typical and most interesting case we say that the behavior of the higher organism depends upon what it has learned by experience. Is the change in the behavior of Stentor in accordance with its past history a phenomenon in any wise similar in character to the learning of a higher organism? In judging of this question we must rely, of course, entirely upon objective evidence; — upon what can be actually observed. When this is done, it is hard to discover any ground for making a distinction in principle between the two cases. The essential point seems to be that after experience the organism reacts in a more effective way than before. The change in reaction is regulatory, not merely haphazard. And this is as clearly the case in Stentor as in the higher organism. It is true that, so far as we can see, the behavior of Stentor shows in only a rudimentary way phenomena that become exceedingly striking and complex in higher organisms. Stentor seems to vary its behavior only in accordance with the experience that either (1) the stimulus to which a strong reaction is at first given, does not really interfere with its activities, so that reaction ceases; or (2) that the reaction already given is ineffective, since the interference with its activities continues, so that another reaction is introduced.[1] If the changes in

[1] It is to be noted that nothing is said in this statement as to the Stentor's perceiving these relations. The statement attempts merely a formulation of the observed facts in such a way as to bring out their relation to what we observe in higher organisms.

the behavior of Stentor were not regulatory, becoming more fitted to the existing conditions, a comparison with the behavior of higher animals in learning would be out of place. But since the changes clearly are regulatory, in the one case as in the other, it would be equally out of place to deny their similarity, in this respect at least.

In another important feature the behavior of Stentor falls, so far as our present evidence goes, far below the level of that found in the learning of higher animals. The modification in the behavior induced by experience seems to last but a very short time. Immediately after reacting in one way, which proves ineffective, it reacts in another. But a short time after it apparently reacts in the same way as at first.[1] As a rule, it is evidently to the interest of an organism living under such simple conditions as Stentor to return to the first method of reaction when again stimulated after a period of quiet, for as a rule this first method is effective, and it would be most unfortunate for the Stentor to proceed to the extremity of abandoning its tube without a trial of simpler reactions. But the difference between behavior which is modified only for a few moments after an experience, and that which is permanently modified, is undoubtedly important. The latter would nevertheless be developed from the former by a mere quantitative change, so that the variation in duration does not constitute a difference in essential nature.

We may sum up the results of the present section as follows: The same individual does not always behave in the same way under the same external conditions, but the behavior depends upon the physiological condition of the animal. The reaction to any given stimulus is modified by the past experience of the animal, and the modifications are regulatory, not haphazard, in character. The phenomena are thus similar to those shown in the "learning" of higher organisms, save that the modifications depend upon less complex relations and last a shorter time.

2. The Behavior of Infusoria under Natural Conditions

We have thus far dealt chiefly with the behavior of infusoria under experimental conditions. In experiments the conditions are usually

---

[1] This matter cannot be considered definitely settled. It is exceedingly difficult in practice to devise and carry out experiments which shall actually determine the length of time that the modified behavior lasts. A thorough, definitely planned investigation should be directed precisely upon this point. Hodge and Aikins (1895) report that Vorticella, which at first took yeast as food, later rejected the yeast, and that for "several hours" it refused to take the yeast again. But unfortunately no further details are given. We do not know whether the Vorticella was injured and took no food at all, or what other conditions were present, so that we can build little upon this observation.

made as simple as possible. All sources of stimulation save one are excluded, in order that we may discover the precise effects of that one. In our account of Paramecium we have seen that when more than one source of stimulation is present, the behavior is determined by all the existing conditions, so that often the behavior cannot be characterized as a precise reaction to a definite stimulus. That this is true also for other infusoria we have seen in a number of instances, particularly in our account of the contact reaction. It would be possible to add many other examples to these, making a special chapter on "Reactions to Two or More Stimuli," but this would add no new principle to what we have already brought out. The general statement may be made, that to account for the way an infusorian behaves at a given time, it is as a rule not sufficient to take into account a single source of stimulation, but all the conditions must be considered.

We shall now look at certain features of the behavior of infusoria under the conditions that are supplied by the environment, in all their variety and complexity. We wish to see how the natural "wild" organism behaves. Our account cannot be exhaustive, for the natural history of the thousands of species of infusoria remains largely to be worked out. We shall merely examine certain typical features of the behavior, devoting especial attention to the food reactions.

In our chapter on the "Action System" we have seen some of the chief variations in the natural behavior of infusoria. We have there seen that the infusoria can be divided, according to their methods of life, into three main groups: those that are attached, those that creep over surfaces, and those that swim freely. The behavior in these different groups necessarily differs much. Yet, as we have seen, every possible gradation exists from one group to another, and even the same individual may at different periods represent each different group. The behavior is simplest and least varied in the free-swimming organisms; more varied in those which habitually creep along a surface; most complex in those which live attached. The reason for this seems to be as follows: In the open water the conditions are exceedingly simple. The free-swimming organism may escape an injurious stimulus simply by swimming away. In the fixed organism, on the other hand, the conditions are more complex. At any moment both the solid and the free fluid are acting on the organism. For a fixed animal to obtain food and escape injurious conditions, varied devices are necessary. It cannot at once solve any difficulty by departing, as the free organism can. We find, then, that such fixed organisms have developed varied reaction methods (see the preceding chapter).

There is much variation in the complexity of behavior even among species living under similar conditions. Some of the free-swimming species are very supple, changing form continually. Such is the case, for example, with *Lacrymaria olor*, which stretches its long neck in every direction, shortens it until it has almost disappeared, reëxtends it, and seems to explore thoroughly the surrounding region. Such an organism has, of course, much better opportunity for effective behavior by the method of trial than has such a rigid form as Paramecium.

Similar differences are found among the creeping infusoria, and among the fixed species. Some fixed infusoria contract frequently, while others contract only rarely. In some cases the contraction occurs at regular intervals, even when there is no indication of an external stimulus. This is the case with Vorticella. There is no evidence that in infusoria periods of rest, comparable with the sleep of higher animals, are alternated with periods of activity. Hodge and Aikins (1895) kept a single Vorticella continuously under observation for twenty-one hours, besides intermittent study for a number of days. They found that there was no period of inactivity. During five days the cilia were in continuous motion, food was continuously taken, and contractions were repeated at brief intervals.

A number of fixed infusoria live, like *Stentor rœselii*, in tubes, some gelatinous, some membranous in character. As a rule these tubes are formed in a very simple manner. The material of which they are composed is secreted by the outer surface of the animal. In the repeated contractions and extensions of the body this material is worked off, in the form of a sheath. The tube may become thicker by the secretion of more material on the surface of the animal. It often grows in length, either as the animal becomes longer or as it migrates farther out toward the open end of the tube. In the secreted material, which is often transparent, all sorts of foreign substances may become embedded, in the following way: They are carried as particles to the oral disk by the cilia. Thence they pass backward over the surface of the body, till they reach the gelatinous substance of the tube, where they become embedded. Thus in most cases the formation of the tube seems a direct consequence of the secretion of the mucus-like substance over the body of the animal, taken in connection with the usual movements. The intervention of any special type of behavior directed toward the end of forming the tube seems unnecessary. But in some cases, as we have seen in our account of Stentor, the tube is formed at the beginning by a definite set of movements, of a character especially fitted to produce such a structure. For details as to different kinds of tubes, and their structure and method of formation, reference may be made to Bütschli's great work on the infusoria (1889).

A set of phenomena that is deserving of careful study for its implications as to the nature of behavior is that involved in the activities preliminary to conjugation. It is possible that the organisms are in a modified physiological condition at this time, behaving differently from usual. Critical observations on this subject, of such a nature that we can use them for our present purpose, are too few in number to make possible a unified account of these phenomena. An account of the facts for Paramecium is given on page 102. The field is one deserving of much further work.

3. Food Habits

The food habits of the infusoria are among the most interesting of their activities to the student of animal behavior. As to their food habits, we can with Maupas (1889) divide the infusoria into two classes. The first includes those that bring the food to the mouth by means of a vortex produced by the peristomal cilia; the second those that go about in search of food, seizing upon it with the mouth, like a beast of prey. The former live chiefly upon minute objects, the latter upon larger organisms. There is, of course, no sharp distinction between the two classes. Most of the infusoria with strong vortices move about more or less in search of food, and most of those that seize upon their prey after a search are aided by a more or less pronounced vortex. Thus the roving or searching movements and the vortex are factors common to the food habits of most of the infusoria. The positive contact reaction further plays a most important part in obtaining food.

Those species that depend primarily upon the ciliary vortex for obtaining food usually feed upon bacteria and other minute organisms and upon finely divided organic matter, — bits of decaying plant or animal material. Of this class of organisms Paramecium and Stentor are types. In some, as in Paramecium, the food is limited to most minute bodies, such as bacteria and small algæ. Stentor and others may take larger objects. Other infusoria and even rotifers of a considerable size are often seen embedded in the internal protoplasm of Stentor. Such animals are caught in the strong ciliary vortex, carried to the buccal pouch, which often contracts in such a way as to prevent their escape, and are then taken through the mouth into the internal protoplasm.

How do these organisms succeed in getting the food that is fitted for them? Is there a selection of food, and how is it brought about? Much of the difficulty as to the selection of food is solved by the conditions under which these animals usually live. They are found as a rule in water which contains decaying vegetable or animal matter, and therefore swarms with bacteria. Hence the usually ciliary current brings food continuously, and little selection is necessary. The animals take, within wide limits, all that the ciliary current brings. Bits of soot, India ink, carmine or indigo, chalk granules, and the like are swallowed along with the bacteria, though of course they are useless as food. They are merely passed through the body and ejected along with the indigestible remains of the food. They do no harm, and the animal may continue to take them indefinitely, provided it receives in addition a sufficient amount of real food. If the ciliary currents do not bring food, of course the organisms die after a time. It is well known that infusoria appear suddenly in immense numbers, or disappear with equal rapidity, according as the conditions are favorable or unfavorable.

But the animals do determine for themselves, to a certain extent, what things they shall take as food, and what they shall not. This is not done, so far as can be observed, by a sorting over of the food by the cilia, as the water current carries it to the mouth. It is true that not all the particles in the vortex produced by the cilia pass into the mouth. But this is due to the simple mechanical conditions. The vortex is very extensive, and the mouth is very small, so that only a fraction of the water in the vortex can ever reach the mouth. Hence inevitably a large share of the particles in the vortex are whirled away. But this is true of particles which are valuable for food as well as of those which are not. If Stentor is placed in water containing immense numbers of small algal cells which are useful as food, it is found that as many of these pass through the vortex without being taken as happens in the case of worthless particles of soot or carmine.

Choice of food occurs in a somewhat cruder fashion than through a sorting of the individual particles by the cilia. It takes place through the reaction with which we have become familiar in studying the behavior of the organisms under various stimuli. Thus in Paramecium the rejection of unsuitable food takes place through the avoiding reaction. If the ciliary current brings water containing various chemicals in solution, or if large solid objects are brought to the mouth, or too great a mass of smaller particles, the Paramecium shifts its position in the usual way. It backs more or less, turns toward the aboral side, and moves to another place. The avoiding reaction is in itself always an expression of choice, in so far as it determines the rejection of certain conditions of existence. In Stentor and Vorticella choice of food occurs in a similar manner, though in these fixed infusoria there is, as we have seen, usually more than one way of rejecting unsuitable conditions.

In Stentor the following behavior is at times observed. The animal is outstretched and feeding quietly in the usual way. Many small objects pass into the buccal pouch and are ingested. Suddenly a larger, hard-armored infusorian, Coleps, is drawn into the pouch. At once the ciliary current is reversed and the Coleps is driven out again. Then the current is resumed in the usual direction. Vorticella and other fixed infusoria often reject large objects in the same way. But besides reversing the ciliary current, these organisms may, when the ciliary current brings unsuitable material, bend over into a new position, contract, or leave their place of attachment and swim away. All these reactions have been described in detail in our account of the behavior of Stentor.

Thus the choice of food in all these organisms depends merely upon whether the usual negative or avoiding reactions are or are not given. The avoiding reaction is the expression of such choice as occurs. Looking at the matter from this standpoint, we are forced to conclude that the entire behavior involves choice in almost every detail. The animals, as we have seen, are giving the avoiding reaction in a certain degree, from a slight widening of the spiral course to the powerful backward swimming, almost continuously. The straightforward course is the expression of positive choice or acceptance; the avoiding reacting of negative choice or rejection. No distinction can be made between choice and the usual behavior. Indeed, choice is the essential principle of behavior based on the method of trial.

What happens if the organisms settle down and attach themselves in a region where no food exists? This question seems not to have been specially investigated. But it is known that under most kinds of unfavorable conditions, — conditions which interfere with the normal functions, — the animal, after a time, leaves its place and swims away to a new location. Doubtless this happens also when food is lacking.

We may sum up the food habits of this first class of ciliates as follows: They settle down in a certain region and then bring a current of water to the mouth. The particles in this current are taken as food, without any sorting, so that many that are not useful are ingested along with the others. But if decidedly unsuitable material is brought, then the animal reacts as to other unfavorable stimuli — reversing the current, contracting, shifting position, or finally moving away to a new place. The method of trial of varied movements is at the basis of the behavior here as elsewhere.

The second class of ciliates includes those which move about in search of their food, preying upon larger organisms and seizing them with the mouth. Maupas has well called these the hunter ciliates. The method of taking food in these animals often resembles in many respects that of the species already described. Thus Stylonychia runs about here and there, producing a strong vortex leading to its mouth. This often carries other infusoria, of considerable size, to the mouth. These are then seized and worked gradually back into the internal protoplasm. Some species move about more rapidly and more extensively, while the ciliary vortex is reduced so that it is of little consequence for food getting. On coming in contact with another infusorian the latter is seized by the usually armored mouth; this is opened widely and the prey is swallowed. In this way such infusoria often feed upon other animals almost or quite as large as themselves, the mouth opening widely and the body becoming greatly distended.

An excellent example of one of these hunter ciliates is furnished by Didinium. This animal (Fig. 113) is cask-shaped, with a truncate anterior end, bearing in its centre the mouth on a slight elevation. The body bears but two circles of cilia. By the aid of these, Didinium swims about rapidly, revolving to the right on its long axis and frequently changing its direction. On coming in contact with a solid object it stops, pushes forward against the object the conical projection which bears the mouth, and revolves rapidly on its long axis. The mouth is armed with a number of strong ribs ending in points, which apparently project a little from the cone bearing the mouth. When pushed forward against a soft organism, these points apparently pierce and hold it. The revolution on the long axis has the appearance of a process of boring into the body. The mouth now opens widely and swallows the prey. Paramecium often falls a victim to Didinium in this way (Fig. 113). Sometimes the Didinium is smaller than its prey, forming after the feeding process a mere sac over its surface.

FIG. 113.—Didinium seizing Paramecium. After Balbiani.

The point which interests us at present is that Didinium reacts in the way described not merely to objects which may serve as food, but also to all sorts of solid bodies. In other words, the process is one of the trial of all sorts of conditions. On coming in contact with a solid, Didinium "tries" to pierce and swallow it. If this succeeds, well and good; if it does not, something else is "tried." In a culture containing many specimens of Didinium, the author has seen dozens of individuals reacting in this way to the bottom and sides of the glass vessel, apparently making persevering efforts to pierce the glass. Others "try" water plants, or masses of small algæ, about which many specimens gather at times. Of course they get no food in this way. On coming in contact with each other, the animals react in the same way, often becoming attached to each other, and sometimes forming chains of four or five. But they never succeed in swallowing one another. They often try rotifers in the same way, but the outer integument of these organisms is so tough that Didinium does not succeed in piercing it, and the rotifer escapes. Stentor and Spirostomum are often fastened upon, but usually escape, owing to their large size, great activity, and rather tough outer covering. The reason why Paramecium is usually employed as food rather than other organisms is clearly due to the fact that when the Didinia try these, they usually succeed in piercing and swallowing them, while with most other objects they fail.[1]

Didinium is a type of the hunter ciliates in this respect. The process of food-getting is throughout these species one of trial of all sorts of things. There is no evidence that in some unknown way the infusoria perceive their prey at a distance, nor that they decide beforehand to attack certain objects and leave others unattacked. They simply "prove all things and hold fast to that which is good."

We cannot do better in emphasizing this point than to quote a portion of the words of the veteran investigator Maupas, as given in Binet's "The Psychic Life of Micro-Organisms" (pp. 48, 49) : —

"These hunter infusoria are constantly running about in search of prey; but this constant pursuit is not directed toward any one object more than another. They move rapidly hither and thither, changing their direction every moment, with the part of the body bearing the battery of trichocysts held in advance. When chance has brought them in contact with a victim, they let fly their darts [2] and crush it; at this point of the action they go through certain manœuvres that are prompted by a guiding will. It very seldom happens that the shattered victim remains motionless after direct collision with the mouth of its assailant. The hunter, accordingly, slowly makes his way about the scene of action, turning both right and left in search of his lifeless prey. This search lasts a minute at the most, after which, if not successful in finding his victim, he starts off once more to the chase and resumes his irregular and roving course. These hunters have, in my opinion, no sensory organ whereby they are enabled to determine the presence of prey at a distance; it is only by unceasing and untiring peregrinations both day

---

[1] Balbiani (1873) described Didinium as discharging trichocysts from the mouth region against its prey, thus bringing it down from a distance. This account has not been confirmed by other observers, and the writer has never seen anything of the sort in the innumerable cases of food-taking in Didinium which he has observed. It can hardly be doubted that the trichocysts represented in Balbiani's figure (our Fig. 113) really come from the injured Paramecium, and not from the Didinium.

[2] This use of the trichocysts has not been confirmed by other writers and was not absolutely observed by Maupas himself.

and night that they succeed in providing themselves with sustenance. When prey abounds, the collisions are frequent, their quest profitable, and sustenance easy; when scarce, the encounters are correspondingly less frequent, the animal fasts and keeps his Lent. The *Lagynus crassicollis*, accordingly, never sees its victim from a distance and in no case directs its movements more toward one object of prey than toward another. It roams about at random, now to the right and now to the left, impelled merely by its predatory instinct — an instinct developed by its peculiar organic construction, which dooms it to this incessant vagrancy to satisfy the requirements of alimentation."

It is evident that these words of Maupas are an excellent description of behavior based on the general method of trial of all sorts of conditions though varied movements, and they bring out clearly the essential principles in the food reactions of infusoria. The same method of behavior is found, as we have seen, throughout almost the whole circle of activities in these organisms; the food reactions epitomize the entire behavior.

LITERATURE X

A. Modifiability of behavior in infusoria: JENNINGS, 1902, 1904 *d*; HODGE AND AIKINS, 1895.

B. Food habits of infusoria: MAUPAS, in Binet, 1889; BALBIANI, 1873.

PART II

BEHAVIOR OF THE LOWER METAZOA

CHAPTER XI

INTRODUCTION AND BEHAVIOR OF CŒLENTERATA

INTRODUCTION

WHILE unicellular forms are the very lowest organisms, an account limited to their behavior alone might give us a one-sided view of the principles of behavior in the lower organisms. The Metazoa differ from the Protozoa structurally in the important facts that their bodies are made of many cells and that they have a nervous system. Does the behavior of such organisms differ essentially from that of the Protozoa? Have we been dealing in our study of unicellular organisms with a peculiar group, whose behavior is of a character essentially different from that of other animals? How far do the general principles to be deduced from the behavior of Protozoa hold for animals in general? To answer these questions is the province of the following chapters.

We shall take up in detail the behavior of only one of the lowest groups of Metazoa — the cœlenterates. This will be followed by a chapter on some of the main features of behavior in other invertebrates. A general analysis of behavior in both Protozoa and the lower Metazoa is found in the third part of the book.

BEHAVIOR OF CŒLENTERATA

The Cœlenterata or Cnidaria form, perhaps, the lowest of the larger groups of Metazoa. This group includes the fresh-water Hydra, hydroids, sea anemones, corals, and jellyfishes or medusæ. The behavior of the corals and of hydroids has been comparatively little studied, so that the present account will be limited mainly to Hydra, the sea anemones, and medusæ.

All of these animals are made up of many cells, of many different kinds, and usually arranged in three more or less irregular layers. Of special interest from the standpoint of behavior are the nerve cells. In Hydra these consist of comparatively few, small cells with long, branched processes, scattered among the ectoderm and entoderm cells. They apparently serve to connect the other cells. In the sea anemones the nerve cells are more numerous than in Hydra, but are likewise scattered throughout the body, in both ectoderm and entoderm. They are somewhat more numerous in the neighborhood of the mouth than elsewhere. In Medusæ the nervous system is more concentrated. The cells and fibres form two rings about the edge of the body: one lies just beneath the ectoderm of the exumbrella, the other beneath that of the subumbrella. These rings are interconnected by scattered fibres. A plexus of nerve fibres covers the entire concave surface of the subumbrella and manubrium, beneath the ectoderm. This plexus is compared by Romanes as regards texture to a sheet of muslin. Nerve cells and fibres are found also in the tentacles, but are not known on the convex surface of the exumbrella. The two marginal nerve rings are often spoken of as the "central nervous system" in medusæ.

1. Action System. Spontaneous Activities

In the cœlenterates we take up animals with action systems differing much from those of the organisms we have hitherto studied. The chief movements are due to contractions and extensions of parts of the body and tentacles, produced by contractions of the muscle fibres. The body is flexible, and being radially symmetrical may contract or bend with equal ease in any direction.

Under natural conditions, Hydra and the sea anemone are usually attached and at rest, while the medusa may be in movement. Let us examine the behavior under such conditions, when no observable stimulus is acting on them, aside from the usual conditions of existence.

If we observe an undisturbed green Hydra attached to a water plant or the side of a glass vessel, we find that it usually does not remain still, but keeps up a sort of rhythmic activity. After remaining in a certain position for a short time it contracts, then bends to a new position, and reëxtends (Fig. 114). In this new position it remains for one or two minutes, then it again contracts, changes its position, and again extends. This continues, the changes of position occurring every one or two minutes. In this way the animal thoroughly explores the region about its place of attachment and largely increases its chances of obtaining food. This motion seems to take place more frequently in hungry individuals, while in well-fed specimens it may not occur.

Thus contractions take place without any present outward stimulus;

the movements are due to internal changes of some sort, like those of Vorticella. The same behavior may be produced, as we shall see later, by external stimuli. In the yellow Hydra such movements do not occur —at least not with such frequency.

FIG. 115. — Diagram of different positions taken by Hydra, as seen from above. After Wagner.

FIG. 114. — Spontaneous changes of positions in an undisturbed Hydra. Side view. The extended animal (1) contracts (2), bends to a new position (3), and then extends (4).

This is apparently correlated with the fact that the yellow Hydra has very long tentacles, which lie in coils all about it, so that exploratory movements are not necessary in order to reach such food as may be found in the neighborhood.

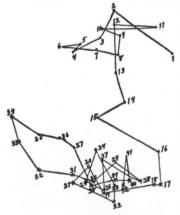

FIG. 116. — Path followed by a green Hydra that was left for some days undisturbed on the bottom of a clean glass dish. After Wagner (1905).

If a green Hydra is left for long periods undisturbed, it does not remain attached in the same position, but moves about from place to place. The movements often take place in random directions, — the animal starting first in one direction, then in another. Figure 116 shows the movements of a green Hydra, which was left alone for some days in the bottom of a large, clean glass dish, the light coming from a window at the right. This movement is probably brought about by hunger — the animals taking a new position when food becomes scarce. Hydra may move about in several different ways. In the commonest method the animal places its free end against the substratum, releases its foot, draws the latter forward, reattaches it, and repeats the process, thus looping along like a measuring worm (Fig. 117). In other cases it attaches itself by its tentacles, releases its foot, and uses the tentacles like legs. A still different form of locomotion has been described, in which the animal is said to glide along on its foot; how this is brought about is not known.

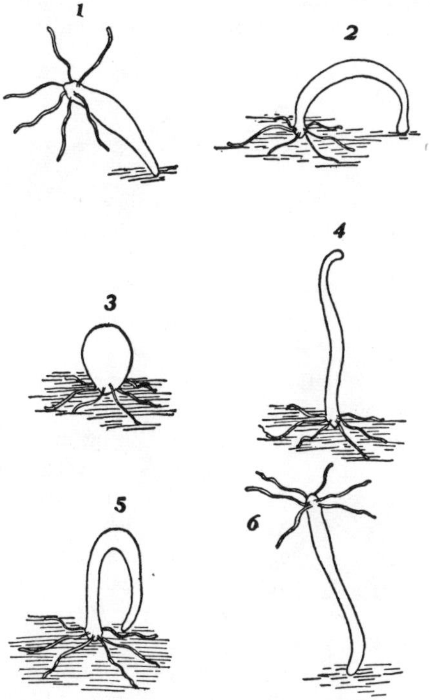

In sea anemones, rhythmical contractions of the undisturbed animal have apparently not been described. But Loeb (1891, p. 59) finds that Cerianthus if not fed will after a time leave its place in the sand and creep about, finally establishing itself in a new place. The common sea anemone Metridium moves about frequently from place to place on the sides or bottom of the aquarium, and so far as can bé observed, this seems often due simply to hunger or other internal conditions; it occurs under apparently uniform external conditions. A common method of movement in sea anemones is to glide about on the foot, — the lower surface of the foot sending out extensions and moving in a manner similar to that of the foot of mollusks. There are doubtless other methods of locomotion.

Fig. 117. — Hydra looping along like a leech. After Wagner (1905). 1-6, Successive positions.

The spontaneous contraction and change of position which plays a subordinate part in fixed forms has become the rule in medusæ. They are commonly found swimming about by means of rhythmical contractions. Since there are no corresponding changes in external conditions, these contractions must be due to internal changes. The internal changes need not of course be themselves of a rhythmical character. They may take place steadily, inducing a contraction only when a result of a certain intensity has been reached (see Loeb, 1900, p. 21).

In the small medusa Gonionemus there is under natural conditions a cycle of activity that is of great interest; it has been well described by Yerkes (1902, *a* and *b*, 1903, 1904) and Perkins (1903). At times the animal is found attached by certain adhesive pads on its tentacles to

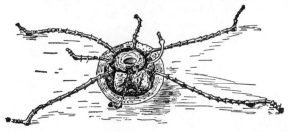

FIG. 118. — Young Gonionemus resting on the bottom, with the opening of the bell upward. After Perkins (1903).

the vegetation of the bottom or to other surfaces (Fig. 118). Leaving its attachment, it swims upward to the upper surface of the water, the convex surface of the bell being upward, and the tentacles contracted

FIG. 119. — Gonionemus swimming upward with contracted tentacles. After Perkins (1903).

(Fig. 119). Reaching the upper surface it turns over "and floats downward with bell relaxed and inverted, and tentacles extending far out horizontally in a wide snare of stinging threads which carries certain destruction to creatures even larger than the jellyfish itself (Fig. 120)" (Perkins, 1903, p. 753). Reaching the bottom, it swims again to the top and repeats the process. It may thus continue this process of "fishing," as Perkins calls it, all day long. It is chiefly in this way that it captures its food.

This cycle of spontaneous activities is in some respects similar to that of the green Hydra described above, though much more complex. Both illustrate the fact that complex movements and changes of movement may occur from internal causes, without any change in the environment.

2. CONDITIONS REQUIRED FOR RETAINING A GIVEN POSITION; RIGHTING REACTIONS, ETC.

Hydra and the sea anemones tend to retain a certain position; we usually find them at rest with foot attached and head free. This usual position is often said to be due to a reaction to gravity or to contact, or to some other simple stimulus. It will be found instructive to examine the different conditions on which depends whether the animal shall or shall not retain a given position in which it finds itself. It will be found that the matter is not an entirely simple one.

Let us take first the case of Hydra. Suppose the animal to be placed on a horizontal surface with head downward and foot upward. It does not retain this position, but bends the body, placing the foot against the bottom, releases its head, and straightens upward. This is what is commonly called the "righting" reaction. In Hydra it is not due to a tendency to keep the body in a certain position with reference to gravity, for the animal may remain attached to the bottom, with head projecting upward, or to the surface film, with head projecting downward, or to a perpendicular surface, with the body transverse or oblique to the direction of gravity. There is even apparently a certain tendency to direct the head downward. Thus out of 100 green Hydras attached to a perpendicular surface, 96 had the head lower than the foot, 3 were horizontal, and 1 had the head directed upward. It is thus clear that the righting reaction of a Hydra which has been inverted on the bottom cannot be due to any unusual relation to the direction of gravity.

To what, then, is the reaction due? Evidently there is a tendency to keep the foot in contact with a surface, for the body is bent till the foot comes in contact. But this is not all; the reaction does not stop at this point. There is likewise a tendency to keep the head free, for it is released. But still this is not all, for now the body is straightened; then the tentacles are spread out symmetrically in various directions.

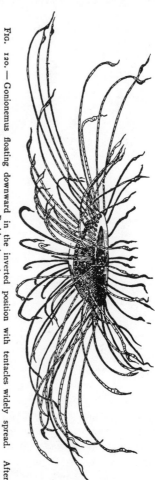

FIG. 120. — Gonionemus floating downward in the inverted position with tentacles widely spread. After Perkins (1903).

It is clear that the reaction is directed toward getting the organism into its usual position, which might perhaps be called the "normal" one; this normal position has various factors, — attachment of foot, freedom of head, comparative straightness of the body, and tentacles outspread.

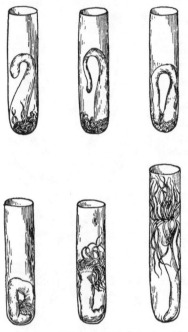

This is, of course, exactly the position which is most favorable for obtaining food.

Suppose now that our Hydra has reached this position, and all the conditions remain constant; is this sufficient? We find that it is not. If the conditions remain so constant that no food is obtained, the Hydra becomes restless and changes the position of its body repeatedly, though still retaining its attachment by the foot. But later even this is given up, and the animal, of its own internal impulse, quite reverses the position attained through the "righting reaction." It now bends its body, attaches its head, and releases its foot, thus bringing it back into the inverted position.

Is this because the irritability of head and foot have become reversed, so that the head now tends to remain attached, the foot free? Apparently not, for no sooner has the organism taken the inverted position than it draws its foot forward and now performs the "righting reaction" again, so that it stands once more on its foot. These alternations of behavior are repeated, and we find that by this means the animal is moving from place to place, as in Fig. 117.

FIG. 121. — Process by which Cerianthus rights itself when inverted in a tube. The figures are taken at intervals during the course of one hour. After Loeb (1891).

It seems clearly impossible to refer each of these acts or the whole behavior to any particular present external stimulus. Through hunger the Hydra is driven to move to another region, and these different opposite acts are the means by which another region is reached. Each step in the behavior is partly determined by the preceding step, partly by the general condition of hunger. The same behavior is often seen, as we shall see later, under continued injurious stimulation of different kinds.

In speaking of righting reactions, it is often said that the organism is *forced* by the different irritabilities of diverse parts of the body to take a certain orientation with reference to gravity or to the surface of contact (see, for example, Loeb, 1900, p. 184). The facts just brought out show that we can in Hydra consider this orientation forced only in the general sense that all things which occur may be considered forced. Man takes sometimes a sitting position, sometimes a standing one, sometimes a reclining one, depending upon his "physiological state" and past history, and the facts are quite parallel for Hydra. So far as objective evidence shows, the behavior is not forced in Hydra in any other sense than it is in man. The animal

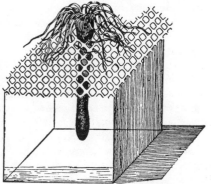

Fig. 122. — Position taken by Cerianthus after it has been placed on its side on a wire mesh. After Loeb (1891).

takes that position which seems best adapted to the requirements of its physiological processes; these requirements vary from time to time.

In the sea anemone Cerianthus the conditions for retaining a certain position are somewhat more complex than in Hydra, according to the account given by Loeb (1891). The animal is usually found in an upright position, occupying a mucus-lined tube in the sand. If placed head down-ward in a test-tube, it rights itself in the same way as Hydra, freeing the head, bringing the foot into contact, and straightening the body (Fig. 121). But in this animal, gravity clearly plays a part in the behavior. Loeb placed the animal on its side on a wire screen of large mesh. Thereupon it bends its foot down through the meshes, lifts up its head, and takes its usual position in line with gravity (Fig. 122). If now the screen is turned over, the animal again directs its head upward, its foot downward — as a human being under similar circumstances would do if possible. It may thus weave itself in and out through the meshes (Fig. 123).

Fig. 123. — Cerianthus which has woven itself through a meshwork, as a result of repeatedly inverting the latter. After Loeb (1891).

But to be in line with gravity, with head free, is not the only requirement for Cerianthus. Loeb found that it would not remain indefinitely in this position on the wire screen, as it does in the sand. After a day or so it pulls its foot out of the wire and seeks a new abode. Only when it can get the surface of its body in contact with something, as is the case when it is embedded in the sand in its natural habitat, is it at rest. If this condition is fulfilled, the requirement of the usual position in line with gravity may be neglected. Loeb found that when the animal is placed in a test-tube, so that its body is in contact with the sides, it remains here indefinitely, even *though the tube is placed in a horizontal position* (Loeb, 1891, p. 54). The head is bent upward, but the body remains transverse to the direction of gravity. Similarly, the anemone Sagartia may ofttimes take a position on the surface film with head down, although usually it maintains an upright position (Torrey, 1904).

But even the usual position in line with gravity, and with sides in contact, does not satisfy Cerianthus indefinitely, if left quite undisturbed. If it secures no food, it again leaves its place and seeks another region.

Thus that the animal may remain quiet in a given position a considerable number of conditions should be fulfilled, constituting altogether what we may call the "normal" state of the animal. The conditions are the following: (1) the foot should be in contact; (2) the head should be free; (3) the body should be straight; (4) the axis of the body should be in line with gravity, with the head above; (5) the general body surface should be in contact; (6) food should be received at intervals.

If these conditions are largely unfulfilled, the animal becomes restless, moves about, and finds a new position. But no one of these conditions is an absolute requirement at all times, unless it be that of having the head free. In the wire screen (Fig. 122) the animal remains for a day or so if in the required position with reference to gravity, even though foot and body surface are not in contact. In the horizontal test-tube it remains with foot and surface in contact, though the body is not straight nor in line with gravity. If all conditions are fulfilled save that of food, the animal remains for a time, then finally moves away.

Clearly, the holding of any given position depends, not on the relation of the body to any one or two sources of stimulation, but on the proper maintenance of the natural physiological processes of the organism. The animal does not always maintain a certain position with relation to gravity, nor does it always keep its body straight, nor its foot in contact, nor its body surface in contact. It does not at all times receive food. It may remain for considerable periods with one or more conditions lacking. It tends on the whole to take such a position as is most favorable to the unimpeded course of the normal physiological processes. Certain usually required conditions may be dispensed with, provided other favorable ones are present. The behavior represents a compromise of the various needs imposed upon the animal by its physiological processes.

In the sea anemone *Antholoba reticulata*, according to Bürger (1903), the requirements for retaining a given position are extraordinary. This animal is usually found attached to the backs of crabs; it is thus carried about, and finds much opportunity for obtaining nourishment. If removed from the crab's back, the animals attach themselves to the stony bottom and spread the tentacles. But after four or five days they release their hold on the bottom and invert themselves, directing the foot upward. Now when a crab's limb comes in contact with the foot, the latter attaches itself and folds about the limb, so that the anemone is dragged about by the crab. It now, in the course of several hours, climbs up the crab's leg to its back, where it establishes itself. The sea anemone thus by its own activity attains the extraordinary situation where it is usually found. The whole train of action is like that shown in the complicated and adaptive instincts of higher animals.

3. General Reaction to Intense Stimuli

The most characteristic reaction of the cœlenterates to intense stimuli of all sorts is a contraction of the whole body. In Hydra and the sea anemones the body is thus shortened and thickened, becoming more nearly spherical. The animals thus shrink close to the substratum and present less surface than before to the stimulating agent. In the medusæ the sudden contraction of course carries the animal away from the stimulating object. The first contraction is usually repeated many times, thus inaugurating a period of swimming by which the animal may be widely removed from the stimulus. Such contractions occur in response both to general stimulation and to local stimulation, if the latter is very intense.

Under most circumstances the contraction of Hydra or the sea anemone of course tends to remove the organism from any source of danger, rendering it for example less likely to be seized by a predatory animal. But the reaction takes place in the same way under circumstances in which it is of no defensive value. If the foot of the attached Hydra is strongly stimulated, the animal contracts as usual; the contraction is then of course *toward* the source of stimulation, not away from it. If the entire vessel containing the animals is heated to 30 degrees, the Hydras contract, though this of course does not tend to remove them from the high temperature. It is clear that for all sorts of stimuli that are unfavorable these animals have a certain reaction which is usually regulatory (beneficial); they give this reaction whatever the nature of the unfavorable stimulus, even under circumstances where it is not regulatory. This is an illustration of a characteristic general trait of behavior in lower animals; their reactions are commonly not specific, but general in character. As we shall see later, this contraction is not the final recourse of the stimulated cœlenterate. If stimulation continues, the animal usually sets in operation other activities, which remove it from the stimulating agent.

4. LOCALIZED REACTIONS

Hydra. — In Hydra, intense stimuli restricted to a small spot on the body or a tentacle usually produce contraction at that point, some-

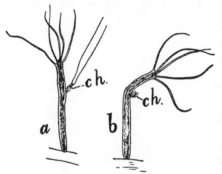

times spreading much or little, sometimes not at all. This reaction is produced by many sorts of stimuli. If the contraction remains precisely localized, as it sometimes does, the body or tentacle bends sharply at the point stimulated.

A precisely localized chemical stimulus is produced in the following way. A fine capillary glass rod is dampened and its tip is dipped in some powdered chemical. Methylene blue

FIG. 124. — A chemical (*ch.*) is brought against a certain spot on one side of a Hydra (*a*). Thereupon this spot contracts, bending the Hydra toward the side stimulated (*b*).

or methyl green is convenient to use, since the distribution of the chemical in the water is easily seen by means of the color. The point of this fine rod, covered with the chemical, is brought close to the body of a Hydra. The chemical diffuses and reaches a small area on the body. Local stimulation by heat may be produced with the simple apparatus devised by Mast (1903). A glass tube is drawn out at its middle to capillary size, then bent so as to form a loop. The two ends are passed through a cork for support, and to them are attached rubber tubes. In this way water of any desired temperature may be passed through the fine tube, and this may be brought close against the body of the animal at any desired point.

When the strong chemical or the heat reaches a certain spot on the body, this spot at once contracts, so that the body makes a knee-shaped bend at this point (Fig. 124). Such a bending is produced by most strong chemicals; strong acids placed in a capillary tube, the tip of which is applied to the body, show it clearly. As a result of the bend the head of the animal becomes directed toward the chemical or the heated region, and is therefore strongly stimulated, so that the Hydra now contracts as a whole. Thus the result of the bending is to carry the most sensitive part of the animal into the injurious agent, where it is still further injured. This reaction is produced only by strong, injurious agents, and is really an incidental result of the local injury produced. The point injured remains contracted for a long time after the stimulating agent has ceased to act. The Hydra may contract completely, so that the bend disappears, but on extension the bend is still found at the injured spot. It is evident that this bending reaction is not a regulatory one, and it is apparently never shown in nature, since the conditions necessary for its production are practically never present. It is a product of the laboratory. As we shall see later, after reaction in this manner, Hydra usually sets in operation other reactions, which do act in a regulatory way.

Sea Anemones. — Intense local stimulation of the column in the sea anemones usually produces a contraction of the entire body, or a movement of tentacles on the side stimulated, in the way described later. In Sagartia (Torrey, 1904, p. 208), stimulation of the edge of the foot induces a local contraction of the foot and base of the column, with discharge of acontia — the defensive weapons of the animal.

Local stimulation of the tentacles causes in the different sea anemones various reactions. Often slight local stimulation causes the tentacles to wave about; this and similar phenomena will be described in connection with the food reactions. In most sea anemones local stimulation of the tentacles, especially if intense, causes them to shorten by contraction, or to collapse and become very slender. This is followed in many cases by a contraction of the whole body. In Aiptasia an immediate contraction of the entire body follows even a slight stimulation of the tip of one of the long tentacles.

Medusæ. — In medusæ, intense stimulation of one side of the bell causes immediate contraction of that side, accompanied by a less marked contraction of the remainder of the bell. The stronger contraction on the side stimulated turns the animal away from that side, and its subsequent locomotion removes it at once from the stimulating agent. Thus the appropriate direction of movement is here determined in the simplest way — by contraction of the part stimulated. Such effects are produced by mechanical and chemical stimulations, by heat, by electricity, and apparently by light. In Hydra, as we have seen, identically the same reaction has the opposite effect, subjecting the animal still further to the action of the stimulating agent; other reactions must supervene before the animal is removed from the stimulus. Intense stimulation of the tentacles of the medusa or of the margin of the bell induces, in Gonionemus, a direct contraction of the tentacles.

FIG. 125.—The medusa *Tiaropsis indicans*, applying its manubrium to a point on the margin which has been stimulated. *x, y, z*, cuts made for experimental purposes. After Romanes (1885).

When the margin or under surface of the medusa bell is locally stimulated, the manubrium behaves in a manner that is of great interest. This has been described by Romanes (1885) in the medusa *Tiaropsis indicans*. If the margin or under surface of the bell is sharply stimulated with a needle, the manubrium at once bends over and applies its tip to the point stimulated (Fig. 125). The reaction is thus very precisely localized. How does it happen that the manubrium is able to locate exactly the point touched, and to bend at once in that direction?

In answer to this question, Loeb presents a very simple explanation, which deserves attention, as it is a type of many of the recent hypotheses put forward to explain the behavior of organisms. According to Loeb, this behavior is due simply to the spreading out of the local contraction caused by the stimulus. "Every localized stimulus leads to an increase in the muscular tension on all sides, which is most intense near the stimulated spot. Now if we decompose each of the lines of increase of tension (aa', ab', ac', ad', ae', Fig. 126) radiating from the stimulated spot, into a meridional component aa', dd', bb', etc., and an equatorial component, it is evident that the

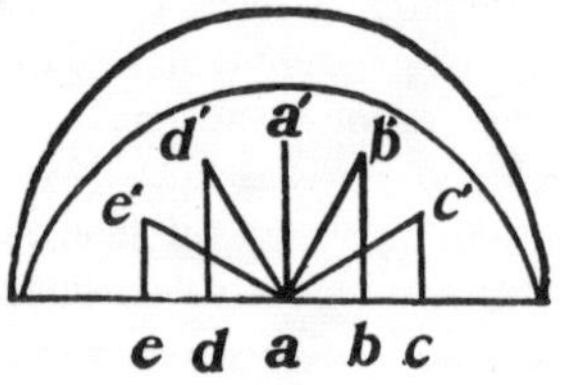

FIG. 126.—Diagram to illustrate Loeb's explanation of the localization of the reaction of the manubrium. After Loeb (1900).

latter can have no influence on the manubrium. Only the meridional components can have an influence, and of these the one passing through the stimulated spot is the largest. This fact must necessarily cause a bending of the manubrium toward the stimulated spot" (Loeb, 1900, p. 32).

This explanation represents the behavior as of the simplest character — a mere spreading of a local contraction from the point stimulated. But is this view adequate to explain the facts? In the protozoa we have found that such local action is as a rule not adequate; that the organism tests the environment; and the behavior at a given moment depends on the success or failure of a previous trial. Is there anything of this kind in the medusa, or does Loeb's simple explanation exhaust the matter?

This question is clearly answered by the experiments of Romanes. He found that if a cut is made parallel to the margin, as at x, Fig. 125, and a point lying below this cut is stimulated, the manubrium is no longer able to locate precisely the stimulated point. It bends, but no longer directly to the point stimulated. This, according to Loeb, is exactly what we should expect. The cut interrupts certain of the lines of tension, so that they no longer pull the manubrium to the precise spot. His explanation, he holds, "also shows why an incision parallel to the margin of the umbrella makes an exact localization impossible and only allows uncertain movements toward the stimulated quadrant" (1900, p. 32). It is easy to see that the manubrium, on Loeb's theory of decomposition of the lines of tension, would be pulled over in the general direction of the stimulated spot, but might not strike it exactly.

Is this what happens? Let us examine the facts as set forth in Romanes' own words: "Although in the experiment just described the manubrium is no longer able to *localize* the seat of stimulation in the bell, it nevertheless continues able to perceive, so to speak, that stimulation is being applied in the bell somewhere, for every time any portion of tissue below the cut a is irritated, the manubrium actively dodges about from one part of the bell to another, applying its extremity now to this place and now to that one, as if seeking in vain for the offending body. If the stimulation is persistent, the manubrium will every now and then pause for a few seconds, as if trying to decide from which direction the stimulus is proceeding, and will then suddenly move over and apply its extremity, perhaps to the point that is opposite the one which it is endeavoring to find. It will then suddenly leave this point and try another, and so on, as long as the stimulation is continued" (Romanes, 1885, p. 112–113).

From Romanes' description it is evident that the manubrium under these circumstances may not even move in the general direction of the point stimulated; he says expressly that it may move toward the opposite point, or toward any other point. At times, he says, a manubrium moves from point to point, "without being able in the least degree to localize the seat of irritation." The considerations adduced by Loeb do not explain these facts; and his theory is quite inadequate to account for the behavior. Contraction occurs, not merely as a direct spreading from the point stimulated, but now in one place, now in another, including even a region directly opposite that stimulated. The manubrium, having reacted once, does not cease, but in some way recognizes its failure and tries again. In other words, failure changes its physiological state, so that now it bends in a new direction. The whole account given by Romanes is as vivid a description of the method of reaction by the production of varied movements subjecting the organism successively to different conditions, as it would be possible to imagine under these circumstances.

It would be most interesting to determine whether the animal may thus by trial finally discover the irritated spot, and later through repetition come to bend toward it directly, as it did before the cut was made.

5. The Rejecting Reaction of Sea Anemones

In some sea anemones the presence of masses of waste matter on the disk leads to the performance of activities which result in the removal of the waste matter; this behavior we may call the rejecting reaction. Such behavior is well seen in the large sea anemone *Stoichactis helianthus*, found in the West Indies. This animal has a flat or concave disk 10 to 15 cm. in diameter, covered closely with tentacles about 8 mm. in length. If a quantity of dead plankton, or a mass of sand, or other waste matter, is placed on the disk, the animal sets in operation measures which remove it. Food placed on the disk of a specimen that is not hungry produces the same result. The behavior under such circumstances is complex, and the removal of the waste matter may be accomplished in more than one way.

The tentacles of that region of the disk bearing the waste body collapse, becoming thin and slender and lying flat against the disk. The disk surface in this region begins to stretch, separating the collapsed tentacles widely. As a result the waste mass is left on a smooth, exposed surface, the tentacles here having practically disappeared, while elsewhere they form a close investment. Thus the waste is left fully exposed to the action of the waves or currents, and the slightest disturbance in the water washes it off. Under natural conditions this must result in an immediate removal of the mass of débris. If this does not occur at once, often the region on which the débris is resting begins to swell, becoming a strongly convex, smooth elevation, thus rendering the washing away of the mass still easier.

But if the débris is not removed by the reaction just described, then new activities set in. If the waste body is near one edge of the disk, this edge usually begins to sink, while at the same time the tentacles between the edge and the waste mass collapse and practically efface themselves. Thus the mass slides downward off the disk. If this does not occur at once, after a time the region lying behind the mass begins to swell; it often forms in this way a high, rounded elevation. The waste mass is now on a steep slope, and is bound soon to slide over the edge. Sometimes by a continuation of these processes the entire disk comes to take a strongly inclined position, with the side bearing the débris below. Often one portion of the edge after another is lowered successively till all of the waste matter is removed and the disk is thoroughly cleaned. The disk then resumes its horizontal position, with nearly flat or slightly concave surface.

Sometimes the edge bearing the débris cannot be lowered, owing to the fact that it is almost against an elevation in the irregular rock to which the anemone is attached. In this case (after perhaps an attempt to bend this edge downward) the part between this edge and the débris swells and rises, rolling the mass toward the centre, while at the same time the region beyond the débris sinks down. In this way the waste matter is rolled across the disk to the opposite side, and dropped over the edge. The process is slow, often requiring fifteen minutes to half an hour.

This whole reaction is characterized by great flexibility and variability. The débris sets in operation certain activities; if these do not put an end to the stimulation, other activities are induced, till one is successful. This is an excellent illustration of the general characteristics of behavior in the lower organisms.

6. Locomotor Reactions in Hydra and Sea Anemones

After contracting in response to stimulation, if the stimulus still continues, Hydra and the sea anemones usually set in operation other activities, having a more radical effect in separating the animal from the source of stimulation. We have examined certain cases of this character in the foregoing section on the rejecting reaction. We shall here consider such reactions as tend to remove the animal, or cause it to take a new position.

Hydra. — After contracting in response to stimulation, Hydra usually bends over into a new position and soon extends again in a new direction, just as happens in its spontaneous contractions (Fig. 114). This may be repeated many times, the animal occupying successively many different positions.

In bending thus into a new position in response to a one-sided stimulus, does Hydra bend directly away from the source of stimulation? Wagner (1905) and Mast (1903) have answered this question experimentally. Wagner tried stimulating one side of the body mechanically, while Mast raised or lowered the temperature of one side. Both authors agree as to the following results: The direction of extension after contraction bears no definite relation to the side from which the stimulus came; the animal is just as likely to extend toward the source of stimulation as in any other direction. In other words, when stimulated, Hydra merely changes its position, without special relation to the localization of the stimulating agent. The direction of bending and extension is determined by internal factors. If the stimulus is repeated, contraction occurs again, and the animal extends in still another direction. The analogy of these relations with those shown by the infusoria is evident; the latter when stimulated usually merely change the direction of movement, without regard to the direction from which the stimulus came. In the infusoria the internal factors (structural in character) which determine the direction have been determined; this has not yet been done for Hydra.

But repeated or strong continued contraction, with extension in a new direction, is not the final recourse of Hydra under strong stimulation. If the stimulation continues, the animal finally bends over, places its head against the surface to which it is attached, releases its foot, and moves away from the spot where it has been subjected to such objectionable experiences. The locomotion is usually of the sort illustrated in Fig. 117. This reaction has been observed by Wagner (1905) under mechanical stimulation, by Mast (1903) under stimulation by heat, and by the present author under stimulation by chemicals. In all cases it was found that the direction toward which the animal moves bears no definite relation to the direction from which the stimulus comes. Wagner stimulated one side repeatedly by striking it with a rod, and found that the animal was as likely to move toward that side as in any other direction. The experiments of Mast are particularly interesting in this connection. Mast placed a considerable number of Hydras in a flat-bottomed trough, and heated one end. At about 31 degrees C. the animals began to release their foothold and move about from place to place. But they were as likely to move toward the heated end as away from it. The results of a series of such experiments are shown in Fig. 127. In this figure are represented not only the movements of locomotion, but also the different directions in which the animal extended after contracting. The diagram shows clearly that both sets of movements are quite without definite relation to the direction from which the heat comes; their direction evidently depends on internal factors. When it experiences the high temperature, the animal merely changes its position, in a way determined by its structure or other internal factors. If the high temperature still continues, it changes position again, and thus continues till the high temperature ceases or the animal dies. The behavior resembles essentially that of infusoria under similar conditions. The reaction is very ineffective under the conditions shown in Fig. 127, owing to the slowness of the movements of Hydra. Most of the animals in the heated region finally die. But if the animals moved rapidly and far at each change of position, then those that moved away from the heated side would escape, and those that moved in the wrong direction the first time would, after one or two changes of direction, likewise get out of the heated region. The reaction would be of precisely the same character as that of the infusoria. But the action system of Hydra is evidently adapted only for meeting changed conditions over a very limited area, such as may be escaped by a slight, slow movement.

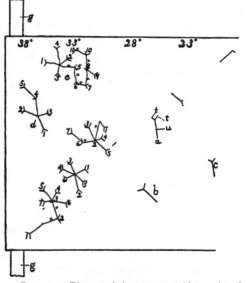

FIG. 127. — Diagram of the movements of a number of Hydras when the trough containing them was heated at the end to the left. Each of the small diagrams represents the movements of a single Hydra. The figures 1, 2, 3, etc., show the successive different directions in which the Hydra extended while remaining attached. The cross (×) between two numbers indicates that here the animal released its foothold and moved in the direction shown to a new point of attachment. After Mast (1903).

When the changed conditions cover too large an area, the Hydra can only "try" its usual reaction; if this fails, it must die.

A decrease of temperature does not cause Hydra to change position. As the temperature becomes lower, the animal merely becomes more sluggish, contracting more slowly and at longer intervals, till finally, near the freezing point, movement almost ceases (Mast).

As we have seen on page 194, an internal condition — hunger — may induce the same locomotor reactions as are produced by continued external stimulation. This is a matter which we shall take up again in the account of food reactions.

Sea Anemones. — In some sea anemones, as in Hydra, repeated strong stimulation causes the animal first to contract, then to bend into new positions, and finally to move away. Each of these reactions may be repeated several times before the succeeding one occurs. There are certain features of this behavior that are of much interest, since they lead to results analogous to habit formation in higher animals. The facts have been most carefully studied in *Aiptasia annulata.*

Aiptasia is a rather slender, somewhat elongated actinian living in crevices beneath and between stones. If stimulated by touching the disk or tentacles with a rod, it contracts strongly. It then extends in the same direction as before. When it is fully extended we repeat the stimulus. The animal responds in the same way as at first. This continues usually for about ten or fifteen stimulations, the animal extending each time in the same direction as at first. But at length, when stimulated anew, the polyp contracts, bends over to one side, and extends in a new direction. As the stimuli are continued, the animal repeats for a number of times the contraction and extension in the new direction, then finally turns and tries a still different position.

This change of position may be repeated many times. But in the course of time the reaction becomes changed in a still different manner. The anemone releases its foothold and moves to a new region. This same reaction is produced in Cerianthus, as we have seen, by hunger.

Aiptasia frequently extends in most awkward turns, the body taking and retaining an irregular and even crooked form. This is evidently due to its life in irregular crevices and crannies. In order that its disk may protrude into the open water, it is compelled to extend in the irregular ways mentioned, and to retain the crooked shapes thus produced. When removed from its natural habitat, it still retains these irregularities of form and action, so that a collection of Aiptasias shows all sorts of right-angled and zigzag shapes. It would appear that these irregularities must have arisen as a result of the way in which the animal extends in its natural surroundings. From this it would appear that a method of extension frequently repeated must in the course of time become stereotyped, forming what we are accustomed to call in higher animals a habit.

If this is the case, then it should be possible to produce new stereotyped reaction forms, by so arranging the conditions that the animal shall be compelled to extend always in a certain way (differing from its former way), and to retain the form thus induced. In some specimens this result is obtained with the greatest ease, and in a very simple manner. Thus, in a certain case, an individual attached to a plane horizontal glass surface was bent in extension far over to the left. Stimulating it repeatedly, it contracted at each stimulation, then bent, in extending, again to the left. But after some fifteen stimulations it turned away, and bent over to the right. Now when stimulated it contracted as before, then bent regularly, in extending, over to the right. It seemed to have acquired a new method of behaving, bending to the right instead of to the left.

Close examination showed that the cause of this phenomenon is as follows: When it contracts in response to stimulation, it does not regain a completely symmetrical structure, but remains a little more contracted on the side that is concave in extension. In extending anew, this side still remains a little more contracted than the opposite one, so the animal takes a curved form, concave toward the same side as in its previous extension. In other words, the structure conditioning the curved form is not completely lost even when the animal contracts, and it becomes evident again on a new extension.

Thus in Aiptasia the formation of a stereotyped method of action depends upon very simple conditions. Yet there can hardly be a doubt that the permanent individual peculiarities of form and action found under natural conditions, as mentioned above, have risen in exactly this way. It thus plays the part taken by what is called habit formation in higher animals.

The facts set forth in the present section show clearly that the cœlenterates do not always react in the same way to the same external stimulus. Internal conditions of the organism, as determined by past stimuli received, past reactions given, and various other factors, are of equal importance with external conditions in determining behavior. We shall see many further illustrations of this fact in the reactions toward food.

7. ACCLIMATIZATION TO STIMULI

Besides the changes in behavior under constant stimuli that we have described in the last section, there are certain others which may perhaps be classed as acclimatization to stimulation. In sea anemones a light stimulus that is not injurious may cause at first a marked reaction, then on repetition produce no reaction at all, or a very slight one. Thus, a drop of water is allowed to fall from a height of 30 cm. on the surface of the water just above the outspread disk of *Aiptasia annulata*. The animal at once contracts completely. After the animal has expanded, another drop is allowed to fall in the same way. As a rule, there is no response to this or to succeeding drops. Sometimes there is a reaction to the first two or even three drops, but usually reaction ceases after the first one.

Sometimes a slight reaction of a different character supervenes after the stimulus has been repeated many times. The animal begins to shrink slowly away from the region where the drops are falling, so that in the course of time the disk has been withdrawn much farther below the surface, though no decided reaction has occurred to any one stimulus.

8. Reactions to Certain Classes of Stimuli

In the foregoing sections we have taken up reactions to mechanical stimuli, heat and cold, and chemicals; we shall have occasion to consider some of these further in the account of food reactions. There are certain other classes of external stimuli which may play a part in determining behavior in these animals; these we will take up separately.

A. *Reactions to Electricity*

Induction shocks have been much employed in experimental work on contraction in cœlenterates. The results of such stimulation do not

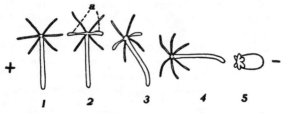

Fig. 128. — Reaction of an attached Hydra to a constant electric current of moderate intensity. 1–5, successive stages in the reaction. After Pearl (1901).

differ greatly from those produced by other forms of stimulation (mechanical, etc.), local or general contractions occurring in dependence on the strength of the current. These may be followed by locomotor movements.

The effects of the constant electric current are more peculiar and of greater interest. They have been studied in Hydra by Pearl (1901); in the medusa *Polyorchis penicillata* by Bancroft (1904).

Hydra. — In Hydra the constant current causes local bendings of the body similar to those produced by sharply localized chemical and thermal stimuli. If a weak current is passed through the water transversely to the Hydra, the animal contracts on the anode side, at a point a little above the foot, thus bending the body (Fig. 128). At the same time or a little before, the tentacles which were in line with the current contract (Fig. 128, *a*). Sometimes, further, there is a contraction on the anode side just below the base of the tentacles. As a result of the contraction on the anode side, the Hydra bends toward the anode. As soon as it comes into a position with the anterior end directed toward the anode, the entire body contracts, since a Hydra in this position is stimulated more than in any other (Fig. 128, 5). In a stronger current the complete contraction takes place first, then the animal slowly bends over toward the anode. If, as sometimes happens, the foot is free while the head is attached, the bending takes place as usual on the anode side.

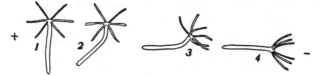

FIG. 129. — Successive stages in the reaction of a Hydra to the electric current when the foot is unattached. The foot becomes directed toward the anode. After Pearl (1901).

The result is necessarily that the foot becomes directed toward the anode, so that in this case the orientation of the animal is the reverse of that found in the specimens attached by the foot (Fig. 129). This result shows clearly that the orientation to the electric current is due to the direct local contractions caused by the current on the anode side, and is not due to an attempt on the part of the animal by anything like a process of trial to come into a certain definite position.

In a Hydra placed transversely to the current, the tentacles contract in a peculiar way. A weak current causes only the tentacles which are in line with the current to contract, and of these, that extending toward the cathode contracts more quickly and more completely than

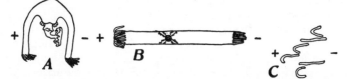

FIG. 130. — Effects of the constant electric current on pieces of Polyorchis. After Bancroft (1904). *A*, meridional strip passing through the manubrium. *B*, similar strip stretched out in line with the current. *C*, isolated tentacles.

that directed toward the anode (Fig. 128, *a*). If the Hydra is lying parallel with the current, the body contracts much more readily when the anterior end is directed toward the anode than when it is directed toward the cathode. In either of these positions the tentacles usually remain extended, and somewhat inclined toward the cathode (Figs. 128 and 129). But if a very strong current is used, both body and tentacles contract strongly. Pieces of the animal react in essentially the same way as the entire organism, and young buds (with tentacles) react in the same way as adults, but are more sensitive to the current (Pearl, 1901).

Medusa. — If strips of various shapes are cut from the medusa Polyorchis, and subjected to the action of the constant current, the tentacles and manubrium bend toward the cathode (Fig. 130, *a*, *b*). This takes place even with isolated tentacles (Fig. 130, *c*). If the current is long continued, such isolated tentacles partially relax, then contract again. This is repeated, so that an irregular rhythmic contraction is produced by the constant current.

B. *Reactions to Gravity*

The position of the body and the direction of locomotion are partly determined in some of the Cœlenterata by gravity. There is great diversity among different members of the group in this respect. In some, gravity is an almost constant determining factor in the behavior. In others it plays only an incidental part, affecting the behavior under certain circumstances, while in still other cases it seems to have no effect on the movements whatsoever.

We have already seen that the position taken by Cerianthus is partly determined by gravity. The sea anemone Sagartia, according to Torrey (1904), usually moves upward when this is possible, and at the same time it tends to keep its body in line with gravity, with the disk above. If while moving on the floor of the aquarium it reaches the perpendicular side, it at once begins to ascend. Since Sagartia creeps by movements of its foot, remaining in the upright position, its ascent on a vertical surface involves bringing the body into an oblique position, in place of the usual perpendicular one. Thus its tendency to creep upward interferes with its tendency to keep its body in line with gravity, and the former prevails. Sagartia may also creep on the under side of the surface film, with head down, so that it is by no means a rigid requirement that the head shall be above. Doubtless many other sea anemones will show a tendency to keep the body in a certain position with reference to gravity.

In the hydroid Corymorpha, according to Torrey (1904 *a*), there is a decided tendency to take a position with the head (or oral end) upward. When placed in an inverted or oblique or horizontal position, Corymorpha rights itself by a bending of the body, which is due, according to Torrey, not to muscular contraction, as in the sea anemones, but to a change in the turgidity of the large axial entoderm cells. Those on the lower side become more turgid, increasing in volume and thus bending the stem directly upward. Either the entire animal or a piece of the stem, without head or foot, reacts in this manner. Thus the reaction is in this animal comparable to the reaction to gravity in a plant.

But in many species of fixed cœlenterates gravity clearly has little or nothing to do with the usual position. Metridium, Aiptasia, *Stoichactis helianthus*, *Condylactis passiflora*, and many others are found occupying all sorts of positions with reference to gravity, and the same is true of Hydra and various hydroids.

In some medusæ the movement is partly guided by gravity. Gonionemus, as we have seen, swims in its "fishing" movements upward to the surface. Yerkes (1903) found that this occurs in the same way when the light comes from below, so that the guiding factor is apparently gravity. This reaction to gravity is of course not constant; it occurs only at intervals and under certain circumstances.

Careful examination will probably show that gravity plays a part in certain episodes of the behavior of most of these animals, even though it may not affect their usual position or direction of motion. Thus, gravity plays a part in the "rejecting reaction" of the actinian Stoichactis, described in Section 5 of the present chapter. The situation "waste-matter-on-the-disk-not-removed-by-the-first-reaction" is responded to by taking such a position with reference to gravity as results in removing the waste; then the reaction to gravity ceases. Similar transitory reactions to gravity, seeming to serve definite ends, are found in many other animals. Thus, in the hermit crab, according to Bohn (1903), we have such a case. While investigating a shell which it may adopt as a home if fitting, this animal takes a certain position with reference to gravity; namely, with body on the steepest slope of the shell, and head downward. It then turns the shell over (the position mentioned being the most favorable one for this action), and ceases to react with reference to gravity. Other cases of the same sort will be described for the flatworm Convoluta (Chapter XII). Gravity has, of course, many diverse effects on the substance of organisms, and in almost no case has its precise action in directing movements been determined. When an animal is inverted, this may cause a redistribution of the constituents of the body or of the separate cells. Such a redistribution would probably interfere with the usual physiological processes, and might therefore act as a stimulus to a change of position. Again, in freely moving organisms, gravity causes differences in the ease of movement in different directions, and such differences may well determine the direction of motion. Again, a change in the usual position with reference to gravity may induce unusual strains in various parts of the body, or may shift the weight of the body to parts unaccustomed to bearing it; and these effects might serve as stimuli to cause the animal to take another position. This possibility will be vividly realized by any one who undertakes to rest with a limb doubled in some unusual position beneath him. Again, certain movements with reference to gravity may produce results involving a change of the conditions affecting the organism, and since it is a well-established fact that the results of behavior partly determine future behavior, this fact may determine movements with reference to gravity. There seems to be no *a priori* reason why each of the relations above mentioned, as well as various others, may not induce reaction in one organism or another, and it seems not difficult to find probable examples of all. We have been assured by various writers that the reaction to gravity must be explained in the same way in all cases, but this is evidently said rather in the capacity of a seer or prophet, than in the capacity of a man of science whose conclusions are inductions from observation and experiment.

C. *Reactions to Light*

Many of the sea anemones and medusæ do not react to light, so far as known. In other cases a reaction to light is very marked. The relation of the behavior to light is in certain cases exceedingly complex, and very instructive, as showing the numerous factors on which behavior depends. We shall take up especially the reactions of Hydra, and of the medusa Gonionemus.

(1) *Reaction to Light in Hydra*

The behavior of Hydra with relation to light has been studied especially by Wilson (1891). Both the green and the brown Hydra are usually found at the lighted side of the vessel containing them. If they are at first scattered, they will in a day or two be found to have moved to the lighted side. If at the side of the dish next the window there are attached light and dark strips of glass, the Hydras collect in the light strips. If different colored lights are used, by placing strips of glass of different colors on the lighted side of the vessel, the Hydras collect in the blue light, while all other colors (except perhaps green, which seems slightly effective) act like darkness. The animals gather in the blue even in preference to the white light, which of course contains all the blue rays. As to the way in which the reaction to light takes place, the following facts were brought out by Wilson. A change from light to dark, or from blue or white light to one of the colors which acts like darkness, causes the animal to become restless and move about. The motion seems undirected, but as soon as the animal comes into the blue or white light, it becomes less restless, and remains. The behavior is thus far, then, like the reaction to heat; the animal when not lighted simply moves about in various directions, till one of its movements brings it into light. Whether the animal when moving draws back or stops on coming to the boundary of the light, where it would pass into the darkness, as Euglena does, has not been determined. But when the vessel is lighted from one side, the animal moves toward the source of light, and the movement is no longer an irregular wandering, but according to Wilson (1891, p. 432) is fairly direct. This is like the reaction of Euglena, and it seems possible that in Hydra the reaction is produced in the same manner as in that organism. If this is true, there is a tendency for the moving animal to keep its anterior end directed toward the light, due to the fact that when it turns this end away, the change to relative obscurity at the anterior end causes further movement, till the light again falls on the anterior end. The movements should be studied further to determine this point. Fixed Hydras do not maintain any particular orientation with reference to the light rays, but change their position frequently, in the way illustrated in Fig. 114. The green Hydra moves to the lighted side of the vessel more rapidly than the yellow Hydra. This is probably due to the generally more rapid movements of the green species.

In a powerful light the reaction of Hydra, like that of most other positive organisms, becomes reversed. The animals collect in the shadow of leaves or on the bottom. They have not been observed to move directly away from the source of light (Wilson, 1891), so that the reaction is probably an irregular wandering based on the method of trial.

Hertel (1904) found that both the green and the colorless Hydra react by contraction when subjected to powerful ultra-violet light. These rays killed the colorless Hydra in about one minute, while *Hydra viridis* resisted their action for six to eight minutes.

The gathering of Hydras in lighted areas and the movement toward a source of moderate light are of much benefit to the animals in obtaining food. Hydra preys upon small crustacea and other minute animals, and these gather as a rule at the lighted side of the vessel. By taking a position on this side, the Hydras find themselves in the midst of a dense swarm of organisms and are able to capture much food. When in such situations one frequently finds them gorged with prey. In other parts of the vessel they would have almost no opportunity of obtaining food (Wilson, 1891).

(2) *Reactions to Light in Gonionemus*

The relation of the behavior of the medusa Gonionemus to light, as studied by Yerkes (1902 *a*, 1903), is exceedingly complex; it can by no means be expressed by any simple formula. In examining the matter it will be well to consider first the relation of the light to the amount of activity shown by the animal; then the nature of the activities in constant lights of various intensities; then the effects of changes of illumination.

In ordinary daylight, Gonionemus continues its usual activities, swimming about by rhythmical contractions, and pursuing its usual occupation of "fishing" (p. 192). It is not clear that the direction of its movements has any relation to the direction of the rays of light, so long as all conditions remain uniform. If the light comes from below instead of above, Yerkes (1903) found that Gonionemus continues to swim to the top and float to the bottom, as before.

If the light is cut off, the medusa usually comes to rest after one to five minutes. By covering the vessel containing them, it is thus possible to bring the animals to rest for experimental purposes. In continued darkness the animal is much less active than in the light.

In strong sunlight the animal becomes very active. At first it swims toward the source of light, thus rising under natural conditions to the surface of the water. Later its reaction changes; it stops coming to the surface, begins to avoid the light, and swims toward the bottom. It may now persistently strike against the bottom in its efforts to swim away from the source of light. Sometimes in a strong light it places the more sensitive subumbrellar surface against the bottom and comes to rest. At times its activities become, under the action of direct sunlight, uncoördinated; it moves upward in its contraction, downward in its expansion.

In a moderate light coming from one side the behavior of Gonionemus is at times very peculiar. When the conditions are quite uniform, as we have seen, its movements often show no relation to the direction of such a light. But when the light first begins to act, as when a jar containing medusæ is placed near a window, they at first swim toward the source of light. The medusæ thus gather at the lighted side of the vessel. But after a time, if undisturbed, they cease to react to light, and may scatter throughout the vessel. If there are regions of light and shade, the animals now usually gather in the shaded region. But if they are again disturbed in some way, as by stirring up the water, they swim toward the light again, — later scattering as before, when the conditions become uniform.

Thus the reaction of the animal depends on its physiological state; when excited it moves toward the light, otherwise it is indifferent or gathers in the shade. In the flatworms we find a parallel condition of affairs, but with the relations reversed. It is not unlikely that the tendency of the medusa to go toward the light when disturbed is related to its usual method of life, and has a functional value. The animal when at rest is commonly attached to the vegetation of the bottom. When disturbed by a large animal foraging among the plants, it would move toward the light, hence out into the free water and upward, thus escaping the enemy.

Thus far we have considered the behavior under light of constant intensity. Let us now see the effects of sudden changes in intensity of illumination. Here we find again that the effect of a given change depends on the state of the animal. If the medusa is at rest on the bottom, a sudden marked increase in the intensity of the light usually causes a sudden contraction of the bell. As a result the animal, of course, swims away from its first position. Sometimes, however, an increase of light merely causes an animal that is at rest with the sensitive concave surface up to turn over, so as to bring the sensitive surface against the bottom, where it is little affected by the light. In a case described by Yerkes, increase of light caused regularly this turn with bell up, while decrease caused a return to the "bell down" position.

A decrease of light usually has no effect on a resting Gonionemus. But sometimes it causes contraction, so that the medusa swims away. In such specimens an increase of light usually causes no reaction. Sometimes, however, a given specimen reacts both to increase and decrease of illumination.

Thus the reaction of a resting medusa to a change of illumination is variable, depending on the individual. Doubtless in a given individual it varies with the physiological state and past history of the animal.

In the swimming Gonionemus, usually both an increase and a decrease of light cause the animal to expand, cease swimming, and sink to the bottom. Here it usually remains for a time, then resumes activity.

If a vessel containing a number of the medusæ is divided by a line *x–x* into two regions, one brightly illuminated, the other shaded, the animals usually behave as follows: A specimen swimming about in the light region crosses in its course the line *x–x*, passing into the shade. It at once ceases swimming and sinks to the bottom. Here it remains for a short time, then continues to swim about in the shaded region.

If a specimen swimming in the shaded region crosses the line *x–x* into the light, it likewise sinks to the bottom and remains quiet for a time. Now, upon resuming activity, it swims in such a way as to pass back into the shade. Yerkes is convinced, from analogy with the effects of other stimuli, that this is due to a stronger contraction on the side most intensely lighted — that farthest from the shadow. This would, of course, turn the medusa back into the shade.

Thus in the course of time practically all the medusæ in the vessel will be found in the shaded region.

In the behavior of Gonionemus with relation to light there are evidently a number of paradoxical facts. The medusa swims toward the source of light, yet tends to gather in shaded regions. It goes at first toward a source of strong light, later reverses this reaction. It moves toward the source of light when excited, but becomes indifferent when undisturbed. Different individuals react differently to the same conditions, and the same individual reacts differently at different times. We have here an excellent illustration of the fact that the reactions of organisms, even to simple agents, depend on a multiplicity of factors. If we could study the medusa in the natural conditions under which it lives, and if we knew thoroughly the physiological processes taking place within it, we should doubtless find all these peculiarities explained, and should probably discover that its reactions are regulatory. When we carry such an animal to the laboratory and experiment upon it there, it is like removing an organ from the body and studying it in a dissecting dish. We cannot understand its activities without knowing their relations to the rest of the body — to the environmental conditions.

9. Behavior of Cœlenterates with Relation to Food

The behavior of organisms is largely determined by the relation of the environment to their internal physiological processes. In no field is this so striking as in the relation of behavior to the obtaining of material for carrying on the processes of metabolism. Under this point of view come the reactions of organisms with reference to food, and to the gases necessary for respiration. These reactions in the Cœlenterata we shall take up now.

A. Food and Respiratory Reactions in Hydra

Hydras are usually found in the upper parts of a vessel of water, near the surface. This is not due to a reaction to gravity, but rather to the relative quantity of oxygen in different parts of the water. If an experiment is arranged in such a way that the lower surface of the vessel is free and in contact with air, while the upper is not, the Hydras tend to gather near the lower surface (Wilson, 1891). Collecting in oxygenated regions is probably brought about through a process of trial, the organisms wandering irregularly till they come into oxygenated regions and there remaining. If the water is allowed to become very foul, all the Hydras soon collect at the very upper surface, often in contact with the surface film itself.

Let us now examine the usual behavior of Hydra in obtaining food, as described by Wagner (1905). As we have seen, the undisturbed green Hydra changes its position at intervals, thus in the course of time exploring thoroughly all the region about it. The tentacles of the green Hydra are comparatively short, so that such exploring movements are needed. In the colorless Hydras the tentacles are often excessively long and slender, lying in coils on the bottom, and almost filling the surrounding waters with a network of fine threads. They may reach three or four inches in length. In these species changes of position are less frequent, the great length of the tentacles rendering this unnecessary. When a small animal comes in contact with one of the tentacles, in a typical case a somewhat complicated reaction occurs. The nematocysts of the region with which the animal comes in contact are shot out, causing the organism to cease its movements.

FIG. 131.—Hydra endeavoring to swallow a large annelid. Camera drawing.

The tentacle is viscid and clings to the animal. Now the tentacle is bent toward the mouth. At the same time the other tentacles bend in the same direction. If the animal is a large one and is inclined to struggle, the other tentacles seize it, and many nematocysts are shot out and pierce it, so that the organism may become quite covered with these structures. An insect larva which was rescued from a Hydra at this stage is shown in Fig. 132, *B*. Meanwhile, the mouth becomes widely opened, sometimes before the prey comes in contact with it. When the food reaches the mouth, the tentacles usually release it and are folded slightly back, while the edges of the mouth, or "lips," actively work up over the food, till it is enveloped and passes into the cavity of the body. In this way a Hydra often takes organisms much larger than itself. Figure 131 shows such a case, where a Hydra endeavored to swallow an annelid that was, at a moderate estimate, fifty times its own bulk. The mouth and body were immensely distended, and the worm was about half enveloped. The Hydra seemed then to have reached its utmost limit, and the process stopped.

We now wish to analyze this complicated behavior, determining as far as possible the nature and causes of the different factors which make it up. We may ask first, What is the cause of the discharge of the nematocysts?

Near each nematocyst there is a projecting point, the cnidocil (Fig. 132, *cl*). This has often been compared to a trigger; touching the cnidocil is said to cause discharge of the nematocyst. That is, it is supposed that a mechanical stimulus is the cause of the discharge. But experiment does not bear out this supposition. Hydra may be rubbed roughly with a needle, without causing discharge of the nematocysts.

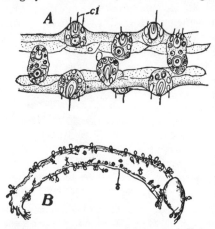

FIG. 132.—Nematocysts and their action in Hydra. *A*, portion of a tentacle, showing the batteries of nematocysts; *cl*., cnidocils. *B*, insect larva covered with nematocysts as a result of capture by Hydra.

Hard organisms, such as Ostracods, may strike against it or run over its surface, brushing against many cnidocils, yet no nematocysts are discharged. On the other hand, various chemicals readily cause discharge of the nematocysts; a solution of methylene blue or methyl green, for example, produces this effect in a marked degree. Apparently, then, some chemical stimulus must be associated with the mechanical stimulus in order to cause discharge of the nematocysts. Chemical stimuli of one sort or another will doubtless usually be received from the organisms which serve as prey.

To what is the remainder of the behavior due? One thing which must be noticed first is that the food reaction depends upon the physiological condition of the animal. Not all Hydras react to suitable food, but only those which have not been recently fed. It is, of course, not surprising that only hungry Hydras should eat. Yet this brings out the important point that the behavior is not an invariable reflex, but depends on the physiological state of the organism.

When the animal eats, are the determining factors of the reaction mechanical stimuli or chemical stimuli? Experiment shows that mechanical stimuli alone do not induce the food reaction. If bits of filter paper, or ostracods with a hard shell, are brought in contact with the tentacles or the mouth of a hungry Hydra, they are not swallowed. But if the filter paper is soaked in meat juice, or if the ostracod is crushed, then they are readily swallowed. A chemical stimulation is a necessary factor in producing the reaction. But under usual conditions the chemical alone — the meat juice — will not produce the food reaction. There must be a combination of chemical stimuli (of the proper character) and of mechanical stimuli before the reaction is induced.

But when the Hydra is very hungry — when it has starved for a long time — then a suitable chemical stimulus acting alone will produce the food reaction. Placed in a solution of extract of beef the very hungry Hydra opens its mouth widely and takes in the fluid. What seems very remarkable is that a solution of quinine produces this effect as well as does extract of beef (Wagner, 1905).

Thus the food reaction is throughout dependent upon the physiological condition of the Hydra. Hydras that are not hungry will not eat at all; moderately hungry specimens will take the solid food (chemical and mechanical stimuli); very hungry ones take liquid food (chemical stimulus alone). Hungry Hydras show still further modifications in their behavior, compared with those that are not hungry. As we have previously seen, they frequently contract and change to a new position and even move about from place to place. Wilson (1891) records a remarkable cycle of behavior in hungry yellow Hydras. Hydras usually remain, as we have seen, in the upper layers of the water, on account of the oxygen there found. But when the crustacea on which the animals feed have become very scarce, so that little food is obtained, Hydra detaches itself, and with tentacles outspread sinks slowly to the bottom. Here it feeds upon the débris composed of dead organic matter which collects at the bottom, often gorging itself with this material. It then moves toward the light, and at the lighted side again upward to the surface. Here it remains for a time, then sinks again and feeds upon the material at the bottom. This cycle may be repeated indefinitely, requiring usually some days for its completion.

B. *Food Reactions in Medusæ*

The food reactions have been studied most carefully in Gonionemus. In this animal, as we have seen, there is a definite set of "fishing" movements, having the function of obtaining food. These movements are of course not direct reactions to food, but are, so far as food is concerned, spontaneous movements of the animal. If food is brought near a resting medusa, this sets the animal to moving. If a piece of fish is placed at one side of the medusa, it does not move directly toward the food, according to Yerkes (1902 *a*). After a few seconds the tentacles nearest the food begin to move about irregularly, and this gives them a chance to find the food if it is very near.   If they do not find it, "there soon follows a general contraction or series of contractions of the bell, which may take the animal either toward or away from the source of the stimulus."   Thus the medusa is induced by the presence of food to swim about, and it usually in this way sooner or later comes in contact with the food (Yerkes, 1902 *a*, p. 438).   The behavior is throughout not a definitely directed action, but an excellent example of the method of trial — of what we call *searching*, in higher animals.

When the tentacles actually come in contact with food, they contract and twist about each other in such a way as to hold it.   The group of contracting tentacles then bends toward the mouth, and that portion of the margin of the bell bearing them contracts, drawing them nearer the mouth.   The manubrium bends toward the food, placing the mouth against it, and the food is enveloped by the lips and swallowed.

What are the determining factors in this behavior?   Doubtless, as in Hydra, internal conditions play a part in determining the reaction to food bodies, but this matter has not been studied in the medusa.   As to external factors, Yerkes (1902 *a*) has brought out the following: In Gonionemus the entire food reaction may be produced by chemicals alone.   If with a pipette a strong infusion of fish meat is applied to the tentacles, they twist and contract, bending toward the mouth, while the manubrium as usual bends toward the tentacles stimulated.   Solutions of common inorganic chemicals do not produce this result; the tentacles merely contract from them, remaining straight.   If the infusion of fish meat is made very weak, the animal begins the food reaction, contracting and twisting the tentacles; but the reaction goes no farther.   In rare cases Yerkes (1902 *a*, p. 439) found that the animal begins the food reaction when a very weak inorganic chemical, such as an acid, is applied to it.   But this quickly ceases, before it has gone far.   The medusa in such cases makes what we call in higher animals a mistake, but changes its behavior as soon as it discovers the mistake.

Mechanical stimuli of a certain sort may likewise produce the food reaction.   With regard to this we find in Gonionemus certain peculiar and most suggestive relations.

If the tentacles come in contact with some quiet object, or are touched with a rod or a needle, they merely contract, remaining straight, as when they are affected by inorganic chemicals.   The response is clearly a negative reaction, not a food reaction.   But if the tentacles are touched in a peculiar way, by drawing the rod quickly across them, they behave differently.   They quickly react and twist, just as when they touch a piece of meat.   Then they bend toward the mouth, the margin bearing them contracting inward as usual, while the manubrium bends toward them. Finding no food, the swallowing movements of the manubrium do not occur. Thus "motile touch," as Yerkes (1902 *a*) calls it, causes the food reaction, while the touch of an object that is at rest causes only a negative reaction. This reaction to a moving object shows clearly the adaptation of the behavior to the natural conditions of life. Usually, when something moves quickly along the tentacles of a medusa, this will be a fish or other small animal, well fitted to serve as food. So the medusa reacts to such a moving thing in such a way as to seize it and bear it to its mouth. If the object turns out not to be good for food, as is rarely the case, there is of course no harm done, and it may be rejected. If the medusa comes in contact with an object that is not moving, this will probably be a stone or plant or other object not fit for food, hence the animal makes no attempt to take it. The behavior is based, at it were, on the probability that any given case will correspond to the usual condition. Movement serves to the medusa as a sign of something living and fit for food, just as it does to hunters among higher animals and even among men.[1] It is a most interesting fact that the positive reaction to a moving object is more rapid than to a quiet one, even though the latter is actually food, while the former is not. The reaction time for a moving object was found by Yerkes (1902 *a*, p. 440) to be about 0.30 to 0.35 seconds, while the reaction time for quiet objects or food is 0.40 to 0.50. This is again directly adapted to usual conditions; to a moving animal reaction must be rapid, or it is useless. One can hardly do otherwise than hold that this specialized reaction to moving objects, so appropriate to the natural conditions of the animal, is not a primitive reflex, but must have been historically developed in some way, and that it would not occur if it were not in the long run beneficial.

C. *Food Reactions in Sea Anemones*

In sea anemones the dependence of the reactions toward food and other agents on the physiological state of the animal, particularly as determined by the progress of metabolism, is very striking.

Finding Food. — Sea anemones remain for the most part quiet, with disk and tentacles outspread, depending for food largely on the accidental contact of moving organisms with these organs. But there are

---

[1] In this, as in other cases, such expressions as "serves as a sign" of course does not affirm a mental sign, concerning which we have no knowledge in animals outside of the self. It signifies merely that movement does, as a matter of fact, cause a reaction which is appropriate to something usually accompanying the motion, so that the behavior is objectively identical with that due in higher animals and man to a stimulus that serves as a sign.

certain active movements which assist in procuring food. In most sea anemones light stimulation of the tentacles, however produced, causes these organs to wave back and forth, just as happens in medusæ; this increases the chances of coming in contact with food. In Sagartia, according to Torrey, the presence of food near one side of the animal, resulting in weak chemical stimulation, gives rise to more definite movements. Part of the tentacles bend toward the food, contracting on the side most strongly stimulated, while others bend toward the mouth. The animal may at times bend its body toward the food, thus securing it. The tendency of the tentacles to bend toward the mouth, as if carrying food, when stimulated in almost any way, is very striking in many cœlenterates. In Sagartia the tentacles when touched bend first toward the side stimulated, then toward the mouth. In the hydroid Corymorpha, according to Torrey (1904 a), the tentacles when thus stimulated bend only toward the mouth. This bending toward the mouth of course serves the function of carrying food, and it seems to have become the reaction to all sorts of stimuli, on the chance, as it were, that it will serve this function, in the given case. The plan of the behavior is that of trial of a reaction that is beneficial under most circumstances.

The Taking of Food. — In the actual taking of food the behavior varies greatly in different sea anemones. In some species ciliary movement plays the chief part in the process,[1] though assisted by muscular contractions. In others, bodily movements brought about by muscles are the main factors. Two or three examples will illustrate the principal variations in this matter.

The common *Metridium marginatum* of the east coast of the United States is an example of the species in which ciliary movement is perhaps the chief agent in food-taking. Under usual conditions the tentacles are pointed away from the mouth, and are covered with cilia, which beat toward the tip of the tentacle. Thus small particles falling on the tentacles are carried outward by the cilia and removed from the animal. But if the particle is something fit for food, the behavior is changed. When a bit of crab's flesh is dropped among the tentacles, they contract on the side touched, thus grasping the flesh. They then bend inward, arching over with tips toward the mouth. The cilia, continuing to strike toward the tip, now of course carry the food toward the mouth instead of away from it. In time the meat drops from the tip of the tentacles into or near the mouth.

The inner surface of the œsophagus, or tube into which the mouth leads, is covered with cilia, which beat outward (save in the two grooves at the angles, known as the siphonoglyphes). They thus bear outward

---

[1] For details regarding this for many different species, see Carlgren, 1905.

any indifferent particles which may fall in the œsophagus. But when a piece of meat is dropped into the mouth, the cilia at once reverse, now beating inward. They thus carry the food into the digestive cavity of the animal. Meanwhile, the muscles surrounding the mouth, and those of the œsophageal tube, contract in such a way as to produce swallowing movements, which aid in ingesting the food. These swallowing movements may begin while the food is still held by the tentacles, showing that the stimulation from the food has been transmitted.

In *Aiptasia annulata* there are cilia which act in the same manner as in Metridium, but the chief rôle in food-taking is played by movements of the tentacles and œsophagus. If a small object comes in contact with a tentacle, it adheres to the surface, and the tentacle contracts strongly, the entire animal usually contracting at the same time. Then the tentacle bends over and places the food with considerable precision on the mouth. The adjacent tentacles likewise bend over and are applied to the food body, holding it down against the mouth. The latter then opens, the lips seizing the food, while the tentacles may release it and bend away. The swallowing of the food is mainly due to the activities of the lips and œsophagus. In this animal a bit of food may be completely enclosed within ten seconds of the time it touches a tentacle.

In the large sea anemone *Stoichactis helianthus*, cilia seem to play no part in the taking of food. In this animal the disk may be 10 to 15 cm. in diameter. If a piece of crab meat is placed on the disk of a hungry specimen, the tentacles immediately surrounding it begin suddenly to wave back and forth. This movement stops for a few seconds, then begins again. All the tentacles that come in contact with the food bend over against it and shrink, so as to hold it down against the disk. Now that portion of the disk bearing the food begins to sink inward, the mouth begins to open, and the walls of the œsophagus protrude from the mouth as large bladderlike lobes. The region between the mouth and the food contracts, the tentacles which it bears collapsing and almost completely effacing themselves. By this contraction the mouth and food are caused to approach each other, the intervening region almost disappearing. The œsophageal lobes increase in size, becoming 3 or 4 cm. long and half as thick; they extend toward the food, finally reaching it. The mouth may, in the way described, be transferred from the centre of a disk 10 cm. in diameter to within 1 cm. of the edge. Now the œsophageal lobes extend over the food, while the tentacles progressively withdraw from it, till the food is lying on the contracted part of the disk, completely covered by the œsophageal lobes. Now that part of the disk below the food withdraws, by an extension and displacement of the mouth, till there is nothing beneath the food body, and it is pressed by the œsophageal lobes into the internal cavity. The lobes then withdraw and the mouth closes.

The determining factors in the food reaction are partly internal, partly external, the variations of the former playing perhaps the most important part. Many of the sea anemones are voracious, taking food until the body forms a distended sac. But in most species, if not all, the behavior changes decidedly as the animal becomes less hungry, and after a time it refuses to take food, even removing it if the food is applied to the disk. The changes in reaction as hunger decreases seem less marked in those species in which the food is taken mainly by ciliary action.

Specimens that have not been fed for a long period frequently swallow indifferent bodies, such as pellets of paper, grains of sand, and the like. This has been observed in Aiptasia (Jennings, 1905 a), Sagartia (Torrey, 1904), Metridium (Allabach, 1905), and in a number of Mediterranean anemones (Nagel, 1892). In Stoichactis the taking of such indifferent bodies is rare, but sometimes occurs. In Sagartia and Metridium such indifferent bodies cause a reversal of the beat of the œsophageal cilia, just as is occasioned by actual food. All together, it is clear that in hungry specimens of various sea anemones mechanical stimuli acting alone may cause the food reaction.

In some cases chemical stimuli acting alone produce the food reaction. If filtered crab juice is applied to the tentacles of Metridium, they arch over toward the mouth. If the juice reaches the mouth, the cilia of the œsophagus are reversed, striking inward, just as when a piece of meat is present. The swallowing movements of the œsophagus may likewise take place under chemical stimulation. Parker (1905) has lately found that certain inorganic chemicals, containing potassium, will cause the cilia to reverse and beat inward; this is the case for example, with KCl and KNO_3. But the reversal which takes place under the action of meat juice is not due to the potassium salts which it contains, for it requires a concentration of the potassium salt to produce this result that is much greater than that existing in meat juice. In Adamsia, according to Nagel (1892), the tentacles react to sugar in the same way as to meat juice; this is not true for Metridium and Sagartia.

As sea anemones become less hungry they usually cease to react to such indifferent bodies as grains of sand, pellets of paper, etc., though they still take crab meat readily. In Metridium and Sagartia bits of paper no longer cause the reversal of the œsophageal cilia, by which particles are carried to the mouth, while crab meat still produces this effect. In *Aiptasia annulata* the tentacles no longer carry pellets of paper to the mouth, but bend backward along the column and drop them. In the Stoichactis that is not very hungry such indifferent bodies are removed by the rejecting reaction described on page 202.

As the sea anemones become still less hungry the reaction to even such food bodies as pieces of crab meat becomes changed. The reaction gradually becomes slower and less precise. In a hungry specimen of Aiptasia the food reaction is rapid, often requiring but ten or fifteen seconds. But after several pieces of meat have been taken, the reaction occupies a much longer period. The tentacles touched by the food may not react for several seconds, then they bend in a languid way toward the centre of the disk, while the adjacent tentacles may not react at all. The food body is not placed so accurately on the mouth as before. At a later stage food applied to the tentacles induces no reaction at all, or a withdrawal of the tentacles, while if it is applied directly to the mouth it is very slowly swallowed. In Stoichactis at this stage food is often carried toward the mouth, then after or even before it reaches the mouth the reaction is reversed and the food is rejected. If two pieces of meat are applied at once to the disk of Stoichactis when in this condition, one may be swallowed while the other is rejected. Often in Aiptasia one piece may be rejected, while the immediately following piece is swallowed. The animal seems in a condition of most unstable equilibrium, so that the reactions are most inconstant and variable. No one could suppose, in studying the behavior of a sea anemone in this condition, that the behavior of such organisms is made up of invariable reflexes, always occurring in the same way under the same external conditions.

As the animal becomes satiated, the food reaction ceases completely. Pieces of crab meat placed on the disk of a Stoichactis in this state are removed by the rejecting reaction already described. Aiptasia either does not react at all when food is applied to the tentacles, or the tentacles contract and bend backward — a negative reaction.

Some anemones are exceedingly voracious, seeming to take food as long as it is mechanically possible for them to do so. This seems to be the case, for example, with Metridium, where the changes in reaction as the animal becomes filled with food are almost lacking. It may feed till the body cavity becomes so completely filled as to cause disturbance of function. As a result the entire mass of food is sometimes disgorged undigested. After this has occurred, Metridium will often take food as before. But in most sea anemones the taking of food ceases before any such disturbance has been produced.

The rejection of food is not determined merely by the mechanical fulness of the digestive cavity, but is evidently due to the effects of food on the internal processes. An Aiptasia (species undetermined), studied by the present author, continued to take filter paper till the body was a swollen sack, and pieces of the paper were repeatedly disgorged. But new pieces, and even those that had just been disgorged, were readily swallowed when applied to the disk. But when specimens of this Aiptasia were fed considerable quantities of meat, they refused to take either more meat or paper.

The reactions of well-fed sea anemones differ in many other ways from those of hungry specimens. They are much less inclined to react to stimuli of all sorts. A disturbance in the water, or a touch with a needle, that would produce a strong contraction in the hungry animal, often causes no reaction whatever in the satiated specimen. A much stronger solution of any given chemical is required to produce contraction than in the well-fed individual. If we should attempt to determine the strength of a given chemical that caused contraction in Aiptasia, we should get totally different results, according as we employed specimens that were very hungry, or only moderately hungry, or thoroughly satiated.

Another factor influencing the food reactions of the sea anemone is fatigue, and the effects due to this are easily mistaken for phenomena of a different character. If the tentacles of a certain region of the disk of Metridium are given many pieces of food, one after the other, they refuse after a time to take the food, though the other tentacles will still take food readily. In taking food very large quantities of mucus are produced, and it is not surprising that many rapid repetitions of this process exhaust the tentacles. If they are allowed to rest five to ten minutes, they usually take food as at first.

As the fatigue comes on, the tentacles first cease to react to weak stimuli, such as are produced by plain paper, or paper soaked in meat juice; later to strong stimuli, such as that produced by meat. If meat and paper are given in alternation, the tentacles will thus at first take both; then they come to refuse the paper, while the meat is still taken. Later they come to refuse the meat also.

The reaction to food varies also with certain other conditions. In Metridium and Aiptasia the following is often observed: A specimen refuses to take bits of filter paper, though it still takes meat. After it has thus refused paper, two or three pieces of meat are given in succession, and taken readily. Now the bit of paper is placed again on the disk, and it too is swallowed. Clearly, the uninterrupted taking of a number of pieces of meat changes the physiological condition in some way, preparing the animal for the taking of any object with which it comes in contact. One cannot fail to note the parallelism with what occurs in higher animals under similar conditions.

10. INDEPENDENCE AND CORRELATION OF BEHAVIOR OF DIFFERENT
PARTS OF THE BODY

There is a general agreement among those who have studied the
behavior of cœlenterates that the different parts of the body show remarkable independence in their reactions. The tentacles of the sea
anemones and medusæ react to most stimuli in essentially the same
manner when cut off from the body as when attached. The isolated
tentacles of Gonionemus react to meat juice by contracting and twisting, as in the usual food reaction, while to inorganic chemicals they react
by a straight contraction, as in the negative reaction of the medusa
(Yerkes, 1902 *b*, p. 183). In Sagartia (Torrey, 1904) and Metridium
(Parker, 1896) the separate tentacles react to meat juice by bending
toward the side which formerly looked toward the mouth. Thus
each tentacle must contain within itself the apparatus necessary for its
usual reactions.

The fact that the tentacles have their own reactions independently of
the rest of the body is illustrated in a curious way in Loeb's experiment
on heteromorphosis in Cerianthus (Loeb, 1891). He succeeded in
causing tentacles to develop at one side of the animal, forming a group
not associated with a mouth. These tentacles reacted to food as usual,
seizing upon it, and bending over with it in the direction in which,
under normal conditions, a mouth would be found. Here it was pressed
down for a time, then released.

Like the tentacles, other parts of the body may react independently.
Yerkes (1902 *b*) cut off the manubrium of Gonionemus and pinned it
by its base to the bottom of a dissecting dish. It now bent toward food,
seized upon and swallowed it, just as in the uninjured medusa. Many
experiments with similar results are described in the work of Romanes
(1885). Parker (1896) isolated a small bit of the ciliated epithelium
of the œsophagus of Metridium. He found that this reacted to meat
juice by a reversal of the ciliary stroke, just as happens in the uninjured
animal. In Actinia, Loeb (1891) found that if the head is cut off, the
lower part of the animal will take food through the œsophageal opening.
If the animal is cut in two, even the open lower end of the upper half
will take food, just as will the mouth.

For experiments of this kind, the bell of the medusa has become,
through the work of Romanes (1885), a classical object. Separating
the margin of the bell, containing the chief portion of the nervous system, from the central part, has been a favorite experiment. Romanes
found that in the Hydromedusæ the margin continues to beat rhythmically, while the centre usually ceases its spontaneous movement. But

this was not due to any actual inability of the centre to initiate movement, for Romanes found that when it was stimulated in various ways,
it contracts rhythmically.   This occurred in the centre of the bell of
Sarsia when placed in certain chemicals, notably in weak acids, and
in a glycerine solution (Romanes, 1885, pp. 190–197).   Rhythmical
contractions have likewise been observed by Loeb (1900 a) in the isolated centre of Gonionemus when placed in a pure solution of sodium
chloride.   Thus it is clear that not only the margin, containing the
greater part of the nervous system, but also the centre of the bell, has
the power of contracting rhythmically.

These and many other experiments have shown that each part of
the body has in the cœlenterates certain characteristic ways of reacting
to stimuli, and that it may react in these ways even when separated
from the rest of the body.   Its reactions may be determined within
itself.   But from this the conclusion cannot be drawn that the behavior
of these animals consists entirely of the separate and independent reactions of these parts to external stimuli.   While each part *may* react
independently, each may also react with reference to influences coming
from other parts of the body.   Thus, the tentacles may react, not only
to external stimuli directly impinging upon them, but also, in many
cœlenterates at least, to stimuli that are transmitted from other parts.
A strong stimulus on the body or on a single tentacle causes a contraction of many tentacles.   In some cases this contraction of the other
tentacles appears to be due to a direct spreading of the muscular contraction.   One fibre pulls on another, setting it in action, until the pull
reaches the base of the tentacle.   This pull then acts as a direct stimulus, causing the tentacle to contract, in the same way that would occur
if it were mechanically stimulated from outside.   This is the way in
which Torrey conceives of the matter in Sagartia.   If this is the correct
explanation, there is of course nothing comparable to nervous transmission — passage of a wave of stimulation independently of a wave of
contraction — in these cases.

In *Aiptasia annulata*, on the other hand, a light stimulus on the tip
of one of the long tentacles induces a sudden quick contraction of the
entire body.   This contraction appears to the eye to take place over
the entire body at once, and it is so rapid as to suggest strongly the operation of a conducting nervous system.   The well-known experiments of
Romanes (1885, p. 76) demonstrated completely that in medusæ there
is such a wave of stimulation independent of a wave of contraction,
and that this wave of stimulation coming from other parts of the body
causes the tentacles to contract.   By cutting off the margin of Aurelia
in the form of a long strip and stimulating one end, he could cause a

wave of stimulation to pass to the opposite end. This wave of stimulation was followed, if the stimulus was intense, by a wave of contraction; if the stimulus was weak, the wave of stimulation passed alone.
This wave caused the tentacles along the margin to contract as it
reached them.

Furthermore, we have seen above that the reaction of the tentacles
or of other parts of the body to a given stimulus depends upon the general physiological state of the body, as determined by the progress of
metabolism. Certain tentacles may, through the activity of totally
different tentacles, in another region of the body, in supplying material
for the metabolic processes, come to react to a given stimulus in a manner entirely different from their former reactions.

The tentacles are therefore not to be compared exclusively to independent organisms associated in a group, but they form parts of a
unified organism. While they may react when isolated, they react also
under the influence of other parts of the body. We have of course the
same condition of affairs in the muscles and various other organs of
vertebrates. They may react when isolated, but, like the tentacles
of the medusa, they likewise react to influences coming from other parts
of the organism.

The same is true for the manubrium and for other parts of the body.
While the isolated manubrium of Gaonionemus may react by bending
toward food, it shows the same reaction when certain of the tentacles
are stimulated by an object moving rapidly across them. The varied
reactions of the manubrium to influences affecting other parts of the body
are shown most clearly in the experiments of Romanes described on
page 201. In Hydra, when the tentacles have seized food, the mouth
often begins to open long before the food has reached it. In Metridium,
according to Parker, when the tentacles are touched by food, the
œsophagus frequently shows peristaltic contractions, and the sphincter of
the mouth closes. It is clear that there is a definite coördination and
unity in the behavior, brought about by a transmission of stimuli from
one part of the body to another. The difference between these organisms
and higher animals is in this respect only one of degree. In the cœlenterates a large share of the behavior is due to the independent reactions
of the different organs to the external stimuli, and the transmission of
influences from one part of the body to another takes place slowly and
without such precision as we find in higher animals.

The part played by the nervous system in unifying the body we
need not take up here, as it has been thoroughly analyzed in the brilliant work of Romanes (1885), and has been further discussed by Loeb
(1900). The essential conclusion to be drawn from the experimental

results seems to be as follows: The nervous system forms a region in which the physiological changes resulting in activity take place more readily and rapidly than in other parts of the protoplasm. These changes occur in the nervous system more readily both as a result of the action of external stimuli, and under the influence of changes in neighboring parts of the body. Hence parts containing the nervous system are more sensitive to external stimulation than other parts of the body, and they serve to transmit stimulation more readily. Furthermore, the spontaneous changes occurring in the protoplasm, which result in the production of rhythmical contractions, are more pronounced and rapid in the nervous system than elsewhere, so that the rhythmical contractions usually begin in parts containing nerve cells. But the difference between nerve cells and other cells is only quantitative in character. The peculiar properties of the nerve cells are properties of protoplasm in general, but somewhat accentuated.

11.   Some General Features of Behavior in Cœlenterates

Comparing the behavior of this low group of multicellular animals with that of the Protozoa, we find no radical difference between the two. In the cœlenterates there are certain cells — the nerve cells — in which the physiological changes accompanying and conditioning behavior are specially pronounced, but this produces no essential difference in the character of the behavior itself. As in the Protozoa, so here, we find behavior based largely on the process of performing continued or varied movements which subject the organism to different conditions of the environment, with selection of some and rejection of others. We find the same changes in behavior under a continued intense stimulus, determined by changes in the physiological condition of the animal. We find at the same time many reaction movements of a fixed character, dependent largely on the structure of the organism, as we do in bacteria and infusoria. Many of these specific responses to specific stimuli are so definitely adapted to the precise conditions under which the organism lives that we can hardly resist the conclusion that they have been developed in some way under the influence of these conditions, as a result of the fact that they are beneficial to the organism. Such, for example, is the quick though complicated grasping and feeding reaction by which Gonionemus responds to a moving object. Possibly such determinate reactions have arisen through fixation of movements which were originally reached by a process of trial, — a possibility to which we shall return in our general analysis of behavior.

In the Cœlenterata we find also, as in Amœba, a certain number of responses due to the simple, direct reaction (by contraction) of the part affected by a local stimulus. Where such simple and perhaps primitive reactions are advantageous to the organism, they are preserved as important factors in behavior, as in the negative reactions of medusæ. Where they are not advantageous to the organism, they are replaced, supplemented, or followed by more complicated reactions, so that they form a comparatively unimportant feature in the behavior of most of these animals.

In cœlenterates we find the same dependence of behavior on the physiological state of the organism that we found so marked in Protozoa. The same organism does not react always in the same way to the same external conditions. In the present group the dependence of behavior on the progress of the internal physiological processes, particularly those of metabolism, stands out strongly. The animal in which material for the metabolic processes is abundant differs radically in its behavior from the hungry specimen. The reaction to a given stimulus depends not alone on the anatomical structure of the animal and the nature of the stimulus, but also upon the way the internal processes are taking place. We cannot predict how an animal will react to a given condition unless we know the state of its internal physiological processes, and often whether a positive or negative reaction will help or hinder the normal course of these processes. The external processes of behavior are an outgrowth and continuation of the internal processes.

The state of the organism as regards its metabolic processes seems indeed the most important determining factor in its behavior. Certain internal metabolic states drive the animal, without the action of any external agent, to the performance of long trains of activity, of exactly the same character as may also be induced by external stimulation. The state of the metabolic processes likewise determines the general nature and the details of the reactions to external stimuli. It decides whether Hydra shall creep upward to the surface and toward the light, or shall sink to the bottom; how it shall react to chemicals and to solid objects; whether it shall remain quiet in a certain position, or shall reverse this position and undertake a laborious tour of exploration. It decides whether the sea anemone shall react to indifferent bodies, and to food, by the long and complex "food reaction" or the equally long and complex "rejecting reaction." It determines whether Cerianthus shall remain quietly in its tube in the sand, or shall seek a new abode. Innumerable details of behavior are determined in the cœlenterates by this factor.

The same dependence of behavior on the metabolic processes of the organism we have seen in the Protozoa, and especially in the bacteria.

Here, however, the change of behavior of a given individual with a change in these processes needs further investigation; this has been experimentally demonstrated only with reference to respiratory processes in certain green organisms.   In higher animals the dependence of the behavior on the state of metabolism is of course most evident.

This dependence of the reaction to stimuli on the relation of external conditions to internal processes is a fact of capital importance, which may furnish us a key to many phenomena that are obscure from other standpoints.   The processes of metabolism are not the only ones occurring in organisms, and the relation of external conditions to other internal processes may equally determine behavior.   This is perhaps the most fundamental principle for the understanding of the behavior of organisms.

Of a character differing from those just considered are certain other factors which modify behavior in the cœlenterates.   Past stimuli received and past reactions given are, as in the Protozoa, important determining factors in present behavior; they may cause either the cessation of reaction to a given stimulus, or a complete change in the character of the reaction.   Certain simple conditions produce a tendency in the organism to perform more readily an act previously performed (p. 206).   The internal state of the organism may be changed in most varied ways, giving rise to corresponding changes in behavior.   These facts give behavior great complexity, as well as great regulative value, even in so low a group as the one now under consideration.

LITERATURE XI

BEHAVIOR OF CŒLENTERATA

A. Behavior of Hydra: WAGNER, 1905; WILSON, 1891; MARSHALL, 1882; PEARL, 1901; MAST, 1903; TREMBLEY, 1744.

B. Behavior of sea anemones: LOEB, 1891, 1895, 1900; NAGEL, 1892, 1894, 1894 a; PARKER, 1896, 1905, 1905 a; TORREY, 1904; JENNINGS, 1905 a; ALLABACH, 1905; CARLGREN, 1905; BÜRGER, 1905.

C. Behavior of hydroids: TORREY, 1904 a.

D. Behavior of jellyfish: ROMANES, 1885; YERKES, 1902 a, 1902 b, 1903, 1904; PERKINS, 1903; BANCROFT, 1904; LOEB, 1900, 1900 a.

CHAPTER XII

GENERAL FEATURES OF BEHAVIOR IN OTHER LOWER
METAZOA

THE foregoing chapters attempt to give a connected systematic account of behavior in the Protozoa and the Cœlenterata. These may serve as types of the lower organisms. The necessary spatial limits of the present work render impossible a similar treatment of other groups. We must content ourselves therefore with a survey of some of the main features of behavior in some other invertebrates. We shall take into consideration chiefly the lower groups.

1. Definite Reaction Forms ("Reflexes")

In the action systems of most organisms we find certain well-defined reaction forms, or what are often known as reflexes,[1] which make up a large proportion of the behavior. In the groups we have thus far considered, such definite reaction types are seen in the avoiding reactions of infusoria, the definite contractions occurring in response to stimuli in the Protozoa and Cœlenterata, the bending of the tentacles toward the mouth when stimulated by food, in the hydroids and sea anemones, and in many other features of the behavior. It is true, as we have seen, that even these so-called reflexes are usually variable when studied in detail, and their occurrence and combination depend upon a multiplicity of internal as well as external conditions. Yet certain elements of behavior do occur in accordance with a definite type, and this fact is one of much importance. In some lower animals behavior is largely made up of such definite reaction forms. This fact has assumed an overshadowing importance in much recent work on behavior; investigation has taken largely the form of a search for precisely definable reflexes and tropisms, and for conditions under which they occur in the typical way, while other factors in the behavior have been neglected. Since these matters have been so much dwelt upon, we need not take them up in great detail in the present work.

The best-known case of behavior made up largely of such definite

---

[1] The use of this term will be discussed later.

reaction forms is that of the sea urchin, as studied by v. Uexküll (1897, 1897 a, 1899, 1900, 1900 a). The sea urchin differs from most lower animals in bearing large numbers of motor organs scattered over its entire surface. Most prominent of these are the spines, which are movable, and may be used as legs, or as means of defence. Among the spines are certain peculiar jawlike organs known as pedicellariæ (Fig. 133), each borne on a movable stalk. These jaws frequently open and close, seizing foreign objects. The surface of the body between the spines and pedicellariæ is covered with cilia. Finally, the body bears five double rows of tube feet, — fleshy tubular suckers, protruded through rows of holes in the shell. These are important organs of prehension and locomotion. All these different sets of organs are interconnected by a network of nerves, one set lying on the outer surface of the shell, another on the inner surface. These nerves connect with the five radial nerve trunks, which unite to form a ring surrounding the mouth.

FIG. 133. — One of the pedicellariæ from a sea urchin. After v. Uexküll

V. Uexküll finds that each of these organs (omitting the cilia) has a number of definite reactions or reflexes, which it performs in response to definite stimuli. In these reactions each organ may act as an independent individual. If a piece of the shell bearing but a single spine or pedicellaria is removed, this organ reacts to external stimuli in essentially the same way as when connected with the entire animal. These reflexes change with different intensities and qualities of stimuli, and with certain other conditions, and they are different in diverse sorts of pedicellariæ. But each reflex has a very definite character. Thus the sea urchin appears to be made up of a colony of almost independent structures. Each of these structures has reactions of such a character that they perform certain functions that are useful in the life economy of the animal.

Yet these organs are not entirely independent. They are connected by the nervous network in certain definite ways, so that when one of them performs a certain action, others may receive a transmitted stimulus, and may perform the same or a differing action. That is, each organ may receive stimuli not only from the outer world, but also, through the nerves, from other parts of the body. These interconnections are of such a character that they cause the various organs to work in harmony, usually assisting to perform certain necessary functions.

Thus, if débris falls upon the sea urchin, the pedicellariæ seize it, break it into bits, and with the aid of the spines and the cilia remove it from the body. Small animals coming in contact with the sea urchin are seized by the pedicellariæ and held, till they are grasped by the slow-moving tube feet and spines, and by them carried to the mouth and eaten. When the sea urchin is attacked by an enemy, the spines all bend toward the region of attack, presenting a serried array of sharp points to the advancing enemy. In some species this occurs even when a shadow falls upon the animal. The spines present their points to the shaded side, thus arranging for an effective defence in case the animal which has cast the shadow shall advance to an attack. In some sea urchins, poisonous pedicellariæ seize an enemy, usually causing a quick retreat. Further, when the animal is severely stimulated from one side, the reflexes of the spines are so arranged as to carry the animal in the opposite direction. When attacked, the animal is thus effectively defended, while at the same time it flees.

V. Uexküll emphasizes the independence of these organs, the definite character of their reflexes, and the definiteness of the interconnections between them. These qualities give the characteristic stamp to the behavior of the sea urchin. According to v. Uexküll, this animal is a "republic of reflexes." Every reflex is of the same rank, and is independent of the others, save for the definite connections that we have mentioned. There is nothing like a central unity controlling the reflexes, according to v. Uexküll. The sea urchin, he holds, is a bundle of independent organs, and it is only through the arrangement of these organs that a seemingly unified action is produced. "It is only by the synchronous course of the different reflexes that there is simulated a unified action, which really does not exist. It is not that the action is unified, but the movements are ordered, *i.e.* the course of the different reflexes is not the result of a common impulse, but the separate reflex arcs are so constituted and so put together that the simultaneous but independent course of the reflexes in response to an outer stimulus produces a definite general action, just as in animals in which a common centre produces the action" (1899, p. 390). The difference between the behavior of the sea urchin and that of higher animals is concretely expressed by v. Uexküll in the statement that when a dog runs the animal moves its legs; when the sea urchin runs the legs (spines) move the animal.

Yet the fixity of these reactions is by no means absolute, even in the sea urchin. As we shall see in the next section, v. Uexküll discovered a number of definite laws in accordance with which they change, and there is positive evidence of still other modifying factors not easily formulated.

In scarcely any other group of lower animals does there appear to be such a multiplicity of these definite units of reaction as in the sea urchin.   In the starfish the extension and withdrawal of the tube feet, and the extrusion and withdrawal of the stomach in feeding, may be considered examples.   In free-swimming rotifers we find an avoiding reaction similar in all essentials to that of the ciliate infusoria, the animals when stimulated turning toward a structurally defined side.   There is the same variability in this reaction that we find in the infusoria.   In planarians, the earthworm, and many other worms, reactions of a fairly well-defined character are seen in the turning of the head toward certain stimuli and away from others.   These reactions play a large part in the behavior of Planaria, according to Pearl (1903).   Weak stimuli of all sorts affecting one side of the body cause the positive turning; stronger ones, the negative turning.

Such reactions often depend closely on the localization of the stimulus.   This may be illustrated from the behavior of the flatworm just mentioned.   A weak stimulus at the side of the head, near the anterior tip, causes the head to turn only a little toward the side touched.   If the stimulus is farther back, the turning is greater.   In each case the turning is so regulated with reference to the point stimulated as to direct the animal very accurately toward the region from which the stimulus came; this aids it much in finding food.   If something touches the flatworm lightly at the middle of the upper surface of the head, the reaction is much modified.   The head is sharply raised and twisted, so as to direct the anterior tip toward the stimulating object, and in such a way that the ventral surface will first come in contact with this object as the animal moves forward.   Similar regulatory changes occur in the negative reaction.   A strong stimulus at the side of the anterior end causes a quick turning away.   A similar stimulus at one side behind the middle causes no turning away, but only a movement forward.   At intermediate regions there is a combination of the two reactions, the animal gliding forward and at the same time turning away.   The farther back the stimulus is given the greater is the tendency to react by moving forward in place of turning away.   This change of reaction with a change in the point stimulated is of course regulatory.   An intense stimulus at the anterior end is best avoided by turning away, while one near the posterior end is most easily escaped by moving rapidly forward.

In most animals there are found a certain number of these relatively fixed reaction types which are determined by the usual conditions of existence, — gravity, light, temperature changes, contact with solids, etc.   We have examined a considerable number of these in the Protozoa and Cœlenterata.   In such reactions the organism often turns or bends directly toward or away from the source of stimulation, as in the positive and negative reactions of the flatworm. Reactions of this character are commonly spoken of as *tropisms*. In the higher animals and man behavior is, of course, largely determined by the same factors. As we have seen in previous chapters and shall find in the following sections of the present one, in neither lower nor higher animals are the reactions with reference to the general forces of nature of a completely fixed and invariable character.[1]

In more complex animals than those considered in the present volume, definite reaction forms are often combined into complex trains of action which are known as *instincts*. Recent work has shown that in these instincts there is by no means that absolute fixity of behavior that was formerly assumed to exist. A detailed treatment of this matter would take us outside the field of the present work.

In the highest animals and man, definite reaction forms, which may take place in certain organs independently of the rest of the body, are of course found as abundantly as in lower organisms. Such reactions are seen in the reflexes of muscles, etc., which persist even after the muscle has been removed from the body. There is no difference in principle along this line between higher and lower animals. The former possess a much larger number of such definite types of movement, and these doubtless make up fully as large a portion of behavior as in the lower animals.

There are some accounts of behavior in various lower animals in which only these definite reaction forms are described and only those conditions are dealt with in which these appear in the typical way. Such accounts have given rise to a widespread impression that behavior in the lower animals differs from that of higher forms in that it is of a fixed, stereotyped character, occurring invariably in the same way under the same external conditions. This impression is in a high degree erroneous. These definable reaction forms are usually in themselves variable within wide limits, as exemplified in the avoiding reaction of infusoria. But even if this were not true, the criteria for judging as to the fixity or modifiability of behavior are to be derived from the study of the conditions that induce reaction, that determine which of several possible reactions shall occur, and that determine the order and combination of reactions. Such a study shows that in lower as well as in higher animals varied internal conditions and changes are of the greatest importance in determining behavior, the animal by no means behaving always in the same way under the same external conditions. With this aspect of the matter we shall deal in the two following sections.

[1] See, for example, the section on reactions to gravity in cœlenterates, Chapter XI.

2. Reaction by Varied Movements, with Selection from the Resulting Conditions

In the foregoing section we have dealt with the fact that stimulation often causes the performance of actions that are of a definite, typical character, such as are often called reflexes. But this by no means exhausts the problem of behavior, as our account of the matter in unicellular animals and in Cœlenterata has shown us. Indeed, we find it not to be the rule that an animal when stimulated performs a single definite movement, then returns to its original state. On the contrary, stimulation is usually followed by varied movements, and the animal may continue active long after the external agent has ceased to impinge upon it. The continued varied movements subject the organism successively to many different conditions, external and internal. In one of these conditions the animal remains through a cessation of the changes in activity. It may thus be said to select certain conditions through the production under stimulation of varied movements. We have seen many examples of this type of behavior in the groups thus far considered.

Behavior of this character is very general in lower animals. We shall in the present section give a number of examples, taken from diverse classes of invertebrates.

As we have seen, the echinoderms furnish perhaps the best examples of organisms in which the behavior is made up largely of more or less independent "reflexes." Yet in the same group we find that much of the behavior is of the type now under consideration. There is, of course, no opposition between the two, the different "reflexes" forming the variables out of which behavior of the present sort is made up. The pedicellariæ of the sea urchin have, as we have seen, a number of these definite reflexes. When the entire animal is suddenly and strongly stimulated, by mechanical shock, by a chemical, or by light, the pedicellariæ respond, not by a single definite reflex, but by beginning to move about in all directions (v. Uexküll). They seem to feel and scrape the entire surface of the body, seizing anything with which they come in contact, and this behavior may continue for an hour or more after stimulation has ceased. Similar effects are often produced in the spines by a general stimulus. They wave about, their tips describing circles, and this may continue for a long time. Such reactions are seen also in the tube feet. When the sea urchin or starfish is suspended in the water or is placed on its back, the tube feet extend and wave back and forth, as if searching for something to which they might attach themselves.

On a more extensive scale, the "righting" reaction of the starfish is a notable example of behavior that is not stereotyped, but is flexible and variable. The usual course of this reaction is as follows: After the starfish has been placed on its back, it extends its tube feet and moves them about in all directions. At the same time the tips of the arms become twisted, so that some of the tube feet are directed downward. In this way, after a time, some of the feet become attached to the bottom. These begin to pull on the arm to which they belong, turning it farther over and bringing other tube feet into contact with the bottom; these now assist in the process. If two or three adjacent rays become thus attached, the other rays cease their searching, twisting movements, and allow themselves to be turned over by the activities of the tube feet of the attached rays. If two or more opposite rays become attached to the bottom in such a way that they oppose each other, then one releases its hold, and allows the turning to be accomplished by the opposing rays. It is evident that the reaction is an example of the performance of varied movements under stimulation, with selection from the conditions resulting from these movements. Certain features in this reaction are of special interest. At first all the tube feet and rays try to find an attachment. When certain ones have succeeded, this is in some way recognized by those parts whose action would oppose the movement, for these cease their attempts, or even release the hold already attained. In some way the physiological state corresponding to "success" in certain rays is transmitted to the other rays, and they change their behavior accordingly.

Variability and flexibility are the essence of such behavior. This is well illustrated by study of repetitions of the righting reaction in the starfish. It is by no means always the same arm or combination of arms that initiates and finally brings about the turning. The essential point is to get started in some way, then to continue on the basis of the start made. Preyer (1886) studied this behavior in the starfish with great care. He says: "Neither in one [species] nor the other is the method of turning always the same. I have likewise seen *Asterias glacialis*, which was several times in succession turned on its back without change in the outer conditions, right itself sometimes in one manner, sometimes in another. The spirals of the twisted arms do not work each time in corresponding directions, but at first the neighboring arms often oppose each other. But soon the correction takes place, in that the attached feet stop those that are disturbing the turning, and the wrongly twisted radii straighten out again. . . . The variability of form in starfish that are righting themselves is great, and no species rights itself in only one way. . . . But here, too, it is true that no Astropecten rights itself twice in succession in exactly the same way. An adaptation to the surface of attachment always occurs, and according as this is convex, concave, smooth, rough, or inclined, is the turning process made easier or more difficult, and brought about in this manner or that" (1886, pp. 107–108). Sometimes the animal turns a somersault; sometimes it extends all its arms upward, taking the "tulip" form and toppling over on one side — and so on through many variations.

Even the main features of the typical reaction may be omitted or changed, the turning taking place by means quite different from the usual ones. Thus, *Astropecten aurantiacus* usually rights itself by means of its tube feet, but sometimes turns without using the tube feet at all. Lying on its back, it lifts the central disk high, resting on the tips of three or four of the arms. Then it turns two of the arms under, while lifting the others upward, so that it now falls with ventral side down. During this action the tube feet are moved about in a lively way, and when the turning is nearly completed the tube feet of the upper radii which are approaching the substratum are pushed far out, as if preparatory to meeting the bottom.

When a portion of the feet were prevented from acting, by subjecting them to alcohol or other drugs, Preyer found that the starfish righted itself by means of the remaining ones, and by bending and twisting its arms. Pieces of the arms may right themselves, and this again occurs in many different ways.

The thorough study of the movements and reactions of the starfish made by Preyer (1886) shows that the righting reaction is typical of the entire behavior. If the starfish is suspended just below the surface of the water with ventral side up, by threads attached to the tips of its arms, it performs varied movements, until in the course of time it turns over, just as in the usual righting reaction. If a short rubber tube was slipped over one of the arms of a brittle star, to its base, Preyer found that this caused the animal to perform many varied movements, till by one of them the tube was removed. Sometimes the animal merely moved rapidly forward, dragging the arm bearing the tube behind it till the tube was scraped off. Sometimes the animal placed one or two of the other arms against the tube and forced it off. In other cases the covered arm was dropped from the body (as often happens in brittle stars). Again, sometimes the arm bearing the tube was lifted and waved back and forth, till the tube was in this way displaced. Thus Preyer observed five different ways in which the tube was finally removed; as he remarks, "If one method does not help, another is used." It may, of course, be maintained that in all these cases the removal of the tube was in a sense accidental. But this is precisely the essential point in much of the behavior of lower organisms. When stimulated they perform varied movements, till one of these "accidentally" removes the source of stimulation. How this may develop into more directly regulatory reactions we shall consider in the next section.

The same qualities are shown in certain experiments of Preyer in which he attempted to confine the starfish by means of large, flat-headed pins. These were placed in the angles between the rays, close against the disk, and driven into the board on which the starfish lay. They thus held it down without injury. The starfish in the course of time escapes from the pins, but only after much effort. The animals try successively various methods, "now they seek to force themselves through, now to climb over the top, now to push through by turning on one side." In scarcely any two cases does the process of escape occur in the same way, according to Preyer. The behavior is as far as possible from that of invariable reflexes always occurring in the same way under the same external conditions.

One further point mentioned by Preyer is of great interest. He says that when the experiment is repeated with the same individual, the time required for escape becomes less. The number of useless movements, "superfluous twistings, feelings about, and forward and backward motions," becomes less the oftener the individual has been placed in such a situation. If this is true, we have in so low an animal as the starfish regulation through the selection of conditions produced by varied movements passing into a more directly regulatory action; in other words, what is commonly called in higher animals intelligence. There seems to be no reason for doubting Preyer's observations on this point, but on account of their great importance they should be repeated and verified or refuted.

Many other illustrations of behavior of the general character set forth above could be presented from the valuable work of Preyer (1886). It has become the fashion to neglect and even speak slightingly of the work of Preyer on the behavior of the starfish. This seems to be due to the tendency observable in recent scientific literature to represent all such matters as extremely simple and reducible to separate well-known mechanical factors, and to avoid all experiments tending to reveal the fallacy of this view. Preyer was not afraid to open his eyes by properly designed analytical experiments to the complexity and regulatory character of the behavior. Such thorough and detailed studies of animal behavior as that of Preyer on the starfish are rare at the present time; his work stands in this respect in most refreshing contrast with some of the superficial work recently put forth. The excellent work of Romanes (1885) had already, before Preyer, brought out many examples of the style of behavior we have illustrated above.

In many free-swimming Rotifera the chief methods of movement and reaction are similar even in details to those of the free-swimming infusoria, which we have already described. Like the infusoria, these rotifers swim by means of cilia, revolve on the long axis, and swerve toward one side (usually dorsal), as they progress. The cilia produce a current passing from in front to the mouth and ventral side, thus allowing the animals to test the conditions in advance. To most effective stimuli these rotifers react, as do the infusoria, by swerving more than usual toward one side, — usually the dorsal side. Thus the spiral becomes much wider, and the animals are pointed successively in many different directions and subjected to many different conditions. In time they may thus reach conditions which relieve them of the action of the stimulating agent. Thereupon the reaction ceases, so that the animals continue in the direction which has thus been reached. All the general features of the reactions are essentially like those of infusoria, so that we need not enter into details. The reactions to mechanical stimuli, to chemicals, to heat and cold, to light, and to electricity are known to occur in the way just

FIG. 134. — Planaria, dorsal view. After Woodworth.

sketched, in a number of species. Orientation to light and to the electric current takes place ·in the same way as the orientation to light in Euglena and Stentor. It is interesting to observe that in the Rotifera, owing to the concentration of the cilia at one end of the animal, there is no such incoherence and lack of coördination in the reaction to the constant electric current, as is found in infusoria. The rotifer (*Anurœa cochlearis*) becomes oriented with anterior end to the cathode by the same method as in reactions to light and other agents.

In many rotifers the reaction plan just described forms only one feature of the activities, so that the behavior, taken all together, may be exceedingly complex. There is much opportunity for further study of the reactions of this group. But so far as known, much of the behavior may be expressed as follows: When stimulated, the animals perform continued and varied movements, the variations often taking place in a systematic way. These movements necessarily subject the animals to varied conditions, one of which is finally selected, through the fact that it removes the cause of stimulation.

Much of the behavior of the flatworm Planaria (Fig. 134), as studied by Pearl (1903), may be summed up under the same formula set forth in the preceding paragraph. Varied movements which subject the animal to many different conditions, are seen even in the unstimulated specimen. As the flatworm glides along by means of its cilia, the head is held upward (Fig. 135) and moved frequently from side to side, while its margins wave up and down, and are extended and contracted. The flatworm thus seems to "feel its way" with its head. Sometimes these feeling movements become much accentuated, the animal almost or quite stopping, then raising the whole anterior part of the body and waving it about in the water.

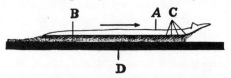

FIG. 135. — Side view of moving Planaria. After Pearl. *A*, body; *B*, mucus; *C*, cilia; *D*, substratum.

These movements of course serve to test the environment on each side; in other words, they subject the sensitive anterior end to varied conditions.

The testing movements are specially marked under certain conditions. When the active planarian is about to come to rest, it stops and moves the anterior end from side to side, touching any object that may be found in the neighborhood. After thus thoroughly testing the surroundings, the muscles relax and the animal comes to rest. When later the animal resumes its active progression, this begins again with the testing movements of the head.

The same testing movements are seen under various sorts of stimulation. On coming to a solid body, the flatworm moves the head about over its surface. If it turns out to be something fit for food, the animal now feeds upon it, otherwise it moves away again. If while a number of specimens of Planaria are moving in a certain direction, the direction of the light is changed so as to fall upon their anterior ends, they usually turn the head from side to side two or three times, then follow up one of these movements by turning the body till it is finally directed away from the light. These testing movements are also seen when the animal begins to dry, and when the water is heated; the worm gives the impression that it is seeking about for other conditions.

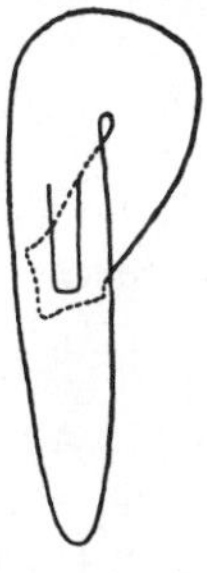

FIG. 136. — Reaction of Planaria to drying. After Pearl.

Other features of the reactions to drying and to temperature changes are of interest from our present standpoint. If the planarian is laid on a glass plate, as soon as the tendency to dry becomes evident the worm curls up closely and thrusts the head under the body (Fig. 136). In this way the exposed surface of the body is made as small as possible, and the sensitive head especially is kept from drying. At intervals the animal straightens out, extends its head as far as possible, and waves it from side to side. If in this way it finds water, it of course moves into it. If it does not find water, it curls up again. After a time, if the drying becomes more decided, the animal attempts to crawl backward. Under natural conditions drying will usually take place at the edge of a pool, and this backward movement carries the animal again into the water. All together, the reaction to drying is not simple and stereotyped, but involves the successive performance of many different activities.

In responses to heat or cold we find a similar train of activities. If the gliding Planaria comes to a region of considerably higher or lower temperature, it waves its head back and forth several times, apparently till it has determined the direction which leads back to the usual temperature, then turns and moves in that direction. Responses of this character usually take place several times before the animal is completely directed toward the region of optimum temperature (see Fig. 137). If the temperature of the water is slowly raised in a uniform manner, so that all parts of the body are similarly affected, then a series of reactions occurs. First the animals become more active, gliding about rapidly, extending the head, and turning it toward one side or the other. The behavior resembles that of specimens showing the "positive" reaction to weak stimuli. As the

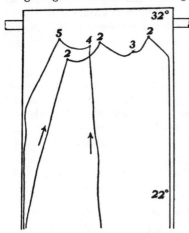

FIG. 137. — Behavior of the flatworm in approaching the heated end of a trough. The lines show the paths followed. At each of the points marked by a round spot, the animal stopped and waved its head to and fro, finally following up one of the trial movements. The figures at these points show the number of trial movements that were made, in each case. After Mast.

temperature rises, the animal begins to contract at intervals, and to turn the head frequently and strongly from side to side, making little progress in advance. The behavior has now the characteristics of the "negative" reaction. As the temperature rises further, the turning ceases, and the animal begins to make rapid, violent contractions, such as occur in "crawling," under other violent stimuli. Later the animal twists its body, as occurs in its righting reaction when placed on its back; it thus forms a spiral of two turns. Finally it behaves in a manner somewhat similar to that shown when it dries. It rolls the two ends under the body, arching the dorsal surface. In this position the animal rolls over on its back and dies.

Thus under a single unlocalized stimulus of gradually increasing intensity, the behavior of the organism passes through a series of stages, closely resembling the reactions given under most diverse conditions. As Mast (1903), to whom these observations are due, expresses it, "the general impression is given that as the thermal stimulus increases, the animal tries, in a sort of 'hit-or-miss' way, every reaction which it has at command in order to get rid of the stimulation."

The "righting reaction" of the flatworm is another example of a response that is not stereotyped in character, but varies greatly. If the animal is turned on its back, it quickly rights itself again. This usually occurs as follows. The animal twists itself into a spiral (Fig. 138, *A*),

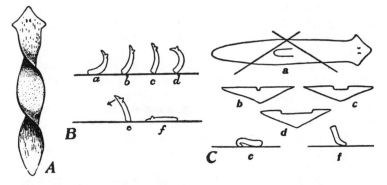

FIG. 138. — Righting reactions in the flatworm. After Pearl. *A*, reaction of entire worm. *B*, righting reaction of short piece from anterior end of worm. *a, b, c, d, e, f*, successive steps in the process. *C*, righting reaction of triangular pieces. *a*, manner in which the piece is cut. *b*, a small portion of the thin edge turns so as to bring the ventral surface in contact with the bottom. *c, d*, this turning increases; by a continuation of the process the whole piece is finally righted. *e, f*, cross sections through the pieces while turning.

thus causing the ventral surface of the head to face the bottom, where it attaches itself. Then the worm creeps forward, bringing successively more and more of its ventral surface in contact with the bottom, proceeding toward the rear. Thus the spiral is unwound, so that after the animal has traversed a short distance, the entire ventral surface is in contact with the bottom, as usual.

But the righting reaction may take place in quite a different way. Pearl (1903) cut the planarian into pieces of such form that it could no

longer twist itself into a spiral. Then some portion of the ventral surface was brought by other means into contact with the bottom, and
from this point the remainder of the surface was pulled into contact. In
small strips from the head region, the posterior ends are turned under,
bringing the ventral side at this point against the bottom, then by pulling from this point, the entire piece was turned over endwise (Fig. 138,
B). In triangular pieces from the middle of the animal, one edge was
turned under, then the remainder righted from this region, by pulling
the rest of the piece over (Fig. 138, C). These modifications bring out
the essentially adaptive character of the behavior. The essential point
seems to be, to get some portion of the ventral surface, by any means
whatever, into contact with the substratum, then by working out from
this point, to bring the whole ventral surface into attachment.

From certain points of view the whole behavior of the flatworm may
be considered a process of testing all sorts of conditions, retaining some
and rejecting others. As we have seen in the section which precedes
the present one, the positive reactions of this animal are not due to any
specific qualities of stimulation. On the contrary, the animal turns
toward weak stimuli of all sorts. Solid bodies, whether fit for food or
not, chemicals of all sorts, including the injurious as well as the beneficial, heat, and cold, all induce, when acting but slightly on one side,
a turning toward the source of stimulation. The flatworm may thus
be said to investigate every slight change occurring in its surroundings.
On reaching a region where the agent in question acts more intensely,
the positive reaction may either continue or be transformed into a negative one. Thus the turning toward food is not due to the specific qualities which make the substance in question fit for food, but is the result
only of this general tendency to move toward all sources of weak stimulation. The flatworm proves all things, holding fast only to that which
is good.

In most if not all other invertebrates there occur many "trial movements" similar to those already described. In many recent accounts
of the behavior of other invertebrates little mention, it is true, will be
found of such movements. This is apparently because attention has
been directed by current theories to other features of the behavior, and
the trial movements have been considered of no consequence. Often
an attentive reading of papers on "tropisms," etc., will reveal paren-
thetical mention of various "disordered" movements, turnings to one
side and the other, and other irregularities, which disturb the even tenor
of the "tropism," and are looked upon for some reason as without significance and not requiring explanation. Further, one often finds in
such papers accounts of movements which are clearly of the "trial"

character, yet are not recognized as such by the author, on the watch only for "tropisms." In the earlier literature of animal behavior, before the prevalence of the recent hard-and-fast theories, one finds the trial movements fully recognized and described in detail. This is the case, for example, in the classical papers of Engelmann on behavior in unicellular organisms, and, as we have seen in detail, for that of Preyer on the starfish. Moebius, in 1873, gave a lecture on behavior in which examples of this fact are found. Thus, he describes the reaction of a large mollusk, Nassa, to chemical stimuli, as shown when a piece of meat is placed in the aquarium containing them, in the following way: They do not orient themselves in the lines of diffusion and travel toward the meat, but move "now to the right, now to the left, like a blind man who guides himself forward by trial with his stick. In this way they discover whether they are coming nearer or going farther away from the point from which the attractive stimulus arises" (Moebius, 1873, p. 9).

Unprejudiced observation of most invertebrates will show that they perform many movements which have no fixed relation to sources of external stimuli, but which do serve to test the surroundings and thus to guide the animal. This the present author has observed, for example, in studies on the leech, on various fresh-water annelids and mollusks, and in less extended observation on many other animals. As Holmes (1905) has recently pointed out, in a most excellent paper, this is really a matter of common observation on all sorts of animals. The fact that such movements are not emphasized by writers on animal behavior is evidently due to their being considered without significance.

In a number of recent papers the importance of trial movements in behavior has been more explicitly recognized. Thus, for the earthworm, the recent papers of Miss Smith (Mrs. Philip P. Calvert) (1902), of Holmes (1905), and of Harper (1905) have set this matter in a clear light. Miss Smith showed that in the reactions of the earthworm *Allolobophora fœtida* to heat and cold, to chemicals, to drying, and to light, "testing movements" play a large part. When stimulated, the earthworm frequently responds by moving the head first in one direction, then in another, often repeating these movements several times. It then finally follows up those movements which decrease the stimulation. Holmes (1905) confirms these results, especially for the reaction of the earthworm to light. His account of the behavior of the earthworm under the action of light coming from one side may be quoted: "It soon developed that what seemed at first a forced orientation, the result of a direct reflex response, is not really such, but that the orientation which occurs and which is often quite definite is brought about in a more indirect manner by a mode of procedure which is in some respects similar to the method of trial and error followed by higher forms" (*l.c.*, p. 99). The precise behavior of the earthworm in becoming oriented to light is described as follows: "As the worm crawls it frequently moves the head from side to side as if feeling its way along. If a strong light is held in front of the worm, it at first responds by a vigorous contraction of the anterior part of the body; it then swings the head from side to side, or draws it back and forth several times, and extends again. If in so doing it encounters a strong stimulus from the light a second time, it draws back and tries once more. If it turns away from the light and then extends the head, it may follow this up by the regular movements of locomotion. As the worm extends the head in crawling it moves it about from side to side, and if it happens to turn it toward the light it usually withdraws it and bends in a different direction. If it bends away from the light and extends, movements of locomotion follow which bring the animal farther away from the source of stimulus" (*l.c.*, p. 100).

Other observers — Parker and Arkin (1901), Adams (1903) — had observed that when the earthworm is lighted from one side, it by no means always turns directly away from that side; Adams, however, showed that it turns more frequently away from the light than toward it, thus indicating that the animal has some direct localizing power. This is confirmed by Harper (1905), who shows that in a strong light the earthworm Perichæta commonly turns directly away from the source of light, though if the light is weak, the "trial movements" are seen. Harper gives many other examples of the performance of varied movements under the action of stimuli in this animal, and brings out some of the internal factors on which some of these depend.

Holmes (1905) found that the leech and the larva of the blowfly react to light in essentially the manner which he had found in the earthworm. For the leech the following account is given: "In its progress the leech frequently raises the anterior part of the body and waves it from side to side as if feeling its way. If the animal turns it in the direction of a strong light, it is quickly withdrawn and extended again, usually in another direction. If the light is less strong, it waves its head back and forth several times and sets it down away from the light; then the caudal end is brought forward, the anterior end extended and swayed about and set down still farther away from the light than before. When the leech becomes negatively oriented, it may crawl away from the light, like the earthworm, in a nearly straight line. The extension, withdrawal, and swaying about of the anterior part of the body enable the animal to locate the direction of least stimulation, and when that is found it begins its regular movements of locomotion. Of a number of random movements in all directions only those are followed up which bring the animal out of the undesirable situation" (*l.c.*, p. 102).

In the case of the blowfly larva, Holmes speaks as follows: "Observations which I have made upon the phototaxis of blowfly larvæ with the problem of orientation especially in mind soon convinced me that the movements of these forms are directed by light through following up those random movements which bring them away from the stimulus. When strong light is thrown on a fly larva from in front, the anterior end of the creature is drawn back, turned toward one side, and extended again. Often the head is moved back and forth several times before it is set down. Then it may set the head down when it is turned away from the light and pull the body around. If the head in moving to and fro comes into strong light, it is often retracted and then extended again in some other direction, or it may be swung back without being withdrawn. If a strong light is thrown upon a larva from one side, it may swing the head either toward or away from the light. If the head is swung toward the light, it may be withdrawn or flexed in the opposite direction, or, more rarely, moved toward the light still more. If it is turned away from the light, the larva usually follows up the movement by locomotion. Frequently the larva deviates considerably from the straight path, but as it continually throws the anterior part of the body about and most frequently follows up the movement which brings it away from the stimulus, its general direction of locomotion is away from the light. In very strong illumination the extension of the anterior part of the body away from the light is followed by a retraction, since in whatever direction it may extend it receives a strong stimulus and the larva writhes about helplessly for some time. Sooner or later, however, it follows up the right movement. Occasionally the larva may crawl for some distance directly toward the light, but after a time its movements carry it in the opposite direction. When once oriented the direction of locomotion of the larvæ is comparatively straight " (*l.c.*, pp. 104–105).

As Holmes points out, these are only examples of a very general condition of affairs in the lower organisms. We cannot do better, in concluding this brief section, than to quote some of Holmes's general remarks, which show that his observations have led him to essentially the same conception of behavior that we have reached in the present work.

"The rôle played by the trial and error method in the behavior of the lower organisms has, as yet, elicited but little comment, owing probably to the fact that attention has been centred more upon other features of their behavior. It may have been considered by some investigators as too obvious for remark, since any one who attentively observes the conduct of almost any of the lower animals for ten minutes can scarcely fail to see the method exemplified. If he were watching a chick pecking at a variety of objects and giving signs of disgust when it had seized a nauseous substance, he would doubtless regard the process as one of trial and error, whatever name he might apply to it. A study of the conduct of much lower organisms would disclose many cases almost equally evident. The lives of most insects, crustaceans, worms, and hosts of lower invertebrate forms, including even the Protozoa, show an amount of busy exploration that in many cases far exceeds that made by any higher animal. Throughout the animal kingdom there is obedience to the Pauline injunction, 'Prove all things, hold fast to that which is good'" (*l.c.*, p. 108).

The well-known behavior of hermit crabs in finding suitable shells in which to live and in changing shells which have become unsuitable shows a systematic application of the method of trial extending to the details of the behavior. This is well brought out in the excellent analysis of this behavior given by Bohn (1903).

Behavior of higher animals based on the selection of the results of varied movements — the "method of trial and error" — plays, as is well known, a large part in recent discussions of that subject. The work of Thorndike (1898) on behavior in the cat, and the books of Lloyd Morgan (1900), in which this matter is dealt with, are, of course, well known, and require no discussion on our part. The fact that behavior of this character plays a large part in higher, as well as in lower, organisms, is of the greatest interest, as showing that this method is one of fundamental and general importance. But with the details in higher animals we are not here concerned.

3. MODIFIABILITY OF BEHAVIOR AND ITS DEPENDENCE ON PHYSIOLOGICAL STATES

In the section preceding the present one we have described many cases of behavior in the lower invertebrates in which the animal, under the action of constant external conditions, passes from one form of behavior to another. All such cases are illustrations of the fact that behavior depends upon internal, physiological conditions, as well as upon external stimuli. Since under the same external conditions the action changes, the animal must itself have changed, otherwise it could not now behave differently from before. It is clear that the continuance of a stimulus, or the performance of a certain action, may change the physiological state of the animal so as to induce new reactions.

In some cases the varied actions performed under stimulation have been spoken of as random movements (Holmes, 1905). The word "random," of course, implies only that these movements are not defined by the position of the stimulus; it does not signify that the movements are undetermined. The principle of cause and effect applies to these movements as well as to others. But the causes lie partly within the animal; each phase of the movement aids in determining the succeeding phase. The earthworm may turn to the right at a given instant merely because it has just before turned to the left. Reactions in which a succeeding phase is determined by a previous one have sometimes been called chain reflexes (Loeb, 1900; Driesch, 1903). If this term is used, it needs to be kept in mind that in most cases the succeeding phase is not invariably and irrevocably called up by the preceding one, as is implied by this term. On the contrary, the relation between the two is extremely variable. One type of action may be repeated many times before the second type comes into play, and the order of the different actions is by no means always the same. Thus the preceding phase is only one factor in deciding what shall be the present action. The latter depends upon the entire physiological state of the organism, which is determined by various factors. Illustrations of this are seen in the righting reaction of the starfish and many other animals; in the series of reactions by which Stentor responds to a mass of carmine grains in the water (p. 174); in that by which Stoichactis gets rid of waste matter lying on the disk (p. 202), and the like.

The diverse physiological states of lower organisms have been little studied. This is partly because it is rarely possible to observe them directly; it is only through their effects upon action that they become evident. Thus the real data of observation are the actions; if we considered these alone, we could only state that a given organism reacts under the same external conditions sometimes in one way, sometimes in another. This would give us nothing definite on which to base a formulation and analysis of behavior, so that we are compelled to assume the existence of changing internal states. This assumption, besides being logically necessary, is, of course, supported by much positive evidence drawn from diverse fields, and there is reason to believe that in time we shall be able to study these states directly. Before we can come to a full understanding of behavior, we shall have to subject the physiological states of organisms to a detailed study and analysis, as to their objective nature, causes, and effects.

The most noticeable and therefore best-known physiological states of lower animals are those which depend upon changes in metabolism. The reactions of the starfish and the planarian to many chemical and mechanical stimuli depend, like those of the sea anemone, on the progress of metabolism. Hungry animals react positively to possible food, while satiated ones react negatively to the same stimuli. This most significant relation is, of course, almost universal in organisms; it shows directly the dependence of behavior on the relation of external agents to internal processes.

V. Uexküll has made precise studies of certain physiological states and of the factors on which they depend, in the sea urchin and a number of other lower animals. In the sea urchin, some of the pedicellariæ will not close in response to a mechanical stimulus, save in case this has been preceded by a chemical stimulus. The latter changes the physiological state of the protoplasm (muscle or nerve), so that it now reacts to a stimulus which before would have had no effect. The spines of the sea urchin usually bend toward a spot on the surface of the body that is mechanically stimulated, as by a needle. But if this stimulus has been preceded by the action of a chemical, the spines now reverse the reaction and bend away from the region stimulated. Many such changes in physiological state are brought by v. Uexküll under the heading of changes in *tonus* of the muscles or nerves. Steady tension, such as is produced in certain muscles by pressing a spine of the sea urchin to one side, decreases the tonus, so that the muscles are no longer so tense as before. Such muscles react more readily to stimuli than do those of higher tonus. Sudden jarring produces the opposite effect, the muscles pull harder and react less readily than before. Decrease of tonus caused by tension is transmitted in some way to neighboring spines, so that after a certain spine has been pressed to one side, all those about it bend in the same direction and react more readily than before. These changes in physiological state play a large part in determining the behavior of the sea urchin under natural conditions.

Besides such changes, there are in the sea urchin others that are less easy to formulate, and that have not been analyzed. V. Uexküll found that the set reflexes of the spines and the changes in tonus mentioned above impose on the sea urchin a behavior that under most conditions seems stereotyped and predictable. This leads the author named to contrast the sea urchin as a "republic of reflexes" with higher animals in which the behavior is unified. But the difference is only one of degree. If the sea urchin is placed on its back, the usual reflexes and their stereotyped interrelations would not restore the animal to the natural position, but merely cause it to walk forward while lying on its back. As a result, we find a physiological state induced that causes a thoroughgoing change in the behavior of the spines. They now move in such a way as to turn the sea urchin again on its ventral surface. As v. Uexküll says, the behavior of the spines is variable and capable of adaptation ("variabel und anpassungsfähig," 1900, p. 98). This adaptation, under unusual conditions, of the movements of the spines to the needs of the organism as a whole, seems to remove all difference in principle between the behavior of the sea urchin and that of higher animals.

Many illustrations of varied physiological states could be given from an analysis of the behavior of the starfish in the righting reaction, and in the various experiments devised by Preyer (see p. 239).

In the flatworm Planaria the work of Pearl (1903) shows that the behavior depends largely upon the physiological state. In this animal the following different states determining behavior may be distinguished:—

1. Conditions of hunger and satiety, determining the reactions to food in a regulatory way.

2. A resting or "sleeping" condition. The animal is often found lying quietly under rocks, the muscles relaxed. In this condition it fails to react to weak stimuli, but strong stimulation induces the negative reaction, followed by continued activity.

3. The condition of normal, undisturbed activity. The animal now responds to weak stimuli of all sorts by the positive reaction, turning toward the side stimulated, while strong stimuli cause the negative reaction.

4. A condition of heightened activity, in which the worm makes many "testing" movements with the head, and reacts positively to most stimuli, whether strong or weak. In this condition the planarian makes the appearance of actively seeking something, and of following up any source of stimulation which it finds.

5. An "excited" condition, produced by stimulating the animal strongly and repeatedly. In this condition the animal moves about violently and reacts negatively to most stimuli to which it reacts at all.

6. Possibly due to an accentuation of the condition last described is a change of reaction observed by Pearl when one side of the head of an excited specimen is stimulated by repeated blows. At first the animal turns farther and farther away from the side stimulated. Then suddenly it jerks strongly backward, and turns far in a direction opposite its previous turning — that is, toward the side stimulated. "The reaction appears as if, after the animal had tried in vain to get away from an uncomfortable stimulus by its ordinary reaction, it finally tries a wild jump in the opposite direction" (Pearl, 1903, p. 580).

The different physiological conditions are determined largely by the history of the individual worm, so that in this sense its behavior may be said to depend on its experience. The dependence of the reactions on the physiological state is in a given specimen very great, so that two individuals often react in opposite ways to the same stimulus. The same individual that reacts to a given stimulus positively may a little later react negatively, and *vice versa*. After long study of Planaria, Pearl concludes that "it is almost an absolute necessity that a person should become familiar, or perhaps better, intimate, with an organism, so that he knows it in something the same way that he knows a person, before he can hope to get even an approximation of the truth regarding its behavior." This remark might be extended to most lower animals.

As we have seen in a previous section (p. 236), the behavior of the flatworm shows certain well-defined reaction types, which might, taken separately, be called reflexes. But when we consider the various factors which determine the production and combination of these reaction types, we cannot consider the behavior of the flatworm as "purely reflex," if we mean by reflexes invariable reactions to the same external stimuli. On the contrary, the behavior is extremely variable in accordance with many conditions, internal as well as external.

A detailed analysis of the behavior of almost any of the lower invertebrates would show as many different physiological conditions on which behavior depends as we find in the flatworm. In the earthworm, for example, the conditions are still more complicated than in the flatworm, so that the same external stimulus, acting with the same intensity, and applied to the same spot on the body, may produce any one of at least six different reactions. The variations of internal state as the animal moves about are what condition the "random movements" described by Holmes in the reactions to light, and by Smith in the reactions to other stimuli (see p. 247).

Of special interest are changes in state that lead to more or less permanent modifications in behavior. These are little known in the lower organisms. Most of the changes of physiological state described in the foregoing paragraphs are not known to last more than a short time. In Vorticella, Hodge and Aikins (1895) state that the modified behavior endured for five hours; this perhaps needs confirmation. In the lowest organisms it is difficult to carry out experiments that shall determine how long modifications last. Perhaps the lowest animal in which an enduring modification of behavior has been demonstrated is the flatworm *Convoluta roscoffensis*. This is one of the lowest of the group, belonging to the division Acœla, which includes the simple forms having no alimentary canal. The behavior of Convoluta, as described by Gamble and Keeble (1903), and by Bohn (1903 *a*), presents many features of the greatest interest; into only a few of these can we enter. Convoluta is a small green worm that lives in immense numbers on the sand of the seacoast of Brittany, just above the water line. It forms thus large green patches. When the tide rises the water covers the region where

Convoluta is found, and the waves would wash the animals away, if their behavior did not prevent. As the water rises and the waves begin to beat on the sand near them, they go downward into the sand, where they are protected. As the water sinks, the animals creep upward and appear again at the surface. These upward and downward movements are reactions with reference to gravity, as is shown by placing the animals on smooth, inclined, or perpendicular surfaces. They go downward as the tide rises, upward as it falls. Bohn (1905) has shown that many littoral mollusks and annelids show similar movements with relation to the tides.

The peculiarly interesting fact concerning this behavior in Convoluta is the following: This periodical alternation of reactions, produced by an environmental factor, becomes so impressed on the organization of the animal that it occurs even when this factor is lacking. The alternation of movement has become habitual. If the worms are removed to an aquarium where the tide no longer acts upon them, they continue to go downward at the period of high tide, upward at the period of low tide. This continues for about two weeks, so that the worms may be carried far away from the shore, and may then be used for a time as tide indicators. But under such conditions the periodicity after a time disappears, showing that it was really due to the external factor, — the tides. This appears to be the lowest known case of what we call in higher animals a habit.

In some of the higher invertebrates, lasting modifications of behavior of a still more complex character may be induced experimentally. This has been accomplished in the Crustacea by Yerkes (1902), Yerkes and Huggins (1903), and Spaulding (1904).

With the crayfish and crab, Yerkes and Huggins (1903) studied the modification of behavior in escaping from danger and in finding water. The crayfish was placed in one end of an inclined pen which opened at the other end into the water. The pen was partly divided by partitions in such a way as to leave two passages leading to the water (Fig. 139). Either of these passages could be closed at its end by a glass plate *G*. The animal was placed at *T* (Fig. 139). In moving away from this region it might enter the blind pocket at *G*, thus not directly reaching the water, or it might go through the other passage straight to the water.

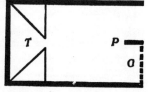

Fig. 139. — Pen used by Yerkes in experimenting with Crustacea. After Yerkes. See text.

After some preliminary experiments without closing either passage, showing that the animals were as likely to pass to the right as to the left, the partition was placed in the right passageway, as in Fig. 139. The crayfish which turned to the left on leaving T escaped at once to the water. But if it turned to the right it passed into the pocket G, and was compelled to explore the region, finally turning to the left and passing the partition P, before it could escape. Three individuals were given sixty trials each in the course of thirty days. In the first ten trials they went just as frequently into the blind passage as toward the water. In the second ten trials, the animals started in 60 per cent of the cases toward the open passage at the left. In the next ten trials this proportion had risen to 75.8 per cent; in the following ten, to 83.3 per cent. In the last ten trials of the sixty, very few mistakes were made. In 90 per cent of all cases they went straight for the open passage. In another series of experiments an individual, after four hundred trials, made only one mistake in fifty trials. Similar results were obtained by Yerkes (1902) in experimenting on the crab *Carcinus granulatus*.

Thus at the beginning of the experiment the animals were as likely to go to the right as to the left, while at the end they went almost invariably to the left. Since the external conditions had not changed, the animals themselves must have changed. Their internal condition now differed in some way from the original condition.

Yerkes and Huggins (1903) endeavored to determine how easily this acquired condition could be modified or destroyed. After the crayfish had learned to go through the open passageway so as to make a mistake in only one case in ten, the experiments were discontinued for two weeks. On the fourteenth day the animals were still inclined to go straight to the open passage, though the habit had become dulled, and they now made mistakes in about three cases out of ten.

In other experiments, after the animals had acquired the habit of escaping through the right passage, the partition G was changed, so as to

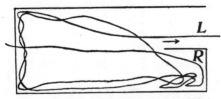

block up this passage, but leave the left one open. At the next trial the animal made a long-continued attempt to escape by the right-hand passageway, following the path shown in Fig. 140. It wandered about for fifteen minutes before discovering the open

FIG. 140.—Path followed by a crayfish which has formed the habit of escaping to the water by the right-hand passageway, when this passage is closed and the left one opened. After Yerkes and Huggins.

way. But in the next trial it turned to the left, and thereafter it turned almost as regularly to the left as it had before turned to the right.

This habit formation took place in the same manner when the floor of the pen was carefully washed out after each trial, showing that the animals were not merely following a path marked by an odor from the previous passage along it. It was evident that the customary direction of turning played a large part in the behavior. When the left passage was closed, the crayfish that had erred into this passage escaped by turning to the right, as indicated by its path in Fig. 141. When after the establishment of this habit, the right passage was closed (Fig. 140), the animal tried persistently to escape from this passage by turning to the right, as it had previously done.

Spaulding (1904) studied the modifiability of behavior in the food reactions of the hermit crab. These animals tend to remain in the lighted parts of the aquarium. They were fed by placing a small, dark screen with a fish beneath it in a certain part of the aquarium. The diffusion of juices from the fish set the crabs to moving about actively, and in the course of time some passed beneath the screen. Here the food was found. At first it took the crabs a long time to find it under these conditions. On the first day only three out of thirty succeeded in fifteen minutes. But by the

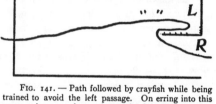

Fig. 141. — Path followed by crayfish while being trained to avoid the left passage. On erring into this passage, it escapes by passing to the right, thus forming the habit of turning to the right. After Yerkes and Huggins.

third day, twenty of the thirty had passed beneath the screen fifteen minutes after it was introduced. At the end of the eighth day, twenty-eight out of the twenty-nine present had passed beneath the screen inside of five minutes. The crabs had become so modified that they went quickly beneath the screen as soon as it was introduced.

Now the experiments were varied by placing in the aquarium the screen alone, without the food. Most of the animals passed beneath it as before. Thus, on the thirteenth day of the experiments, twenty-five specimens out of twenty-seven present had passed under the screen within five minutes. After they had entered they were fed, in order that the association between the screen and food might not be destroyed.

Phenomena of this character are usually spoken of as learning, or as the formation of habits or associations. The facts may be expressed in a purely objective way as follows: When subjected to the stimulus of the screen and the food, the animals reacted to the food by gathering about it — incidentally of course gathering under the screen. After many repetitions of such stimulation, the animals had become changed so that they responded to the dark screen alone by the reaction proper to food. We shall analyze these phenomena more fully in our general discussion of behavior (Chapter XVI).

These processes, by which behavior becomes more or less enduringly modified, are known to play a large part in the behavior of higher invertebrates, such as ants and bees, and in the vertebrates. As investigation progresses, we find analogous processes lower and lower in the animal scale. It was only eight years ago that Bethe (1898) could deny their occurrence even in ants and bees; now they have been fully demonstrated in these and much lower animals. The study of these matters has hardly begun, and it is not too much to say that no experiments have been carried through on the lowest invertebrates that would show this lasting modifiability, even if it exists. We are therefore still in the dark as to how far downward such modifiability extends; time may show it to be a universal property of living things.

The importance of this modifiability for the understanding of behavior is obviously great. Where such modifiability exists, the definite "reflex" is not to be considered a permanent, final element of behavior. On the contrary, it is something developed, and it must differ in individuals with different histories. Two specimens of Convoluta side by side might show at the same moment, one "positive geotropism," the other "negative geotropism," depending on their past history. Whether a hermit crab will pass beneath a dark screen, or will avoid it, is not determined by the permanent properties of its colloidal substance; this can be predicted only by knowing the history of the individual.

The process by which an organism acquires a definite reaction which it before had not is, of course, nothing mystical, but an actual physiological one, whose progress is open to investigation as is that of any other. It needs to be studied and analyzed in the same objective way as the circulation of the blood. The power of changing when acted upon by outer agents, in such a way as to react differently thereafter, is one of the most important properties of living matter, and it is misleading to ignore this property and deal with animals as if their reactions were invariable. How the modifications occur is one of the fundamental problems of physiology. We must remember that even what we call memory, intelligence, and reasoning are composed objectively of certain physiological processes. In other words, as Liebmann has emphasized, there are objective material processes that follow the laws of intelligence, of reasoning, of logic. This is a capital fact. In searching for the laws of life processes we must remember that those just mentioned are as real as any others, and their laws must be provided for in the physics and chemistry of colloids if these are to give us the laws of life processes.

We shall attempt to analyze some of these matters farther in our general discussion of behavior, which forms the remainder of the work.

LITERATURE XII

A. Behavior of echinoderms: UEXKÜLL, 1897, 1897 *a*, 1899, 1900, 1900 *a*; PREYER, 1886; ROMANES, 1885.

B. Behavior of planarians: PEARL, 1903; MAST, 1903.

C. Behavior of Rotifera: JENNINGS, 1904 *b*.

D. Reaction by varied movements in other invertebrates: SMITH, 1902; HOLMES, 1905; BOHN, 1903; MOEBIUS, 1873; HARPER, 1905.

E. Method of trial and error in vertebrates: THORNDIKE, 1898; MORGAN, 1900.

F. Modifiability of behavior in lower animals: BOHN, 1903 *a*, 1905; GAMBLE AND KEEBLE, 1903; JENNINGS, 1904 *d*; YERKES, 1902; YERKES AND HUGGINS, 1903; SPAULDING, 1904.

PART III

ANALYSIS OF BEHAVIOR IN LOWER ORGANISMS, WITH A DISCUSSION OF THEORIES

CHAPTER XIII

COMPARISON OF BEHAVIOR OF UNICELLULAR AND MULTI-CELLULAR ORGANISMS

We have now examined the behavior of a number of Protozoa and of a number of Metazoa. What characteristic differences do we find between the two?

This question is of interest from a number of points of view. The Protozoa consist each of but a single cell, while the Metazoa are composed of many cells, which are differentiated for the performance of different functions. Does this difference in structure correspond to any fundamental difference in behavior? Le Dantec (1895) proposed to distinguish the life manifestations of the Protozoa as " elemental life " from the life of the Metazoa, holding that the two are so different in fundamental character that it is improper to apply the same name to them; this point of view is often met in scientific literature. The life of the Protozoa is considered "as the direct result of the diverse reactions of a small mass of a certain chemical substance in the presence of appropriate substances " (*l.c.*, p. 26), while that of the Metazoa is " the result of the functioning of an extremely complicated machine, in which the reactions of the chemical in question serve as motor power." The former is compared to the burning of the alcohol in an alcohol motor, the latter to the functioning of the motor itself (p. 27). We are interested in the question whether this theoretically fundamental difference shows itself in any way in the phenomena to be observed. Is there any objective evidence in the behavior for the belief that the life of the Protozoa differs fundamentally from that of the Metazoa?

Again, the Metazoa possess a nervous system, while the Protozoa have none. To the specific properties of the nervous system many of the manifestations of behavior in higher animals have been attributed. This system is often considered an essential prerequisite for certain fundamental features of behavior. Do we find a striking difference in the behavior of organisms after a nervous system has been developed? What can animals do *without* a nervous system? A comparison of organisms with and those without this system should give us evidence as to the real nature of the functions of the latter, and will perhaps prevent us from overestimating its importance.

We will sum up briefly in a number of paragraphs the resemblances and differences between the behavior of animals with and without a nervous system.

1. First, we find that in organisms consisting of but a single cell, and having no nervous system, the behavior is regulated by all the different classes of conditions which regulate the behavior of higher animals. In other words, unicellular organisms react to all classes of stimuli to which higher animals react.[1] All classes of stimuli which may affect the nervous system or sense organs may likewise affect protoplasm without these organs. Even the naked protoplasm of Amœba responds to all classes of stimuli to which any animal responds. The nervous system and sense organs are therefore not necessary for the reception of any particular classes of stimulations.

2. The reactions produced in unicellular organisms by stimuli are not the direct physical or chemical effects of the agents acting upon them, but are indirect reactions, produced through the release of certain forces already present in the organism. In this respect the reactions are comparable with those of higher animals. This is true for Amœba as well as for more differentiated Protozoa.

3. In the Protozoa, as in the Metazoa, the structure of the organism plays a large part in determining the nature of the behavior. There are only certain acts which the organism can perform, and these are conditioned by its organization; by one of these acts it must respond to any stimulus. If the behavior of the Metazoa is comparable in this respect to the action of a machine, the same comparison can be made for the behavior of the Protozoa.

4. Spontaneous action — that is, activity and changes in activity induced without external stimulation — takes place in the Protozoa as well as in the Metazoa. Both Vorticella and Hydra, as we have seen, spontaneously contract at rather regular intervals, even when the external conditions remain uniform. Continued activity is the normal state of affairs in Paramecium and most other infusoria. The idea that spontaneous activity is found only in higher animals is a totally erroneous one; action is as spontaneous in the Protozoa as in man.

---

[1] Considering auditory stimulation as merely a special case of mechanical stimulation.

5. In unicellular organisms, without a nervous system, certain parts of the body may be more sensitive than the remainder, forming thus a region comparable to a sense organ in a higher animal. Whether such a part may become more sensitive to one form of stimulation, while insensitive to others, as in higher organisms, seems not to have been determined.

6. Conduction occurs in organisms without a nervous system. This is, of course, seen in the fact that a stimulus limited to one part of the body may cause a contraction of the entire body, or a reversal of cilia over the entire body surface. A strongly marked case is the contraction of the stalk in Vorticella, when only the margin of the bell is stimulated.

7. Summation of stimuli occurs in Protozoa as in Metazoa. This is shown most clearly in Statkewitsch's experiments with induction shocks (p. 83). Weak induction shocks have no effect until frequently repeated.

8. In the unicellular animal, as in that composed of many cells, the reaction may change or become reversed as the intensity of the stimulus increases, though the quality of the stimulus remains the same. Such a change in reaction has sometimes been claimed as a specific property of the nervous system. The protozoans Amœba and Stentor, as well as the metazoan Planaria, move toward sources of weak mechanical stimulation, away from sources of strong stimulation.

9. In the Protozoa, as in the Metazoa, the reaction may change while the stimulus remains the same. That is, the animal may respond at first by a certain reaction; later, while the stimulus remains the same, by other reactions. This has been shown in detail in the account of Stentor (Chapter X). The change may consist in either a cessation of the reaction, or in a complete alteration of its character. These changes are, as a rule, by no means due to fatigue, but are regulatory in character. The behavior thus depends on the past history of the organism. For such modifications of behavior a nervous system is then unnecessary.

10. In the Protozoa, as in the Metazoa, the reactions are not invariable reflexes, depending only on the external stimulus and the anatomical structure of the organism. The reaction to a given stimulus depends upon the physiological condition of the organism. In Stentor we could distinguish at least five different conditions, each with its characteristic reaction to the given stimulus.

11. In unicellular as well as multicellular animals we find two chief general classes of reactions, which may be designated positive and negative. The positive reaction tends to retain the organism in contact with the stimulus, the negative to remove it from the stimulus. In many classes of stimuli we can distinguish an optimum condition. A change

leading from the optimum produces a negative reaction, while a change leading toward the optimum produces no reaction, or a positive one. The optimum from this standpoint usually corresponds, in a broad way, to the optimum for the general interests of the organism. These relations hold equally for Protozoa and Metazoa.

12. In both the Protozoa and the Metazoa that we have studied, the behavior is based to a considerable degree on the selection of certain conditions through the production under stimulation of varied movements (see Chapter XII). This shows itself in two characteristic types. In the one case the organism when subjected to a change leading away from the optimum responds by a movement that subjects it successively to many different conditions, finally remaining in that one which is nearest the optimum. This form of reaction is strongly developed in Paramecium. In the second type, which may be considered a development of the first, the organism first responds by one reaction, then by another, continuing at intervals to change its response until one of the reactions frees it from the stimulation. This way of behaving is well seen in Stentor. Both methods of reaction may be expressed as follows: When the organism is subjected to an irritating condition, it tries many different conditions or many different ways of ridding itself of this condition, till one is found which is successful.

All together, there is no evidence of the existence of differences of fundamental character between the behavior of the Protozoa and that of the lower Metazoa. The study of behavior lends no support to the view that the life activities are of an essentially different character in the Protozoa and the Metazoa. The behavior of the Protozoa appears to be no more and no less machinelike than that of the Metazoa; similar principles govern both.

Further, the possession of a nervous system brings with it no observable essential changes in the nature of behavior. We have found no important additional features in the behavior when the nervous system is added. In the lower Metazoa, experiment has shown the nervous system to have two chief functions, — the maintenance of tonus, and the bringing of the parts of the body into relation with each other by serving for conduction. But both these functions are performed in the Protozoa without a nervous system. The body of Paramecium maintains marked tonus, and the different parts of the body work together. A comparison of the behavior of the Protozoa with that of the lower Metazoa lends powerful support to that view of the functions of the nervous system which is so ably maintained by Loeb in his brilliant work on "The Comparative Physiology of the Brain and Comparative Psychology." According to this view we do not find in the nervous system *specific qualities* not found elsewhere in protoplasmic structures. The qualities of the nervous system are the general qualities of protoplasm. Certain of these general qualities have become much accentuated in the protoplasm of the nervous system, while in the remainder of the protoplasm of the metazoan body they are less strongly marked, being partially obscured by differentiations in other directions. Most if not all of the fundamental activities which have been considered peculiar to the nervous system may be demonstrated, as we have seen, in the Protozoa, yet in them no nervous system exists.

These facts show the necessity of guarding against overrating the importance of the nervous system. It is doubtful if the nervous system is to be considered the *exclusive* seat of anything; its properties are accentuations of the general properties of protoplasm. Dogmatic statements as to the part necessarily played by the nervous system in given cases must be looked upon with suspicion unless supported by positive experimental results. If acts objectively identical with "reflex actions" and still more complex types of behavior may exist in the Protozoa without the intervention of a nervous system, it is not impossible that they may occur in the same manner in Metazoa, as Loeb has maintained. Where a nervous system exists, we are not justified in dogmatically referring all phenomena of behavior to it, for other protoplasm exists too, and may still retain some of the characteristics which it had in the Protozoa. In an animal possessing a nervous system we cannot tell without experimentation whether a given reflex action or other reaction depends on the nervous system or not. The possibility always remains open that the remainder of the protoplasm may perform the act in question by its own capabilities, as it does in the Protozoa. In any animal, we are justified in attributing exclusively to the nervous system only those properties which rigid analytical experimentation shows it alone to possess.

CHAPTER XIV

TROPISMS AND THE LOCAL ACTION THEORY OF TROPISMS

A LARGE share of the behavior of lower as well as of higher animals consists of movements either toward or away from certain objects or sources of stimulation. Behavior can thus be largely classified into two great classes: "positive and negative" reactions; movements of "attraction and repulsion," of approach and retreat. To account in a general way for these directed movements certain theories have been proposed, and one of these has become widely accepted. This is the so-called "tropism theory." The word "tropism" has been used in several different senses by different authors, and not always as implying a definite theory (see page 274). But there is a certain theory which is usually implied when tropisms are mentioned; it has become so generally accepted that it is often spoken of as *the* tropism theory. It will perhaps be more accurate to speak of it as the local action theory of tropisms. "Tropisms" has become the key-word for the behavior of lower organisms, and the theory mentioned is supposed to furnish explanation of most of the puzzles found in this field. A theory so generally accepted demands separate special treatment. What is this tropism theory as usually understood in discussions of animal behavior, and how far does it go in helping us to understand the behavior of lower organisms?

According to this tropism theory the primary feature in the directed movements of lower organisms is the position or orientation of the body with respect to the source of stimulation, and this orientation is brought about by the direct local action of the stimulating agent on that part of the body on which it impinges. The essential points in this theory are then two: first, orientation; second, the production of orientation by local action. These points we may consider separately.

(1) By this tropism theory a stimulus is considered to force the animal to take a certain position with respect to the direction from which the stimulus comes; in this position it is said to be oriented. Usually the organism becomes oriented with anterior end either toward or away from the source of stimulation. This is the essential feature in the action of the stimulus. "The essential point in all directive stimulation is

265

therefore the axial orientation of the cell body, and the central point in
the mechanism of this phenomenon, lies in the explanation of this axial
position" (Verworn, "General Physiology," 1899, p. 480). After the
animal has thus become oriented it may move forward in the usual
way. If it does so, it will of course incidentally move toward or away
from the source of stimulation, but this approach or retreat is not an
essential or determining part of the reaction. "The really fundamental
phenomenon which characterizes these directed movements is always
not so much the forward movement as such, as rather a process which
may be called a movement of orientation. The organism places its
axis in a definite localized relation to the stimulus, which may be photic,
thermic, chemical, etc. That is, it places its axis either in the direction
of the stimulation or perpendicular to it (diatropism). In the former
case the 'anterior' end may be directed 'positively,' toward the source of
stimulation, or 'negatively,' away from it. It now appears a matter of
course that if forward motion takes place after such orientation, its
direction will correspond to the direction of the stimulus" (Driesch,
1903, p. 5, translation).

(2) This orientation is produced, according to this tropism theory,
by the direct action of the stimulating agent on the motor organs of that
side of the body on which it impinges. A stimulus striking one side of
the body causes the motor organs of that side to contract or extend or
to move more or less strongly. This, of course, turns the body, till the
stimulus affects both sides equally; then there is no occasion for further
turning, and the animal is oriented. "These tropisms are identical
for animals and plants. The explanation of them depends first on the
specific irritability of certain elements of the body surface, and, second,
upon the relations of symmetry of the body. Symmetrical elements at
the surface of the body have the same irritability; unsymmetrical elements have a different irritability. Those nearer the oral pole possess
an irritability greater than that of those near the aboral pole. These
circumstances force an animal to orient itself toward a source of stimulation in such a way that symmetrical points on the surface of the body
are stimulated equally. In this way the animals are led without will of
their own either toward the source of stimulus or away from it" (Loeb,
1900, p. 7). Holt and Lee (1901, pp. 479–480) bring out this point in
the prevailing theory, as applied to light, as follows: "The light operates,
naturally, on the part of the animal which it reaches. The intensity of
the light determines the sense of the response whether contractile or
expansive, and the place of the response, the part of the body stimulated,
determines the ultimate orientation of the animal."

How the orientation is brought about according to this theory may be illustrated most simply by considering an organism covered with cilia. For this purpose we may employ the accompanying diagrams, based on those given by Verworn (1895, p. 484), but modified to make them clearer. In Fig. 142 a stimulus is supposed to act from the right

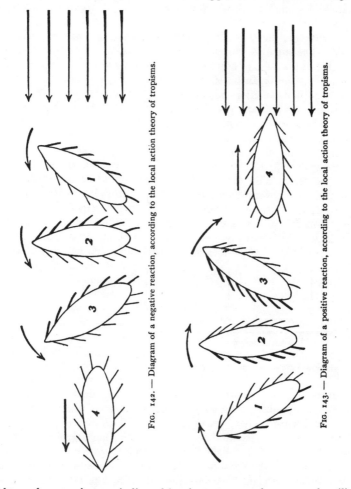

Fig. 142. — Diagram of a negative reaction, according to the local action theory of tropisms.

Fig. 143. — Diagram of a positive reaction, according to the local action theory of tropisms.

side on the organism, as indicated by the arrows, and to cause the cilia of that side to contract more strongly, as is indicated by the heavier shade and greater curving. This must, of course, turn the body to the left, as a boat is turned to the left when the right oar is more strongly pulled. The animal therefore occupies successively the positions 1, 2, 3, and 4. In the position 4 both sides are equally affected by the stimulus, so that there is no cause for further turning. The animal has become oriented and its usual forward movements now take it away from the source of stimulation. We have here a case of negative tropism or taxis.

Figure 143 illustrates the conditions producing positive tropism or taxis. The stimulus, coming from the right side, is supposed to cause the cilia of that side to beat less strongly backward, or to beat forward. As a result the organism is turned to the right, through the positions 1, 2, 3, 4, till its anterior end is directed toward the source of stimulation. Both sides are now affected alike, and there is no cause for further turning. The animal now moving forward in the usual way of course travels toward the source of stimulation.

As an example of the application of the tropism scheme to a muscular organism, we may take Davenport's exposition of the action of light in determining the direction of locomotion of the earthworm. "Represent the worm by an arrow whose head indicates the head end (Fig. 144, A). Let solar rays SS fall upon it horizontally and perpendicularly to its axis. Then the impinging ray strikes it laterally, or, in other words, it is illuminated on one side and not on the other. Since, now, the protoplasm of both sides is attuned to an equal intensity of light, that which is the less illuminated is nearer its optimum intensity. Its protoplasm is in a photo-tonic condition. That which is strongly illuminated has lost its phototonic condition. Only the darkened muscles, then, are

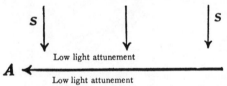

Low light attunement

Low light attunement

Fig. 144. — Diagram to explain a tropism in a muscular organism, such as the earthworm. After Davenport. See text.

capable of normal contraction; the brightly illuminated ones are relaxed. Under these conditions the organism curves toward the darker side; and since its head region is the most sensitive, response begins there. Owing to a continuance of the causes, the organism will continue to turn from the light until both sides are equally illuminated, *i.e.* until it is in the light ray. Subsequent locomotion will carry the organism in a straight line, since the muscles of the two sides now act similarly. Thus orientation of the organism is effected. The same explanation, which is modified from one of Loeb ('93, p. 86), will account, *mutatis mutandis*, for positive phototaxis" (Davenport, 1897, p. 209).

From the relations above set forth, it follows that for determination of the direction of movement in accordance with this tropism theory, a stimulus must act upon one portion of the body differently from or more intensely than on other parts. Without such differential action on different parts of the body there is nothing to cause the animal to turn in one direction or another.

This tropism schema is made by its upholders the basis for the larger part of the directed activities of the lower animals. "Thus the phenomena of positive and negative chemotaxis, thermotaxis, phototaxis, and galvanotaxis, which are so highly interesting and important in all organic life, follow with mechanical necessity as the simple results of differences in biotonus, which are produced by the action of stimuli at two different poles of the free-living cell" (Verworn, 1899, p. 503). Verworn (1899) and Loeb (1900) have developed the theory as a general explanation for all sorts of directed activities, and many authors have accepted it for reactions to particular stimuli. In recent times, Holt and Lee (1901) have applied it in detail to the responses to light, Loeb (1900, p. 186) and Garrey (1900) to chemicals, Loeb (1897) and Verworn (1899) to gravitation, Mendelssohn (1902 *a*) to heat and cold.

In the foregoing chapters we have examined the behavior of a considerable number of lower organisms, of many different kinds. How far does this examination support the above theory? How far is the observed behavior due to orientation produced by the local action of stimuli on the different parts of the body? To what extent does this tropism theory aid us in understanding the behavior of these organisms?

In Amœba there are no permanent body axes; anterior and posterior ends continually interchange places in the rolling movement, and any part may become at any time the advancing portion. Under these conditions the term "orientation" can have little meaning, and we can hardly say that stimulation causes the body to become oriented in a certain way. But stimulation does determine the direction of motion, and anything like orientation that can be distinguished is a result of the direction of motion, not its cause. Under stimulation the direction of movement is changed first, then in consequence the animal takes an elongated form which furnishes the only possible basis for the use of the term "orientation."

In the fact that to produce directed movement, local action of the stimulus on a certain part of the body is necessary, causing local contraction or extension, the conditions in Amœba agree with the fundamental postulates of the tropism theory. The agreement is most precise in the positive reactions, where the part stimulated is the part that extends and determines the direction of movement. In the negative reactions the agreement with the theory is less complete; for while the part that contracts is determined by the region stimulated, the extension and consequent direction of movement are, as a rule, not thus determined.

But while some important features of the behavior of Amœba are thus in agreement with the underlying assumptions of the tropism theory, it is certain that for such organisms alone the theory would never have been proposed. The facts for Amœba can be formulated in a much simpler way than by bringing in the conception of orientation, — a conception derived from organisms with permanent body axes, and fitting only these.

But when we turn to an examination of the behavior of those unicellular organisms having permanent body axes, we find the conditions widely at variance with the assumptions of the local action tropism theory. In the infusoria most of the behavior is quite inconsistent with the theory. The reactions are not determined by the direct action of a localized stimulus in producing greater contraction or extension in that part of the body on which it impinges. The organism responds as a whole, by a reaction involving all parts of the body. It does not necessarily turn directly toward or directly away from the source of stimulation, as would be the case if it reacted in accordance with this tropism theory. The direction of turning is determined by internal factors; the animal turns toward a side which is structurally defined. For inducing directed motion it is not necessary that the stimulus should act differently on different parts of the body. The cause of reaction — that is, of a change in the movements — is usually a change from one condition or intensity to another. Thus the essential point in determining whether reaction shall occur is in most cases the direction of movement — whether this takes the organism (or its most sensitive portion) away from, or toward, the optimum. It is difficult to conceive a type of behavior more completely opposed to the local action theory of tropisms above set forth.

In some cases this method of reaction produces orientation with relation to the direction of some external force, in other cases it does not. The orientation when it occurs is brought about through continued movements that are varied in direction, with final selection of one of these directions. Whether orientation shall or shall not result depends on whether it must result in order that there shall be a cessation of the stimulation which is producing the varied movements. These relations have been set forth in detail in our account of the behavior of Paramecium (Chapter IV, Section 6), so that it is not necessary to take them up here.

To almost all the relations set forth in the preceding paragraphs there is one exception. In the reaction of ciliate infusoria to the electric current we find certain features which agree with the local action tropism theory. These features are so striking and so utterly at variance with everything found in the remainder of the behavior of these organisms that they throw into strong relief the contrast between the usual behavior and the requirements of this tropism schema. Owing to the remarkable cathodic reversal of the cilia (a phenomenon not paralleled under any other conditions), the motor organs of opposite sides or ends of the ciliate infusorian act under the electric current in different ways. The result is behavior partly in accordance with the tropism schema. This furnishes us with a picture of what behavior would be if this schema held throughout. The unity and coördination that are so striking in the remainder of the behavior are here quite lost. Different parts of the motor organs urge the organism in different directions at the same time. The animal seems to be trying to do two opposed things at once (see p. 89). Nothing more ineffective and unpurposive can be imagined than such behavior. But in producing these local effects the electric current is unique among stimuli, and the reaction is as far from the typical behavior of these organisms as can be imagined. The electric current may be used for producing local contractions in man as well as in Paramecium, but such contractions cannot be considered an adequate type of the behavior of mankind. The electric current never acts effectively on the organisms under the natural conditions, so that normally they never show the peculiar behavior produced by it. To all the natural conditions of existence they react in a totally different manner — a manner quite at variance with this tropism schema.

In the bacteria as in the infusoria the behavior is not in accordance with the above-discussed theory of tropisms. The details of the reactions are not so completely known as in the infusoria. But what we know shows that the behavior of these organisms so far as involved in the directed reactions is as follows: When stimulated the bacterium changes its course, moving in some other direction, — a direction determined by its own body structure, and not by the position of the stimulating agent.

Thus we find in the unicellular organisms very little in the behavior that can be interpreted in accordance with this local action theory of tropisms. The latter does not by any means express the fundamental nature of their behavior in directed reactions. These are based chiefly on the performance under stimulation of varied movements, with selection from the resulting conditions, — the "method of trial."

In the symmetrical Metazoa we of course find many cases in which the animal turns directly toward or away from a source of stimulation, without anything in the nature of preliminary trial movements. This is a simple fact of observation, which leaves open the possibility of many different explanations. Is the simple explanation given by the local action theory of tropisms one that is of general applicability to the directed reactions of lower and higher Metazoa? In considering the evidence on this question, we find that even in symmetrical Metazoa the direction of movement with reference to external agents is by no means always brought about by a simple, direct turning. On the contrary, in many of the Metazoa, trial movements are as noticeable and important as in the Protozoa. This we have illustrated in detail for many invertebrates in the section devoted to this subject (Chapter XII, Section 2). For such behavior the local action theory of tropisms fails to give determining factors.

In some cases the turning movements are directly toward or from certain stimuli. But the question here is, whether this turning is produced by the local action of the agent in question on the part of the body against which it impinges, as is asserted by the theory which we are considering, and illustrated in Fig. 144.

In a few instances this is apparently the case. The medusa escapes unfavorable stimulation by contracting most strongly on the side on which the stimulus impinges. In Hydra local stimulation by chemicals, heat, or electricity often produces limited local contraction, causing the animal to bend toward the side stimulated. In various sea anemones the tentacles, and sometimes the body, may bend toward the side stimulated, as this theory demands. Yet this direct contraction plays very little part in the behavior of these animals. In Hydra it is only injurious agents to which the animal responds in this way, and the result is to still further subject the animal to the action of the injurious agent. In order to escape the action of injurious stimuli, Hydra has recourse to behavior of quite a different character, and in its natural life there seems to be no indication that behavior ever occurs in accordance with this theory of direct local action. In sea anemones the direct turning toward the region stimulated is at once supplemented by movements determined in quite a different way, — through the structure of the organism, — the tentacles bending toward the mouth. Without this supplementary reaction the local bending would be of no service. In the hydroid Corymorpha it is only this second method of bending that occurs at all. Throughout the Cœlenterata the part played by trial movements, not directly determined by the position of the stimulating agent, is most striking and important.

In the echinoderms we have, as in Amœba, organisms which are as a rule without a definite body axis, so far as the direction of locomotion goes; there is usually no permanent anterior, posterior, right, or left. Hence a theory like that of tropisms, based primarily on the position or orientation of the body axis with reference to the direction of the stimulating agent, can find little precise application. Yet it is again in this group that we find behavior that is in certain respects at least in accordance with the tropism theory. For locomotion in a certain direction the stimulus must be localized, acting in a different way on the two sides; this is one of the postulates of the tropism theory. Further, a local stimulation may have at least a partially local effect, and this may result in movement in a certain direction. But as v. Uexküll has well pointed out, the elementary factors here are the typical reaction methods ("reflexes") of the individual organs of the body surface. The tropism, if we attempt to apply the concept at all, is a mere collection of these elementary reactions; it is not in any sense itself an elementary factor. In other words, the tropism theory would never have been based on the known behavior of the echinoderms, for the facts, even so far as they agree with the fundamental postulates of the theory, can be formulated more directly and simply in another way. The tropism theory is furnished with an apparatus of relations that finds no application to the starfish and sea urchin.

Furthermore, as we have shown in detail, much of the behavior of these animals is based on the method of trial. In such bilaterally symmetrical animals as the flatworm Planaria we have the most favorable possible conditions for action on this tropism theory, and such animals often do turn directly toward or away from sources of stimulation. But when this occurs, is it due merely to the local contraction or extension of the musculature on the side on which the stimulus impinges, or is it a reaction of the animal as a whole?

This question can be answered only by a thorough study of all the factors in the reaction; such a study is given us for the flatworm by Pearl (1903). The positive reaction of the flatworm — the direct turning toward the source of stimulation — seem to present ideal conditions for explanation on the simple tropism theory. But Pearl, after exhaustive study, concludes that the processes in the reaction are as follows: —

"A light stimulus, when the organism is in a certain definite tonic condition, sets off a reaction involving (1) an equal bilateral contraction of the circular musculature, producing the extension of the body; (2) a contraction of the longitudinal musculature of the side stimulated, producing the turning toward the stimulus (this is the definitive part of the reaction); and (3) contraction of the dorsal longitudinal musculature, producing the raising of the anterior end. In this reaction the sides do not act independently, but there is a delicately balanced and finely co-ordinated reaction of the organism as a whole, depending for its existence on an entirely normal physiological condition" (Pearl, 1903, p. 619).

Similar lack of uniformity and simplicity appears in the remainder of the behavior of the flatworm. In few of the lower metazoa has the movement been so thoroughly analyzed as in Planaria. But there seems to be no reason for thinking that in this simple animal these relations are more complex than in most invertebrates.

The recent thorough studies of Rádl (1903) on reactions to light in many animals have shown clearly the inadequacy of this theory to account for most of the reactions to this agent. Bohn (1905) has likewise been compelled to reject this theory, on the basis of the results of his thorough studies on the behavior of the animals of the seashore. To the writer it appears that most of the recent thorough work on animal behavior points in the same direction.

We must then conclude from our examination of the facts that for the lower organisms taken into consideration in the present work, the local action theory of tropisms is of comparatively little value for interpreting behavior. This theory uses and attempts to make of general application certain elements here and there observable in the behavior of some organisms. But in many organisms even these elements are almost completely lacking, and in no organism that we have taken up does this theory adequately express the nature of behavior. The tropism as applied to animal behavior in the sense we have considered, is not an elementary factor; it is only a more or less artificial construction, made by combining certain elements of behavior and omitting others that are of most essential significance. It makes use of certain simple phenomena that actually exist, but elevates these into a general explanation of directed behavior, for which they are utterly inadequate. The prevalence of this local action theory of tropisms as a general explanation of behavior in lower organisms is based only on an incomplete knowledge and an insufficient analysis of the facts of behavior.

OTHER TERMS EMPLOYED IN ACCOUNTS OF ANIMAL BEHAVIOR

In the foregoing pages we have criticised a certain definite theory of tropisms, this being the theory most commonly implied when the word is used in a precisely defined way. But the term "tropism" is often used in a looser sense. By some writers the word is applied merely to the general phenomenon that the movements of organisms show definite relations to the location of external agents. In this sense the word implies no theory, and is not open to criticism on the basis of observed facts. It is, of course, equally applicable to the behavior of man and that of lower organisms; in this sense the botanist Pfeffer (1904, p. 587) consistently remarks that a man who bends toward a lighted window shows phototropism as does a plant. The use of the word in this purely descriptive sense is often convenient, but we need to keep in mind the fact that the word thus used involves no explanation, and includes phenomena of the most heterogeneous character.

By some writers the word " tropism " is restricted to the bending or inclination of a fixed organism, while the movements of free organisms under the influence of external agents are called *taxis*. This distinction is a purely descriptive one.

Some writers reserve the term " tropism " (or taxis) for those reactions in which the organism takes up a well-defined orientation with relation to the line of action of some external agent. Other reactions, in which orientation is not a feature, are variously designated as *kinesis* (Engelmann, 1882 *a*; Rothert, 1901; Garrey, 1900), as *-pathy* (Davenport, 1897; Yerkes, 1903 *b*; and others), as *-metry* (Strasburger, 1878; Oltmanns, 1892), and by various other names, depending on the method by which the author in question considers them to be brought about. On this basis the reactions of infusoria to water currents, gravity, the electric current, and to light coming from one side would be called tropisms or taxis; while the reactions to chemicals, osmotic pressure, heat and cold, and mechanical stimuli would be designated by some other term.

An immense number of technical terms have been devised for application to the phenomena of behavior in the lower organisms. A systematic exposition of a very complete set of such terms will be found in the paper of Massart (1901). The "Plant Physiology" of Pfeffer (1904) likewise deals extensively with this matter. A proposed new terminology applying to many of the features of behavior is set forth by Beer, Bethe, and v. Uexküll (1899). A number of other references to this matter will be found in the literature list at the end of the present chapter.

As to the value of giving technical names to every distinguishable act that an organism performs, opinions will differ. So far as the names are purely descriptive, expressing nothing more than some observed action of the organism, it is difficult to see any very great advantage in their use. To say that an organism shows phobism (Massart), is merely to say that it moves backward; to say that it reacts by dorsoclinism (Massart), is the same as to say that it reacts by turning toward the dorsal side. To most readers the latter expressions are more intelligible than the former, and they are equally accurate and complete. Such purely descriptive terms embody no results of scientific analysis. Their use is therefore merely a question of convenience or taste on the part of the writer. They are doubtless at times convenient and may perhaps be used to advantage.

So far as the terms involve a certain explanation of the phenomena, their use requires that the writer shall accept that explanation for the phenomena in question, otherwise their use gives rise to misconception. This makes many of the terms unavailable, save in a very restricted degree. The study of behavior seems hardly to have reached as yet the stage where a hard and fast nomenclature can be used to advantage. To the present writer, after a long-continued attempt to use some of the systems of nomenclature devised, descriptions of the facts of behavior in the simplest language possible seems a great gain for clear thinking and unambiguous expression. If investigators on the lower organisms would for a considerable time devote themselves to giving in such simple terms a full account of behavior in all its details, paying special attention to the effect of the movements performed on the relation of the organism to the stimulating agent, this would be a great gain for our understanding of the real nature of behavior, and some theories now maintained would quickly disappear. Less attention to nomenclature and definitions, and more to the study of organisms as units, in their relation to the environment, is at the present time the great need in the study of behavior in lower organisms.

LITERATURE XIV

A. The local action theory of tropisms: LOEB, 1900, 1897; VERWORN, 1895, 1899; DAVENPORT, 1897; DRIESCH, 1903; RÁDL, 1903; HOLT AND LEE, 1901; MENDELSSOHN, 1902 *a*; GARREY, 1900; BOHN, 1905; JENNINGS, 1904 *c*.

B. Nomenclature and classification in behavior: PFEFFER, 1904; MASSART, 1901; BEER, BETHE, AND V. UEXKÜLL, 1899; NAGEL, 1899; CLAPAREDE, 1905; HABERLANDT, 1905; ZIEGLER, 1900; NUEL, 1904; DAVENPORT, 1897; ROTHERT, 1901; ENGELMANN, 1882 *a*; LOEB, 1893, 1900; GARREY, 1900; STRASBURGER, 1878; OLTMANNS, 1892; YERKES, 1903 *b*.

CHAPTER XV

IS THE BEHAVIOR OF THE LOWER ORGANISMS COMPOSED OF REFLEXES?

THE simplest reaction of an organism is the performance of a definite simple act in response to a definite stimulus. Such is the contraction of Vorticella, such the reversal of movement in a bacterium or in Paramecium or the flatworm. A simple responsive action of this sort is commonly known as a reflex. The question has been raised as to whether the behavior of the lower organisms differs from that of higher animals in being purely reflex or not; in other words, whether all their reactions to stimuli are reflexes. For various organisms this question is answered by many authors in the affirmative. In some cases the behavior of animals much higher in the scale than most of those we have considered is characterized as purely reflex. This is v. Uexküll's view for the sea urchin. We must examine briefly the question whether behavior in these lowest organisms is properly characterized as reflex.

What is a reflex? The concept of reflex action has had a complex origin, and as a result it is defined in various ways. One of the phenomena on which the concept is based is the contraction of a muscle when a certain nerve is stimulated. The stimulation is supposed to pass from the nerve to the spinal cord, whence it is reflected back to the muscle; hence the name *reflex*. Some authors hold that the term can be properly used only of acts thus performed by the aid of the nervous system. This would of course exclude reflexes from the behavior of unicellular organisms, and introduce uncertainty in dealing with the lower Metazoa, for in many of these we do not know whether the reactions are throughout mediated by the nervous system or not. But it is more usual to consider the reflex as a certain type of action, without regard to the particular anatomical structures involved. Even where the term is limited to actions produced through the nervous system, some other term is employed to indicate the corresponding type of action in animals without a nervous system, so that the existence of a particular kind of action, indicated usually by the word " reflex," is recognized. Thus, Beer, Bethe, and v. Uexküll (1899) use for reflexes performed without a nervous system the word "antitype." We may then examine the reflex (or antitype) simply as a type of action, without regard to the existence of a nervous system.

A second phenomenon on which the concept of reflex action is based is the following: In ourselves, certain acts are performed unconsciously. These acts have been considered identical with those due to the passage of an impulse from the nerve-ending to the spinal cord, and thence back to the muscle; that is with reflexes. Hence the reflex is often defined as an unconscious or involuntary action: "Such involuntary responses we know as 'reflex' acts" (James, "Psychology," Vol. I, p. 13). "Reflexes are voluntary acts that have become mechanical" (Wundt). This definition of a reflex act as involuntary or unconscious is widely employed. If we accept this definition, there is of course no way by which we can tell whether the reactions of lower animals are reflex or not. By observation we cannot tell whether the reacting organism is conscious, for this would require, as Titchener (1902) says, an objective criterion of the subjective, — an objective criterion of that which is not objective, and this is impossible. It is certainly as dogmatic and unscientific to assert that the actions of organisms are reflex in the sense of unconscious, as to assert the opposite, for we have no knowledge on this point. We can recognize reflex acts, from this point of view, only in ourselves.

A third phenomenon on which the conception of a reflex is based is the supposed uniformity of certain reactions. The muscle responds to all sorts of stimuli by contracting. This uniformity is considered by many authors the essential feature in reflexes. Hobhouse (1901, pp. 28, 29) defines reflexes as "uniform responses to simple stimuli." According to Beer, Bethe, and v. Uexküll (1899, p. 3), reflexes are reactions "always recurring in the same manner." Driesch (1903) says a reflex is "a motor reaction which as a response to a stimulus occurs the first time completely and securely."

This objective definition of a reflex as an invariable reaction to a simple stimulus is the only one which we can really use in determining by means of objective study whether the behavior of animals is reflex in character. Is the behavior of lower organisms composed of reflexes in this sense?

Possibly the best case for an affirmative answer to this question could be made out for the bacteria. Here there is so far as known only one form of motor reaction, — the reversal of movement when stimulated. But even in the bacterium the uniformity is disturbed by the fact that on coming in contact with a solid the organism sometimes comes to rest against it, while at other times it reacts by the reversal of motion. Owing to their minuteness, the behavior of these organisms is less known than that of other unicellular forms, so that it is difficult to make a positive generalization on such a point as the present one.

If we attempt to apply our definition of a reflex to the behavior of the infusoria, — of Paramecium, for example, — we at once get into difficulties. The "avoiding reaction" of Paramecium is sharply limited in many ways, and always takes place in accordance with a definite type. But it is far from being invariable. The reaction is composed of three factors, which may vary more or less independently of each other, in such a way that an absolutely unlimited number of combinations may result, all fitting the generalized type. The possible variations may be summed up as follows: If the animal be taken as a centre about which a sphere is described, with a radius several times the length of the body, then as a result of the avoiding reaction the animal may traverse the peripheral surface of this sphere at *any point*, moving at the time either backward or forward. In other words, the reaction may carry it in any one of the unlimited number of directions leading from its position as a centre. While the direction of turning is absolutely defined by the structure of the animal, yet the combination of this turning with the revolution on the long axis permits the animal to reach any conceivable position with relation to the environment. In other words, Paramecium, in spite of its curious limitations as to method of movement, is as free to vary its relations to the environment in response to a stimulus as an organism of its form and structure could conceivably be. Such behavior does not fall within the concept of a reflex, if the latter is defined as a uniform reaction.

Still less does the behavior of Stentor yield itself to formulation as purely reflex. To the same stimulus, under the same external conditions, this animal may react, as we have seen, in several different ways; its reaction depends upon its physiological condition. The same is true for Hydra and other Cœlenterata, for the echinoderm, the flatworm, and many other invertebrates, as we have set forth in detail in the description of the behavior of these organisms. In the sea anemone we have examples of indecision, parts of the positive reaction being combined with parts of the negative. In all these cases the behavior is far from that sureness and fixity that characterizes the supposed reflex.

Even in Amœba it is difficult to apply the reflex concept to the behavior. So far are the reactions here from being uniform, that we can almost say, on the contrary, that Amœba never does the same thing twice. The behavior is here formless, undefined, not held within narrow bounds by structural conditions, as in the infusoria and in most higher animals; the essential criteria of reflex action seem lacking. It would be very difficult to apply the reflex concept, for example, to the behavior of a floating Amœba in attaining a solid support, as described on page 8, or to the food reaction illustrated in Fig. 21. Further, as we have seen on page 20, Amœba may at different times react in opposite ways to the same stimulus.

Indeed, consideration shows that it is impossible to apply rigidly the conception of a reflex, as an invariable reaction to a definite stimulus, to the behavior of any organism having more than one motor reaction at its command. James ("Psychology," Vol. I, p. 21) and Pearl (1903, p. 704) have given us sketches of what would be the behavior of an organism whose acts were purely reflex. Taking the reaction to food as an example, James says: "The animal will be condemned fatally and irresistibly to snap at it whenever presented, no matter what the circumstances may be; he can no more disobey this prompting than water can refuse to boil when a fire is kindled under the pot. His life will again and again pay the forfeit of his gluttony. Exposure to retaliation, to other enemies, to traps, to poisons, to the dangers of repletion, must be regular parts of his existence. His lack of all thought by which to weigh the danger against the attractiveness of the bait, and of all volition to remain hungry a little while longer, is the direct measure of his lowness in the mental scale" (*l.c.*, p. 21). Such a picture has only to be presented to make us see the impossibility of constructing the entire behavior of an organism out of such irresistible reflexes. For the reactions to dangers and enemies must then be reflexes, as well as the reactions to food, and the two are incompatible. Suppose the food and the danger are present together, as often happens. The organism *cannot* react fatally and irresistibly to both, for the movements required are in opposite directions. It *must* decide to react either with relation to one or to neither, and in either case the fatality and irresistibility of at least one of the reflexes disappears.

If, then, we consider the reflex an invariable reaction to a given stimulus, we cannot hold that behavior in lower organisms is made up of reflexes. Indeed, the fact that stands out most clearly in the behavior is the following: Each stimulus causes as a rule not merely a single definite action that may be called a reflex, but a series of "trial" movements, of the most diverse character, and including at times practically all the movements of which the animal is capable. The reaction to a given stimulus depends on the physiological state of the organism, not alone on its anatomical structure; and physiological states are variable. This is true both for the infusoria and for man.

The attempt to characterize the behavior of the lower organisms as purely reflex has risen from the desire to show that the structural conditions of the organism and the physical and chemical action of the stimulus are sufficient to account for their behavior, without the necessary intervention of consciousness. This is well expressed by v. Uexküll (1897, p. 306) when he says that we are to regard the reflex as "the necessary course of a process that is conditioned by nothing else than the mechanical structure of the organism." Shall we include the physiological state of the organism as part of its mechanical structure? If we answer this question in the negative, then it is clear that the behavior of the lower organisms is not reflex in character. If on the other hand we answer this question in the affirmative, holding that the physiological state is some chemical or physical configuration of the substance of the organism, and therefore to be included in its mechanical structure, then the entire question concerning the reflex character of behavior in a given organism loses its objective character and evaporates into thin air. For in the highest as well as the lowest organism the reactions must be supposed to depend upon the physical and chemical constitution of the organism, unless we are to accept vitalism. And if when we say that the behavior of an organism is reflex in character, we mean only that its behavior depends upon its physical and chemical make-up, we can make no distinction upon this ground between the behavior of lower and higher organisms. This point is indeed well recognized by thoughtful psychologists. "The conception of *all* action as conforming to this [the reflex] type is the fundamental conception of modern nerve physiology," says James ("Principles of Psychology," Vol. I, p. 23). Those who have been most strenuous in attempting to demonstrate that the behavior of certain lower organisms is "purely reflex" in character would probably be the last to hold that in the higher organisms behavior must be explained on essentially different principles. The attempt often made to contrast the behavior of lower organisms as reflex with that of higher organisms as something else, seems therefore a shortsighted and pointless proceeding. What a given organism does under stimulation is limited by its action system, and within these limits is determined largely by its physiological condition at the time stimulation occurs. In the lowest organism the action system confines the variations in behavior within rather narrow limits, and the different physiological conditions distinguishable are few in number; hence the behavior, is less varied than in higher animals. But the difference is one of degree, not of kind. The behavior of Paramecium and the sea urchin is reflex if the behavior of the dog and of man is reflex; objective evidence does not indicate that there is from this point of view any fundamental difference in the cases.

The importance attributed to the concept of reflex action is of course due to the desire to find a simple invariable unit for behavior, comparable to the atom in physics. To obtain such a unit it is necessary to take into consideration as an additional possible variable, the physiological state of the organism. Dr. E. G. Spaulding has suggested the following: We cannot properly say for a given organism "same stimulus, same reaction," as appears to be the usual idea of a reflex. On the other hand we can say "same physiological state, same stimulus, same reaction," and this supplies whatever need there may be for a simple invariable element of behavior. To this element the term "reflex" or an equivalent one might be applied, and we might then maintain that the behavior of all organisms is made up of reflexes. But on this definition the question whether the behavior of a given organism is made up of reflexes is not a problem for objective investigation; but the conception that it *is* thus made up is a postulate, in accordance with which we interpret the results of our observations; and this applies to the highest as well as to the lowest organisms. The assumption that varied physiological states exist is of course one of these interpretations, made to save what is essentially this very postulate, — the principle that like causes always produce like effects.

LITERATURE XV

REFLEXES AND BEHAVIOR

HOBHOUSE, 1901; JAMES, 1901; BEER, BETHE, AND V. UEXKÜLL, 1899; TITCHENER, 1902; DRIESCH, 1903; V. UEXKÜLL, 1897.

CHAPTER XVI

ANALYSIS OF BEHAVIOR IN LOWER ORGANISMS

1. THE CAUSES AND DETERMINING FACTORS OF MOVEMENTS AND REACTIONS

IN the following sections we shall analyze the behavior of the lower organisms described in previous chapters, attempting to determine the essential characteristics of behavior and to bring out the chief factors of which it is made up. We shall take up first the factors causing or determining the movements and reactions, treating first the inner, then the outer, factors. Then we shall consider the movements and reactions themselves, attempting to bring out the features of essential importance. From a synthesis of our results on both sets of factors — the causes and the effects — we shall try to arrive at a general statement of the fundamental character of behavior in the lower organisms.

The external factors in behavior are usually known as stimuli, and their effects on movement as reactions. The term "reaction" has been used in various ways. In our analysis we shall employ the word "reaction" as signifying an actual *change in movement*. The word is sometimes used in a looser sense. For example, the movement toward a source of light is often spoken of as the reaction to light, even though the only observable *change* of movement was that by which orientation was brought about. This looser sense is sometimes unavoidable, either from our ignorance of the facts, or for other reasons; when used in this loose sense in the following, the context will clearly indicate it. Where question might arise, reacton is to be understood as meaning an observable *change of movement*. To avoid ambiguity, the latter phrase will sometimes be used in place of the word "reaction." The following discussion will be intelligible only if this meaning of the word "reaction" is kept in mind.

A. The Internal Factors

(1) *Activity does not require Present External Stimulation.* — A first and essential point for the understanding of behavior is that activity occurs in organisms without present specific external stimulation. The

normal condition of Paramecium is an active one, with its cilia in rapid motion; it is only under special conditions that it can be brought partly to rest. Vorticella, as Hodge and Aikins (1895) showed, is at all times active, never resting. The same is true of most other infusoria and, in perhaps a less marked degree, of many other organisms. Even if external movements are suspended at times, internal activities continue. The organism *is* activity, and its activities may be spontaneous, so far as present external stimuli are concerned.

The spontaneous activity, of course, depends finally on external conditions, in the same sense that the existence of the organism depends on external conditions. The movements are undoubtedly the expression of energy derived from metabolism. The organism continually takes in energy with its food and in other ways, and continually gives off this energy in activities of various sorts. The point of importance is that this activity often depends more largely on the past external conditions through which the energy was stored up than upon present ones. Thus the organism may move without the present action of anything that may be pointed out as a specific external stimulus to this movement.

This fact is of great importance for understanding behavior, and many errors have arisen from its neglect. If we see an organism moving, it is not necessary to assume that some external stimulus now acting is producing this movement. In studying the reactions to present particular stimuli, as light or gravity or a chemical, it is in many cases not necessary to account for the fact of movement, for the movement comes from the discharge of internal energy, and often the organism was moving (though perhaps in another direction) before the stimulus began to act. It is only the change in the movement when the stimulus acts that the present stimulus must account for. In the movement of Paramecium toward the cathode, it is not necessary to assume, as some have done, that a special force (as cataphoric action) is required, to carry the animals. They were moving equally before the electric current began to act; the difference that the stimulus has made is in the *direction* of motion, and it is only this that the stimulus must account for. In the movements of infusoria toward chemicals, some have supposed that an attractive force from the chemical was necessary, actually bearing the organisms along; this is quite superfluous. In general, when an organism moves toward or away from any agent, it is unnecessary to assume that an actually attractive or repellent transporting force is acting upon it. Often — perhaps usually in the lower organisms — movement in a certain direction is due only to the release of inhibition. The organism moves in the given direction because it is moving from internal impulse, and because movement in this direction is not prevented. This possibility must be considered in all cases. Further, when the action of a stimulus actually changes the direction of movement in an organism, persistence in this new direction by no means demands persistence in stimulation. The new direction once attained may be followed, from the internal impulse to movement, merely because there is nothing to change this direction, or because stimulation does occur when this direction is changed, bringing the organism back to it. This is apparently the case, as we have seen, in the reactions of infusoria to gravity, to water currents, and to light coming from a certain direction.

Often, of course, stimulation does rouse an organism to increased activity. But even in this case the activity is due to the release of internal energy. It may, therefore, continue long after the stimulation which inaugurated the release has ceased to act. Such continuance thus does not necessarily imply continued action of the stimulus. In many cases the specific stimulus to action is only the *change* of conditions. Thus, if light or a chemical acts upon an organism, the only stimulus may be the sudden change, even though the organism continues to move after the conditions have become constant. Whether the effective stimulation actually continues, must be determined by experiment; it cannot be simply assumed.

In general, when an organism is moving in a certain way — even when toward or from a certain agent — careful analytical experimentation is necessary to determine whether this movement is due to present stimulation, or to the simple outflow of the stored-up energy of the organism through the channels provided by its structure. In most cases, apparently, the latter is true.

The spontaneous activities of the organism — those not due directly to present specific external stimulation — are, perhaps, the most important factors in its behavior.

(2) *Activity may change without External Cause.* — If we watch a specimen of Vorticella under uniform conditions, we find that its behavior does not remain uniform. At first the animal is outstretched, its cilia bringing a current of water to the mouth. After a certain period its stalk contracts, its peristome folds inward, and its cilia cease moving. Soon it extends and resumes its normal activity. These alternations of different ways of behaving occur at rather regular intervals, though the external conditions remain unchanged. Hydra shows parallel changes of behavior at intervals, under uniform external conditions (p. 189); the medusa contracts at intervals, though there is no change in the outer conditions, and similar examples could be given for many other organisms.

(3) *Changes in Activity depend on Changes in Physiological States.* —
What causes the changes in behavior described in the foregoing paragraph? Since the external conditions have not changed, the animal itself must have changed. The Vorticella which contracts and folds its cilia is in certain respects a different animal from the one that remains extended and keeps its cilia in active motion, otherwise it would not act thus differently. Its internal or physiological condition has been changed. Soon its original condition is restored; it unfolds and behaves as it did at first. In the same way, the physiological condition of the Hydra that stands quiet with outspread arms is different from that of the Hydra which, without external cause, contracts and changes its position. The behavior produced by these differences in physiological condition is the same as that producible by an external stimulus.

Other examples of changes in behavior due to changed physiological states are shown in the different reactions of hungry and of well-fed individuals, which we have seen in so many cases, and in the different reactions of organisms as determined by their respiratory processes.

The precise nature of these internal changes of condition we of course do not know. The expression "physiological states" evidently includes a great many things of heterogeneous character, having merely the common characteristic that they are internal modifications of the living substance resulting in changed behavior. In the lower organisms it is difficult to define the different classes of physiological states in an objective way, though the progress of investigation will doubtless make this possible. Certain fundamental differences in diverse states will be pointed out in the following pages.

(4) *Reactions to External Agents depend on Physiological States.* —
Change of activity is, of course, often produced by external agents. With this point we are to deal later; here what interests us is the fact that in any given organism the reaction to a given external agent depends on the physiological condition of the organism. This principle is of such importance that we must dwell upon it.

First we have the important fact that the reaction to a given stimulus depends upon the progress of the metabolic processes. To a given external condition the nature of the reaction often depends upon whether it favors these metabolic processes. If material for these processes is lacking, the reaction to stimuli is of such a character as to secure such material. In such organisms as the cœlenterates almost the whole character of the behavior, down to the details of the reactions to specific stimuli, depends thus on the condition of the processes of metabolism (see Chapter XI). The behavior of organisms is similarly determined by the course of other internal processes; these are, perhaps, the most important factors determining physiological states.

Of a somewhat different character are the changes in physiological state exemplified in the behavior of Stentor and the flatworm. In Stentor, as we have seen in Chapter X, we can distinguish at least five different physiological states in which the same individual reacts differently to the same conditions. Under stimulation by numerous grains of carmine in the water, the Stentor in condition No. 1 does not react at all. In condition No. 2 it reacts by turning into a new position. In condition No. 3 its reaction is a reversal of the ciliary current. In No. 4 it responds by contracting at brief intervals. In No. 5 the contractions are stronger and the organism remains longer in the contracted condition, finally breaking its attachment to its tube and swimming away. Throughout this entire series of reactions the external conditions remain the same, so that we can attribute the different reactions only to different conditions of the organism.

In the flatworm we have seen in Chapter XII that six different physiological conditions may be distinguished, in each of which the flatworm is a different animal, so far as its reactions to stimuli are concerned. We need not repeat the details regarding these conditions here. Illustrations of the fact that the reaction of the organism depends on its physiological state might be drawn from the behavior of many other animals.

(5) *The Physiological State may be changed by Progressive Internal Processes*, particularly those of metabolism. The well-fed sea anemone or Hydra is a very different animal, so far as its behavior is concerned, from the specimen that has fasted. Under uniform conditions, the sea anemone that is well fed remains quiet; while the individual that has exhausted the material for metabolism toils painfully away on a tour of exploration. The well-fed individual reacts negatively or not at all to that to which the hungry individual reacts positively. The *Paramecium bursaria* that has exhausted its supply of oxygen behaves in one way with regard to light, the individual in which respiration is progressing normally in another way. Innumerable examples illustrating this principle can be found in the behavior of lower and higher organisms. It is hardly too much to say that the progress of the metabolic and other physiological processes is the chief factor in determining the behavior of lower organisms.

(6) *The Physiological State may be changed by the Action of External Agents.* — This follows directly from the behavior of Stentor and the flatworm, to which we have referred in the preceding paragraph. The Stentor in condition No. 1, as we have seen, does not respond to the

stimulus of the carmine grains in the water. The stimulus continues, and after a time the physiological condition changes so that the animal *does* respond. The change in physiological state can then be due only to the action of the stimulus. In the same way the other changes in the physiological condition of Stentor and the flatworm are evidently due largely, at least, to the continued action of the stimulus.

(7) *The Physiological State may be changed by the Activity of the Organism.* — This is demonstrated by the spontaneous changes in the behavior of Vorticella or Hydra, of which we have already spoken. At first the animal is in a certain condition which corresponds to extension and activity. It then passes into a condition which results in contraction. But it does not remain contracted; the contraction itself restores the original condition, so that the animal now again extends and becomes active. Certain of the changes in physiological state seen in Stentor and the flatworm are probably due to the reactions of the organism. Thus, we find that the flatworm, after turning for a long time away from a lateral stimulus, suddenly changes and turns in the opposite direction (p. 253). The change of physiological state conditioning this change of reaction was probably due, not alone to the continuance of the stimulus, but to the previous prolonged turning of the flatworm in a certain direction.

(8) *External Agents cause Reaction by changing the Physiological State of the Organism.* — We have found that external stimuli cause changes in physiological state, and that changes in physiological state induce changes in behavior, — activities of a definite character. It is evident, then, that external agents must change the behavior of organisms by changing their physiological condition. In other words, in a reaction to an external stimulus the course of events is probably as follows: The stimulus causes first a change in the physiological condition of the organ or organism. This, then, causes a change in behavior, which we call a reaction to the stimulus. What the organism reacts to is the change produced within it by the external agent. Hence, if two different external agents induce the same internal change (as by blocking certain processes) they will receive the same reaction.

(9) *The Behavior of the Organism at any Moment depends upon its Physiological State at that Moment.* — This follows immediately from the principles already developed. We have seen that both in "spontaneous" movements and in reactions to stimuli the behavior depends on the physiological condition of the animal. The behavior must then depend, secondarily, not only upon the present external stimulus, but upon all the conditions which affect the physiological states. This point will be developed under the two succeeding heads.

(10) *Physiological States change in Accordance with Certain Laws.* — It is evident that we may distinguish at least two great classes of physiological states, — those depending on the progress of the metabolic processes of the organism, and those otherwise determined. The changes in the metabolic states, as we may call the former, of course depend largely upon the laws of metabolism. In the physiological states not directly dependent on metabolism, but rather upon stimulation and upon the activity of the organism, such as we have seen in Stentor, we find certain fairly well-defined laws of change, of a peculiar character.

In a number of organisms we have found the following phenomenon: Under certain conditions the organism reacts in a certain way. These conditions continuing, the organism changes its first reaction for a second or third or fourth. Later the same external conditions recur, and now the organism at once responds, not by its first reaction, but by its final one. This is illustrated for unicellular organisms by the case of Stentor (Chapter X); for higher Metazoa it is well seen in the behavior of certain Crustacea, as described by Yerkes and Spaulding (Chapter XII). There are certain differences in these two cases that will be taken up later.

How does this state of affairs come about? The "physiological state" is evidently to be looked upon as a dynamic condition, not as a static one. It is a certain way in which bodily processes are taking place, and tends directly to the production of some change. In this respect the "law of dynamogenesis," propounded for ideas of movement in man, applies to it directly (see Baldwin, 1897, p. 167); ideas must indeed be considered, so far as their objective accompaniments are concerned, as certain physiological states in higher organisms. The changes toward which the physiological state tends are of two kinds. First the physiological state (like the idea) tends to produce movement. This movement often results in such a change of conditions as destroys the physiological state under consideration. But in case it does not, then the second tendency of the physiological state shows itself. It tends to resolve itself into another and different state. Condition 1 passes to condition 2, and this again to condition 3. This tendency shows itself even when the external conditions remain uniform.

In this second tendency a most important law manifests itself. When a certain physiological state has been resolved, through the continued action of an external agent or otherwise, into a second physiological state, this resolution becomes easier, so that in the course of time it takes place quickly and spontaneously.

This may be illustrated from the behavior of Stentor, as described in Chapter X as follows: When the organism is stimulated by the flood of carmine grains (or in any other way), this produces immediately a certain physiological state (corresponding to that accompanying a sensation in ourselves); this state we may call A. This state at first produces no reaction. As the carmine continues or is repeated, this state A passes to a second state B, producing a bending to one side. (The two may differ only slightly, but a difference must exist, otherwise B would not produce a reaction while A does not.) After several repetitions of the stimulus, the condition B passes to the condition C, producing a reversal of the cilia, and this finally passes to D, resulting in a contraction of the body. The course of the changes in physiological states may then be represented as follows: —

$$A \longrightarrow B \longrightarrow C \longrightarrow D$$

Now we find that after many repetitions of the stimulation the organism contracts at once as soon as the carmine comes in contact with it. In other words, the first condition A passes at once to the condition D, and this results in immediate contraction.

$$A \longrightarrow D$$

It seems probable that the same series occurs as before, save that conditions B and C are now passed rapidly and in a modified way, so that they do not result in a reaction, but are resolved directly into D. The process would then be represented as follows: —

$$A \longrightarrow B' \longrightarrow C' \longrightarrow D$$

But whatever the intermediate conditions, it is clear that after the state A has become resolved, through pressure of external conditions, into state D, this resolution takes place more readily, occurring at once after the state A is reached.

The same law is illustrated in the experiments of Yerkes and Spaulding on much higher organisms. In the experiments of Spaulding with the hermit crabs (Chapter XII), the introduction of the screen and the diffusion of the juices of the fish cause the animals to move about. In so doing they reach the dark screen, which induces, let us say, the physiological condition A. This leads to no special reaction. But this is followed regularly by contact with food, inducing the physiological condition B, which is concomitant with a positive reaction. The physiological condition A is thus regularly resolved into the condition B. In the course of time this resolution becomes automatic, so that as soon as the condition A is reached it passes at once to B. The positive reaction concomitant with B is therefore given even though the original cause of B is absent.

In the experiments of Yerkes, using the two passages to the water, described in Chapter XII, the following are the conditions. The presence of the investigator or the drying of the animal at T, Fig. 139, acts as a stimulus to cause movement away from T. A turn to the right is accompanied, let us say, by the physiological condition A. This is soon followed by contact with the glass plate G, inducing the condition B, which involves inhibition of movement and a turn in another direction. In the course of time the condition A comes to be resolved immediately into B, so that movement is inhibited at the start. On the other hand, the physiological condition C, concomitant with a turn to the left, is regularly resolved into the condition D, concomitant with reaching the water, and inducing a positive reaction. This resolution becomes automatic, so that the turn to the left is followed at once by forward motion to the water. In these cases the actual number of physiological states that could be distinguished is, of course, greater than what we have set forth above. But this does not alter in any way the general principle involved.

The law of the resolution of physiological states illustrated in the foregoing examples is of the highest importance for the understanding of behavior. With selection from among varied movements, it forms one of the corner-stones for the development of behavior. The law may be expressed briefly as follows: —

The resolution of one physiological state into another becomes easier and more rapid after it has taken place a number of times. Hence the behavior primarily characteristic for the second state comes to follow immediately upon the first state.

The operations of this law are, of course, seen on a vast scale in higher organisms, in the phenomena which we commonly call memory, association, habit formation, and learning. In the lower organisms the manifestations of this law are comparatively little known. This is probably due largely to difficulties of experimentation. Since the law has been demonstrated to hold in unicellular organisms (Stentor and Vorticella), there is much reason to suppose that it is general, and that it will be demonstrated in one form or another for other lower organisms. There seems to be no theoretical reason for supposing it to be limited to higher animals. Very great differences exist among different organisms as to the ease with which the quick resolution of one physiological state into another is established. There are likewise great differences in the permanency of existing connections among the present reaction methods. Hence it does not follow, as Yerkes (1902) has well pointed out, that because a few experiments do not demonstrate this law in a given case, the law, therefore, does not hold. In his experiments with crustaceans,

Yerkes found that a very large number of repetitions were necessary before a given resolution was established.

(11) *Different Factors on which Behavior Depends.* — We have seen that the behavior of the organism at a given moment depends on its physiological state, and that it therefore secondarily depends upon all the factors upon which the physiological state depends. Hence we cannot expect the behavior to be determined alone by the present external stimulus, as is sometimes maintained, for this is only one factor in determining the physiological state. The behavior at a given moment may depend on the following factors, since these all affect the physiological state of the organism: —

1. The present external stimulus.
2. Former stimuli.
3. Former reactions of the organism.
4. Progressive internal changes (due to metabolic processes, etc.).
5. The laws of the resolution of physiological states one into another.

All these factors have been strictly demonstrated by observation and experiment, even in unicellular organisms. Any one of these alone, or any combination of these, may determine the activity at a given moment.

CHAPTER XVII

ANALYSIS OF BEHAVIOR (*Continued*)

B. *The External Factors in Behavior*

(1) As we have seen in the foregoing chapter, external agents produce reactions through the intermediation of changes in the internal physiological condition of the organism. This proposition is, perhaps, a truism, yet it needs to be kept in mind if behavior is to be understood. In the following discussion it will be unnecessary to mention specifically in each case the intermediate step in the process.

(2) The most general external cause of a reaction is a *change* in the conditions affecting the organism. This has been illustrated in detail in the descriptive portions of the present work. In most cases the change which induces a reaction is brought about by the organism's own movements. These cause a change in the relation of the organism to the environment; to these changes the organism reacts. The whole behavior of free-moving organisms is based on the principle that it is the movements of the organism that have brought about stimulation; the regulatory character of the reactions induced is intelligible only on this basis. Reactions due to stimulation produced in this manner are seen when an organism progresses from a cooler to a warmer region, or *vice versâ;* when it moves into or out of a chemical in solution; when it strikes in its course against a hard object; when the unoriented infusorian shows lateral movements while subjected to light coming from one side. In all these cases it is the movement of the organism which causes a change in its relation to the external agent, and this change produces reaction. In most, if not all, cases the change is one in the intensity of some agent acting on the organism.

But an active change in the environmental conditions, not produced by movement of the organism, may likewise produce reaction; this is, of course, most frequently the case in fixed organisms, such as the sea anemone. Responses produced in this way are seen in the reactions of organisms when heated or cooled from outside, or when a chemical or a solid object is brought in contact with them, or when the source of light changes in intensity or position, or when the direction of a water current changes. The general fact is that a change in the environment produces a change in behavior.

A. Change of conditions often produces a change of movement when neither the preceding nor the following condition would, acting continuously, produce any such effect. Thus when Euglena is swimming toward the source of light, if the light is suddenly diminished, the organism reacts by a change in its course; it then returns to its course and continues to swim toward the light as before. Its behavior before and after the change is the same; but at the moment of change there is a reaction. Paramecium may live and behave normally in water at 20 degrees or at 30 degrees, yet a change from one to the other, or a much less marked change, produces a definite reaction. This relation could be illustrated by many cases from the behavior of any of the organisms described in the foregoing pages. Thus change simply *as* change may produce reaction.

To constant conditions, on the other hand, unless differing very greatly from the normal, the organism usually does not react. The Paramecium placed in $\frac{1}{10}$ per cent sodium chloride reacts at first, but soon resumes its normal behavior. Euglena or Stentor when subjected to changes in the illumination of the anterior end react till they come into a position of orientation where these changes cease; they then swim forward in the normal manner. As a general rule, organisms soon become acclimatized to a continuous condition, if it is not too intense. Exceptions to this rule will be considered later.

Of course a change must reach a certain amount before reaction is produced; that is, there is a certain necessary threshold of stimulation. In the best-known cases the amount of the change which produces reaction is proportional to the intensity of the original condition; in other words, the relation of stimulus to reaction follows Weber's law (see pp. 38, 123). That is, it is relative change, not absolute change, that causes reaction.

B. But not every change, even if sufficiently marked, produces reaction. It is usually not change alone that determines reaction, but change in a certain direction. Of two opposite changes, one usually produces a certain reaction, while the other either produces none or brings about a reaction of opposite character. This point is one that is of fundamental importance for an understanding of behavior. It may be illustrated in its simplest aspect from the behavior of the infusoria, where any reaction that is produced is usually of such a character as to remove the organism from the source of stimulation (the "avoiding reaction"). Paramecium at a temperature of 28 degrees reacts thus negatively to a change to a higher temperature, not to the opposite change. Paramecium at 22 degrees reacts to a decrease of temperature, not to an increase. Stentor reacts to an increase of illumination, not to a decrease. Euglena when moderately lighted reacts negatively to a decrease of illumination, not to an increase; if strongly lighted, it shows the opposite relations. Paramecium reacts at passing into an alkaline solution, but not at passing out; it reacts at passing out of a weak acid solution, not at passing in. Hydra at 24 degrees reacts to an increase of 2 degrees in temperature, not to an equivalent decrease. Innumerable instances of this fact could be given from the behavior of the lower organisms.

What decides whether a given change or its opposite shall produce this negative reaction? Examination of the facts brings out the following relations: The organism generally reacts by a change in its behavior when the change is of such a nature as to lead *away from the optimum.* By optimum we mean here the conditions most favorable to the life processes of the organism in question. Changes leading toward this optimum produce in many animals no reaction; the organisms simply continue the activity which has brought about this change. Changes leading away from the optimum produce a negative reaction, by which the organism is removed from the operation of this change. There are undoubtedly some limitations and exceptions to this, and with these we shall have to deal later, but, as we have seen for Paramecium, it is unquestionably the rule. Cases where this rule does not hold are striking because exceptional. Reaction in this manner keeps the infusoria in regions of moderate temperature, prevents them from entering injurious chemical substances, brings green organisms such as Euglena into the light, where their metabolic activities are aided, and in general keeps the organisms in regions where the conditions are favorable. In these organisms the chief cause of reaction to a change is its *interference with the normal life activities,* and the reaction if successful serves to remove the interference.

C. But in many cases changes which favor the normal activities produce reaction. The response is then of such a character as to retain the organism under the conditions producing the change. Such responses we usually call positive reactions. In many cases it is clear that such reactions are determined by a previously existing unfavorable state of metabolism or of other processes. The Hydra or the sea anemone does not react positively to food substances unless metabolism is in such a state as to require more material; and parallel relations exist in the behavior of many if not all organisms. In unicellular organisms definite positive reactions play a comparatively small part, favorable conditions being secured primarily by a negative reaction to less favorable conditions. It is possible that all positive reactions are to be traced to this as the primitive type (see the following chapter). That is, while the negative reaction is impelled by new unfavorable conditions, tending to retain the more favorable old condition, the positive reaction is impelled by the old unfavorable condition, tending to retain the new more favorable one.

(3) Sometimes change of behavior occurs without change in the environment, the external conditions remaining uniform. As a rule, we have found that change of behavior occurs under uniform conditions only when these are decidedly injurious to the organism. If the water containing infusoria or the flatworm is heated to about 37 degrees, the animals react not merely to the *change* in temperature; they continue to react violently, with frequent alternations in the behavior, until they die. Many examples could be given of such reactions. Under uniform conditions a change in behavior also occurs at times owing to internal changes. The commonest cases of this sort are the changes in behavior due to hunger. In almost all cases of reaction under uniform conditions we find that the reaction is due to some interference with the normal life processes. But reactions under uniform conditions play only a small part in the behavior, as compared with reactions to changes.

We have then two main results as regards the external causes of changes in behavior: (1) change alone may produce reaction; (2) interference with the normal life processes or release from such interference may produce reaction. The usual cause of a change in behavior is a combination of both these factors — a change that hinders or helps the normal life processes. In the lowest organisms it is chiefly interfering changes that cause reaction.

(4) *Reactions to Representative Stimuli.* — In the reactions due to change, one further point is of much importance. The organism may react to changes that in themselves neither favor nor interfere with the normal life activities, but which do lead to such favor or interference. The reaction given is then positive or negative in correspondence with the benefit or injury to which the change leads. Thus, Stentor may bend toward a small solid body when touched by it (Fig. 83), this reaction aiding it to procure food, though there is no indication that the touch itself is directly beneficial. Or it may contract away from a light touch, this enabling it to escape from a possible approaching enemy, though the touch itself is not injurious. Euglena reacts negatively when its colorless anterior end alone is shaded, yet it is only when the shadow affects its chlorophyll bodies that it interferes with metabolism. The flatworm may turn toward a weak stimulus of any sort. This leads in the long run to its obtaining food, though sometimes the stimulus does not come from a food body. In such cases the animal

reacts positively merely to the localized change, not to the nature of the change. Certain colorless infusoria, and the white Hydra, react to light in such a way as to gather at the lightest side of the vessel containing them. There is no evidence that the light itself is beneficial to them, but their reaction does aid them in obtaining food, since their prey gathers on the lightest side of the vessel. The collecting of Paramecia in CO_2 can hardly be considered to favor directly the life processes of the animals, but it apparently aids them to obtain food. The sea urchin tends to remain in dark places, and light is apparently injurious to it. Yet it responds to a sudden shadow falling upon it by pointing its spines in the direction from which the shadow comes. This action is defensive, serving to protect it from enemies that in approaching may have cast the shadow. The reaction is produced by the shadow, but it *refers*, in its biological value, to something behind the shadow.

In all these cases the reaction to the change cannot be considered due to any direct injurious or beneficial effect of the actual change itself. The actual change merely *represents* a possible change behind it, which *is* injurious or beneficial. The organism reacts as if to something else than the change actually occurring; the change has the function of a *sign*. We may appropriately call stimuli of this sort *representative* stimuli.

This reaction to representative stimuli is evidently of the greatest value, from the biological standpoint. It enables organisms to flee from injury even before the injury occurs, or to go toward a beneficial agent that is at a distance. Such reactions reach an immense development in higher animals; most of our own reactions, for example, are to such representative stimuli. Only as we react to actual physical pain or pleasure do we share with lower organisms the fundamental reaction to direct injury or benefit. Practically all our reactions to things seen or heard are such reactions to representative stimuli. While such behavior plays a much larger part in higher than in lower organisms, the existence of reactions to representative stimuli even in the low organisms considered in the present work is an evident fact.

How can we account for such reactions? It is perhaps worth while to point out that the operation of the law of the resolution of physiological states, set forth on page 291, would result naturally in the production of such reactions. Let us take as the simplest possible case the reaction of Euglena when its colorless anterior tip is shaded. Since it is only the metabolism of the chlorophyll bodies that is blocked by shade, we cannot suppose that the shading of the colorless tip actually interferes with the life processes. Yet to this change Euglena reacts negatively. We may suppose that the shading of this colorless part induces the indifferent physiological state A, which of itself produces no reaction. But this is invariably followed by the shading of the chlorophyll bodies, interfering with metabolism and inducing the physiological state B, resulting in a negative reaction. Thus the state A is regularly resolved into the state B. In accordance with the law of the resolution of physiological states, this resolution in the course of time becomes spontaneous. A passes at once to B and a negative reaction occurs, even when the colorless anterior tip alone is shaded. In unicellular organisms a condition so reached would naturally continue to succeeding generations, since the organisms in reproducing merely divide.

In the same way the defensive reaction of the sea urchin when shaded could be produced. The condition A, induced by the shade, is usually resolved into the condition B, induced by the attack of an enemy, and resulting in the defensive movement. This resolution in the course of time may then become spontaneous, so that the sea urchin now reacts defensively even when a cloud passes over the sun. This condition could be continued to succeeding generations only if acquired characters are inherited.

Thus through the operation of the law of the resolution of physiological states the following general result will be produced: If a given agent induces a physiological state A, and this is usually followed by a second state B, then in time the given agent will produce at once the response due primarily to B. The organism will have come to react to A as representative of B.

We do not know whether the development of reactions to representative stimuli has actually taken place in this way, or not. But the fact that there is a factor, whose existence is demonstrated, that would produce exactly these results, certainly suggests strongly the probability that they have been at least partly brought about in the way above set forth. If the law of the resolution of physiological states is actually operative throughout behavior, the effect would be to make behavior depend on the results of the animal's own action. This would produce behavior that is regulatory, such as we actually find to exist.

(5) The reaction to a given external stimulus depends, as we have previously seen, on the physiological condition of the organism, not alone on the nature of the external change. The physiological condition depends partly on whether the normal stream of life activities is proceeding uninterruptedly. In certain physiological states, such as hunger, the processes are not proceeding normally. This impels the organism to a change, so that to almost any external stimulus it may react in a way that tends to bring about a change. The hungry sea anemone in this condition reacts positively to all sorts of neutral bodies;

the hungry Hydra reacts positively to chemicals. In certain physiological conditions the flatworm reacts positively to almost any stimulus. At other times the opposite conditions prevail; the animal reacts negatively to the stimulus to which it before reacted positively. In closely related organisms differing in their metabolic processes, the reaction to a given agent depends on the nature of the metabolic processes, tending to retain the conditions favoring these processes. This is especially well illustrated in the bacteria (pp. 36, 39) and in the cœlenterates (pp. 224, 231), but is equally true for other organisms. Thus what the organism does depends on the course of its life processes, and upon the completeness or incompleteness of their performance. In other words, the behavior of the animal under stimulation corresponds to its needs, and is determined by them. This correspondence is of course not always perfect; with this point we can deal after we have considered the nature of the reactions given. But a study of the determining factors of behavior demonstrates that the relation of external conditions to internal processes is the chief factor, and that hence behavior is regulatory in essential nature.

(6) We may sum up the external factors that produce or determine reactions as follows: (1) The organism may react to a *change*, even though neither beneficial nor injurious. (2) Anything that tends to interfere with the normal current of life activities produces reactions of a certain sort ("negative"). (3) Any change that tends to restore or favor the normal life processes *may* produce reactions of a different sort ("positive"). (4) Changes that in themselves neither interfere with nor assist the normal stream of life processes may produce negative or positive reactions, according as they are usually followed by changes that are injurious or beneficial. (5) Whether a given change shall produce reaction or not, often depends on the completeness or incompleteness of the performance of the metabolic processes of the organism under the existing conditions. This makes the behavior fundamentally regulatory.

CHAPTER XVIII

ANALYSIS OF BEHAVIOR (*Continued*)

2. THE NATURE OF THE MOVEMENTS AND REACTIONS

IN the preceding section we have dealt primarily with the causes and conditions of movements and reactions; here we are to deal with the movements and reactions themselves.

A. The Action System

Every organism has certain characteristic ways of acting, which are conditioned largely by its bodily structure, and which limit its action under all sorts of conditions. This perhaps seems a mere truism. Amœba of course cannot swim through the water like Paramecium, and the latter cannot fly through the air nor walk about on dry land. But the behavior of any given lower organism is actually confined in this way within narrower limits than is frequently recognized. Formulæ have at times been proposed to explain the movements of various organisms, when the latter are incapable of performing the movements called for by the formulæ. It is usually possible to determine with some approach to completeness the various movements which a given organism has at command. These form as a rule a coördinated system, which we have called in previous pages the *action system*. The action system of an organism determines to a considerable extent the way it shall behave under given external conditions. Under the same conditions, organisms of different action systems must behave differently, for to any stimulus the response must be by some component of the action system. Thus, Amœba, the bacteria, Paramecium, Hydra, and the flatworm have action systems of different character, and their behavior under given conditions must differ accordingly. This matter has been dealt with in detail in the descriptive portion of the present work, so that we need not dwell upon it here. In studying the behavior of any organism, the first requisite to an understanding is the working out of the action system.[1]

---

[1] The action system corresponds largely to what Pütter (1904) calls the "Symptomatology" of organisms.

B. *Negative Reactions*

In our discussion of the causes of reaction we found that we could classify most stimuli into two groups — those that interfere with the normal life processes, and those that do not. It will be best to consider separately the reactions to these two classes of stimulation, and to take up the reactions to unfavorable stimuli first, since these seem to present the most primitive conditions.

The simplest reaction to unfavorable stimuli is merely a change in the direction or character of the movement. The organism is moving in a certain direction; when subjected to an unfavorable change, it changes its direction of movement. This is the case in Amœba, in bacteria, in infusoria, in rotifera, in the flatworm; indeed, in most free organisms. The mere fact of a change is in itself regulatory or adaptive. The original behavior has brought on the unfavorable change, hence the best thing to do is to change this behavior. If the unfavorable condition still persists, the behavior is changed again; this being continued, the organism is bound to escape from the unfavorable conditions if it is possible to do so. The repeated change in behavior under unfavorable stimulation is very striking in Paramecium, in Stentor, in Hydra, in the flatworm, and elsewhere.

The fundamental principle for this method of reaction is that *a change of behavior under unfavorable conditions is in itself regulatory.* As we have before pointed out, the reactions of organisms are based on the principle, usually correct, that it is the previous behavior of the organism that has brought on the present conditions. Hence if these conditions are unfavorable, a change of behavior is required.

The developments of this method of behavior found in different organisms consist in defining, varying, and systematizing the changes that occur. In Amœba we find perhaps the simplest condition. When this animal in its forward course meets unfavorable conditions it merely goes *in some other direction.* In what direction it will go cannot be predicted from either the structure of the organism or from the localization of the stimulus, for Amœba can move with any part in advance. It is evidently determined by transient internal conditions. In organisms with definite body axes and other structural relations, the change of motion becomes more definite. In bacteria the organism moves after stimulation in the *opposite* direction. In the free-swimming infusoria, as illustrated by Paramecium, and in the free Rotifera, there is an elaborate system of movements which make the reaction effective. The animal stops or reverses the movement which has brought on the unfavorable condition, then swings its anterior end about in a circle as it moves forward, so as to try successively many different directions. The behavior shows the "method of trial" reduced to a system. It would be almost impossible to suggest any modification of this reaction, as exemplified in Paramecium, that would make it better fitted, under the given relations, for meeting all sorts of conditions. In fixed infusoria, such as Stentor, this behavior is modified to adapt it to the fixed life. In the free-swimming animal the organism is subjected to new conditions every time the reaction is repeated, hence there is little occasion to try other methods of behavior. But if the organism is fixed in one place, this is not true; when a given reaction is repeated it merely brings on the same conditions its first performance induced. So different methods are developed. Under unfavorable conditions the organism first turns to one side, then reverses its ciliary current, then contracts, etc. (see p. 174), trying many different changes of behavior. In Hydra, in the starfish, in the flatworm, we have seen this same "method of trial" appearing under various forms. In all these organisms persistent unfavorable stimulation induces first one physiological state, then another, then another, and to each state there corresponds a certain method of behavior.

C. *Selection from the Conditions produced by Varied Movements*

In all this behavior we find the manifestations of a most important principle, one of far-reaching significance for the understanding of behavior. The stimulus does not produce directly a single simple movement (a reflex act), of a character that relieves the organism at once from the stimulating condition. On the contrary, stimulation is followed by many and varied movements, from which the successful motion is selected by the fact that it *is* successful in causing cessation of stimulation. This is the principle of the "selection of overproduced movements," of which much use has justly been made by Spencer, Bain, and especially by Baldwin (1897, 1902), in attempting to explain behavior. It is more accurate to speak of the selection of the proper conditions of the environment through varied movements. It is primarily the proper environmental conditions that are selected; the movements are only a means to that end. From this point of view what we have often called in the foregoing pages the method of trial may be formulated as follows: When stimulated the organism performs movements which subject it to varied conditions. When in this way it reaches a condition that relieves it of stimulation the reacton movement ceases, since there is no further cause for it. The organism may then resume its usual movements. In the case where the reaction consists of changes in direction, as in infusoria, the resumption of the usual forward motion of course carries the organism in a new direction brought about by the reaction.

What movements are produced by the stimulating agent depends on the action system of the organism; it performs the movements that it is accustomed to perform. In some cases these movements are of a rather uniform character, yet are of such a nature as to subject the animal to many changes of the environmental conditions. This is the case, for example in the reactions of such infusoria as Paramecium. In other cases the movements themselves are varied; the organism first reacts in one way, then in another, running thus through a whole series of activities, till one succeeds in ridding the organism of the stimulating condition. This is the method of behavior seen in Stentor and in most higher organisms. In both methods the essential point is the same — the subjection of the organism to varied environmental conditions, until one of these relieves it from the stimulation. This condition is then said to be "selected." In some cases the maintenance of this favorable environmental condition involves continuance of the movement finally resulting from the varied trial movements; in other cases it does not.

Reaction by selection of excess movements depends largely on the fact, previously brought out (p. 283), that the movement itself is not directly produced by the stimulus. The movement is due, as we have seen, to the internal energy of the organism. In the case of free-moving animals like Paramecium, stimulation usually neither increases nor decreases the amount of motion, but merely causes it to change in various ways. Reaction, of course, sometimes does take the form of an increase of motion; this is seen in the increased movements of infusoria under strong chemicals or heat; of Planaria under light, etc. But even in these cases the energy for the motion comes from within and is merely released by the action of the stimulus. It is important to remember, if the behavior is to be understood, that energy, and often impulse to movement, come from within, and that when they are released by the stimulus, this is merely what James has called "trigger action." There is thus no reason to expect that upon stimulation an organism will perform merely a single simple movement (a "reflex action"), and then become quiet. Movement of one sort or another is its natural condition, and after stimulation has ceased it may show movements (the character or direction of which may have been determined by the stimulus) for an indefinite period.

Behavior by selection from the results of varied movements is based on general principles. The reactions are not specific ones, definitely adapted to particular kinds of stimulation, but are responses to *any* stimulation of a certain general character, — namely, to any condition that interferes with the normal course of the life processes. On receiving an unfavorable stimulus that it has never before experienced, the organism behaving on this plan is not at a loss for some method of reacting; it merely responds in the usual way, performing one movement after another, till one of these relieves it of the stimulation, if this is possible.

Of course special circumstances may arise in which this general method of reacting may be ineffective. If dropped into a strong chemical, Paramecium reacts in the usual manner, though this does not help it. If the water containing a flatworm is heated, the animal goes through, one after the other, almost every reaction it has at command, though all are unavailing (p. 245). The difficulty, of course, lies in the fact that under these circumstances nothing the organism can do is of any avail, and a man in similar conditions would be equally helpless. The infusorian and the flatworm, like the man, merely try everything possible before succumbing.

D. "Discrimination"

The effectiveness of reaction by continued varied movements in preserving the organism depends upon several factors. One of these is what is called in higher animals the power of discrimination, — that is, the accuracy with which the tendency to react is adjusted to the injuriousness of the stimulating agent. If an injurious agent resembles in its first action a non-injurious one, so that the animal reacts in the same way toward both, its behavior will not preserve it from injury. Using the more subjective form of expression, if the organism does not discriminate between the first action of injurious and non-injurious agents, it cannot react differently to them, until perhaps the injury has become irremediable. The facts show that in both higher and lower organisms the power of discrimination under weak stimulation is far from perfect. Thus, in the sense in which we have used the term, Paramecium discriminates acids from alkalies and salts, and these again from sugar. But it does not effectively discriminate the first effects of different acid substances, so that it swims into weak carbonic acid, which is harmless, and likewise into weak sulphuric acid and copper sulphate, which kill it. It does not discriminate the first action of a 10 per cent sugar solution from that of water, hence it swims readily into the sugar solution and is killed by the osmotic action. In all these cases it does discriminate and react to the injurious agent when its effect has become marked, but injury has then already occurred and the reaction does not preserve the animal. In regard to these injurious substances Paramecium thus makes what we would call in ourselves a "mistake." The whole scheme of reaction by the selection of the results of varied movements is not a set, perfected, final one, but is a tentative plan, based on the confusing world taken as it comes; it is liable to mistakes, and is capable of development. Progress in this method of behavior takes place largely through increase in the accuracy of discrimination of different stimuli. This may occur through the law of the increased readiness of resolution of physiological states after repetition, in the way that we shall attempt to set forth later (Chapter XIX).

E. *Adaptiveness of Movements*

The second chief factor on which depends the effectiveness of behavior by selection of overproduced movements lies in the relative fitness of the movements to relieve the organism from the unfavorable conditions. This, of course, depends on many things. If a powerful chemical is diffusing from a certain direction, the rapid movements of Paramecium are more likely to save than is the slow motion of Amœba. There are two factors on which the effectiveness of the movements depends, that are worthy of special consideration.

In what we may call the pure method of trial, a most important requirement for effectiveness is that the movements shall be so varied as to give much opportunity for finding other conditions. There are great differences in the behavior of different organisms from this standpoint. This may be illustrated by a comparison of the reactions of Paramecium and Bursaria to heat, as previously described. When a portion of the area containing the organisms is heated, these two infusoria react in accordance with essentially the same plan, yet practically none of the Paramecia are injured, while a large proportion of the Bursariæ are killed. The difference is due chiefly to the fact that Paramecium rapidly repeats its reactions and revolves on its long axis as it turns, so that in a short time it has tried in a really systematic way many different directions, and is practically certain to find one leading away from the heated region, if such exists. Bursaria, on the other hand, changes its direction of movement only at longer intervals, and usually soon ceases to revolve on its long axis as it turns toward the aboral side. This failure to turn on the long axis deprives it of the great advantage of being directed successively in many different directions in the different planes of space. The result is that it is likely to be destroyed by the heat before it has found a direction leading to a cooler region.

F. Localization of Reactions

A second factor that is of great importance in making the movements effective lies in the proper localization of the reactions. An organism that moves directly away from an unfavorable agent (or directly toward a favorable one) has a great advantage over an organism whose movements are not thus accurately directed. There are great differences in different organisms in this respect; some react very precisely with reference to the position of the stimulating agent, while others do not.

How is the relation of the reaction to the localization of the stimulus brought about, and what is the cause of the differences between different organisms in this respect?

In answering this question, we can distinguish three different classes of phenomena. These are the following: —

(1) First we have the simple phenomenon that when a portion of an organism is stimulated this portion may respond by contraction, extension, or other change of movement. If the remainder of the body does not respond, or responds in a different way, this gives at once a reaction localized in a certain way with reference to the place of stimulation. Such local responses we find in Amœba, where the part strongly stimulated contracts, or if stimulated by a food body it extends. The same phenomenon is found in Hydra, in the bending of the body when one side is powerfully stimulated, in the bending of the tentacles of Sagartia toward the point stimulated, and in the local contractions of the medusa and of stimulated points on the body of the flatworm and many other soft-bodied animals. The same thing is seen even in man when the electrode of a battery is applied directly over a muscle; this muscle now contracts. This seems a simple and primitive phenomenon, and as such has been seized upon by the "tropism theory" and made the chief factor in the behavior of lower organisms, and particularly in all directed reactions. As we have shown in our chapter on that theory, this factor plays by no means the extensive part assumed by the theory, and is quite inadequate to account for most of the behavior of lower organisms. Even in the behavior of the organisms mentioned above, where it clearly does play a part, this part is a subordinate one (see Chapter XIV). In many organisms, such as the free infusoria and some rotifers, it is hard to detect any part of the effective behavior that is due to local reaction at the point stimulated. The fact that such local reactions may and do occur in organisms is of course a fact of much importance, but taken by itself it is utterly inadequate as a general explanation of directed reactions.

(2) In many cases we find that the relation of the movement to the source of stimulation is brought about indirectly through selection from among varied movements. The organism tries moving in many directions, till it finds one in which there is no stimulus to further change. In this way it may become oriented very precisely if the conditions require. This is the prevailing method in the infusoria and in various other organisms, as we have seen. It is becoming evident that this method is more common even among higher organisms than has been hitherto set forth. Movements of the head from side to side, such as we find in the flatworm and many other animals, movements of the eyes or other sense organs, such as are common in higher animals, or movements of the body from side to side, as in the swimming of many creatures, give opportunity for determining which movement tends to retain the stimulus, which to get rid of it. In this way they form a basis for the determination of the direction of locomotion through the method of trial. How much part such movements play needs careful study.

(3) In still other cases the reaction shows a definite relation to the localization of the stimulus, yet it is not due to local reaction of the part stimulated, nor is it brought about by trial. If an infusorian is stimulated at the anterior end it swims backward; stimulated at the posterior end it swims forward. Both these movements are reactions of the entire organisms, all the motor organs of the body concurring to produce them; they are not produced by local reactions of the organs at one end or the other. The flatworm turns toward or away from the side stimulated, by reactions involving the muscles of both sides, as well as transverse and dorso-ventral muscles, all at a distance from the point stimulated. If stimulated on the upper surface of the head, a complicated twisting reaction occurs, involving many sets of muscles in various regions (p. 273), by which the ventral surface is made to face the stimulating agent (p. 236). Innumerable instances of this class of reactions could be given; they include perhaps the greater number of the directed movements of organisms.

In these reactions a stimulus at one side or end evidently produces a different reaction from a stimulus at the opposite side or end, though the reaction is not primarily at the point stimulated. Doubtless the stimulus starts a physiological process of some sort at the point upon which it impinges, and this determines in some way the direction in which the organism shall move. This effect in the region directly acted upon corresponds to the "local sign" in human physiological psychology. Behavior thus brought about is of course more effective than that of the two preceding classes, permitting more direct and rapid reaction than the method of trial, and meeting the conditions in an incomparably more adequate way than the simple local reaction of the part stimulated.

Such behavior apparently represents not a primitive condition, but a product of development. How has it been brought about?

It is evident that the operation of the law of the readier resolution of physiological states after repetition, taken in connection with behavior by selection from varied movements, would in course of time produce such reactions. Let us suppose that the original reaction to a stimulus at the anterior end was simply the production of a change resulting in varied movements, according to the principles governing the actual reactions of Paramecium. These varied movements would include forward as well as backward motion. The forward movement would induce still further stimulation, hence it would be changed. The backward movement would give relief from stimulation, hence would not be changed (till internal conditions require). Hence after stimulation at the anterior end the physiological states induced will always be resolved finally into that state corresponding to backward movement. This resolution will in time become spontaneous; the physiological state due to stimulation at the anterior end will pass at once into that producing movement backward. Trial movements will no longer occur, but the organism will respond at once by backward motion. A similar exposition will account, *mutatis mutandis*, for other localized reactions.

Whether this condition has been brought in the way above sketched or not, its existence is evidently a fact of great importance. It is a step forward from the pure "trial movement" condition. Wherever the organism can react in this manner, and this will meet the conditions equally well, we may expect such behavior in place of repeated trials. In higher organisms especially we find this behavior playing a large part. Such organisms could not be expected, for example, to orient to gravity or to light rays by trial movements, as the infusoria do, but rather to turn directly toward or from the source of action of the stimulating agent. This is, of course, known in many cases to be true.

But under many circumstances the reaction by trial is surer, though less rapid, than that depending directly on the localization of the stimulus, so that we find the trial method much used even by higher organisms (see Chapter XII). Further, the more direct reactions due to precise localization are again combined as elementary factors to produce behavior based on the method of trial, as when the flatworm turns toward and "tries" any source of weak stimulation, accepting or rejecting it finally, according as it proves fit for food or not. Thus we have behavior rising to a higher degree of complexity, — the method of trial in the second or third degree, as it were. Examples of this character are abundant.

G. *Positive Reactions*

We have thus far dealt primarily with reactions to environmental conditions that interfere with the normal life processes. We find that these induce changes in behavior, subjecting the organism to new conditions, the more favorable one of which is selected. This gives us a basis for the understanding of reactions toward conditions which favor the normal life processes, — that is, positive reactions.

In conditions that are completely favorable — so that all the life processes are taking place without lack or hindrance — there is of course no need for a change in behavior, for definite reactions of any sort. The most natural behavior on reaching such conditions, and that which is actually found as a general rule among lower organisms, is a continuation of the activities already in progress. These activities have resulted in favorable conditions, hence it is natural to keep them up; there is no cause for a change. This we find strikingly exemplified in bacteria, infusoria, rotifers, and many other organisms under most classes of stimuli. A change in behavior takes place only when the activities tend to remove the organism from the favorable conditions. Unfavorable conditions cause a change in behavior; favorable conditions cause none. It is perhaps a general rule in organisms, high or low, that continued completely favorable conditions do not lead to definite reactions. Of course while the external conditions remain the same, the internal processes may change in such a way that these conditions are no longer favorable, and now the behavior may change.

But when the organism is not completely enveloped by favorable conditions, but is on the boundary, if we may so express it, between favorable and unfavorable ones, then there is often a definite change in the behavior leading toward the favorable conditions, — a positive reaction. To understand such reactions, we may start from the fact that unfavorable internal conditions (as well as external ones) cause a change of behavior. The Hydra or sea anemone whose metabolic processes are interfered with by lack of material, exchanges its usual behavior for activities of a totally different character, setting forth on a tour of exploration. It is a general fact that the hungry animal sets in operation trains of activity differing from the usual ones. Interference with respiration or with other internal processes has similar effects. An increase of temperature above that favorable for the physiological processes likewise starts violent activities. Indeed, it is a general rule that changes of internal condition unfavorable to the physiological processes set in operation marked changes in behavior.

But the activities thus induced are in themselves undirected, save by structural conditions. There is nothing in the cause that produces them, taken by itself, to specifically direct them with reference to external things. Let us suppose, however, that certain of these movements lead to a condition which relieves the interference with the internal processes. The cause for a change of behavior is now removed, hence the organism continues its present movement — continues in the direction, we will say, that has led to the favorable conditions. But perhaps later — sometimes at the very next instant — this same movement may tend to remove the organism from the favorable conditions — as when a heated Paramecium passes across a small area of cool water, or a hungry organism comes against food. Thereupon the cause for a change — interference with the life processes — is again set in operation, and this movement changes to another. Thus the animal changes all behavior that leads away from the favorable condition, and continues that which tends to retain it, so that we get what we call a positive reaction. The change of behavior is due primarily in each case to the unfavorable condition, internal or external — perhaps in last analysis always internal.

Behavior of this character is seen with diagrammatic clearness in the free-swimming infusoria. These animals continue their movements so long as they lead to favorable conditions, changing at once such movements as lead away. They thus retain favorable conditions by avoiding unfavorable ones; the positive reaction is seen to be a secondary result of negative ones.

In the infusoria we have then the most elementary condition of the positive reaction. Let us now examine a more pronounced type of positive reaction, — movement directly toward the favorable condition. Amœba flows toward and follows a food body with which it comes in contact, as illustrated in Fig. 19, p. 14. Take, for example, its action at 3 in this figure. It moves forward with broad front, part of the movement taking it toward the food, part away. On coming in contact with the food, all movement is changed which takes it away, only that being retained which keeps the animal in contact with the food. We have here then, as in infusoria, a case of selection from varied movements, the central point being the changing of all motion that leads to less favorable conditions.

This is, perhaps, the fundamental condition of affairs, from which all positive reactions are derived. The animal moves (partly or entirely from internal impulse, as we have seen), but changes all movements that lead to less favorable conditions. It therefore moves toward the favorable conditions. In many higher animals, even, this behavior is seen in the random movements by which food is sought, by the aid of the chemical stimulation which it sends forth. The movements leading to loss of the favorable stimulation are changed, the others continued, till the food is found (see p. 247).

But many animals have developed, in some way, as we have seen in the account of the negative reactions, the power of localizing their reactions precisely, so as to move in a certain definite way with relation to the position of the source of stimulation. Let us suppose that such an organism is reached by a favorable stimulus on one side — food, or the optimum temperature. It has the power of turning *directly* toward this favorable condition — and this, of course, is what happens in many higher organisms. There is the same reason to think that this condition is not primitive that we saw in the case of negative reactions. It may, perhaps, be conceived as derived from behavior through selection of overproduced movements in the way set forth on page 308. The precise reactions shown in the actual taking of food are perhaps derivable in the same way.

In those animals whose positive reactions are precisely defined and localized, there is, of course, the same evidence that the impulse to change of behavior comes from within and is due to lack or hindrance of the physiological processes, that we find elsewhere. If the metabolic processes lack material for proper action, the medusa or sea anemone changes its behavior and moves about, even though there is nothing present to which it can react positively. When some object is reached, whether there shall be a positive reaction or not depends again on the state of the metabolic processes. If their state is bad, the animal reacts positively to almost anything; if fair, the animal reacts positively to substances that will improve them; if they are in a completely satisfactory condition, the animal does not react positively even to good food.

Thus with all conditions absolutely favorable there will be no reaction, either positive or negative. At the boundary between favorable and unfavorable conditions, the animal moves in such a way as to retain the favorable conditions. This is primitively due to selection from varied movements — all movement leading to less favorable conditions being changed. The "negative reactions" thus seem to furnish in a certain sense the primitive building stones from which the derived positive reactions are constructed. By development of the power of precise localization of reactions, the derivation of the positive reaction in this manner is in higher animals obscured. The fundamental fact for both positive and negative reactions is that interference with the physiological processes of the organism causes a change of behavior.

3. Résumé of the Fundamental Features of Behavior

We have considered in the three foregoing chapters, first the determining factors of movements, and second the movements themselves. Let us now attempt to put together the most important points in both, so as to reach a general characterization of behavior.

The three most significant features of behavior appear to be (1) the determination of the nature of reactions by the relation of external conditions to the internal physiological processes, and particularly the general principle that interference with these processes causes a change in behavior; (2) reaction by varied or overproduced movements, with selection from the varied conditions resulting from these movements — or, in brief, reaction by selection of overproduced movements; (3) the law of the readier resolution of physiological states after repetition. The first of these phenomena produces the regulatory character of behavior. The second and third furnish the mainsprings for the development of behavior, the second being constructive, the third conservative.

The activity of organisms we found to be spontaneous, in the sense that it is due to internal energy, which may be set in operation and even changed in its action without present external stimuli. In reactions this energy is merely released by present external stimuli. What form the activity shall take is limited by the action system, and within these limits is determined by the physiological state of the organism. Physiological states depend on many factors. The two primary classes of states depend on whether the internal life processes are proceeding uninterruptedly in the usual way. Interference with these processes produces a physiological state of a certain character ("negative"), while release from interference or assistance to those processes produces a different state ("positive"). Within or beside these contrasted primary classes, many subsidiary variations of physiological condition are possible, each with its corresponding method of behavior; at least five of these have been distinguished in a unicellular organism. Any change, external or internal, may modify the physiological state, and hence the behavior.

The effects of external agents depends largely on their relation to the normal course of the life processes — whether aiding or interfering, or neither. A primary fact is that interference with the life processes produces progressive changes in physiological state, inducing repeated changes in behavior. This is in itself regulatory, tending to relieve the interference, whether due to internal or external causes; it is a process of finding a reaction fitted to produce a more favorable condition. When through such changes a fitting reaction is found, the changes in physiological state and hence of behavior cease, since there is no further cause for change. In the same way a fitting reaction to a beneficial change, or one releasing from interference, may be found. This fitting reaction then tends to be preserved, by the law of the resolution of physiological states, in accordance with which the physiological state inducing this reaction is reached more readily after repetition. Thus the production of varied movements by stimulation is the progressive factor in behavior, while the law of the resolution of physiological states is the conservative factor, tending to retain fitting reactions once attained.

Through the law of the resolution of physiological states behavior tends to pass from the pure "trial" condition to a more defined state. The operation of this law tends to produce reactions precisely localized with reference to the position of the stimulating agent; increased appropriate reaction to the first weak effects of injurious or beneficial stimuli; and appropriate reactions to representative stimuli, according as they are followed by injurious or beneficial stimuli. In higher organisms such defining of the reactions has gone far; much of the behavior consists of derived reactions. There are in such organisms doubtless other factors producing derived reactions, besides the law just mentioned. These are treated in our chapter on the "Development of Behavior."

Thus through the production of varied movements by stimulation the organism finds the best method of behavior, and through the law of the resolution of physiological states it tends to retain this method as long as it is the best method. Through the same process it of course tends to lose this method when it is no longer adapted to the conditions. Thus behavior is regulatory in essential character; it is the process by which the organism tends to find conditions favorable to its life processes and to retain them, and it contains within itself the conditions for its own more efficient development.

CHAPTER XIX

DEVELOPMENT OF BEHAVIOR

IT is not the primary purpose of the present work to treat the problems of development, but rather to give an analysis of behavior as we now find it. But the results of this analysis furnish a certain amount of evidence as to how development may have occurred; this it will be well to set forth briefly. We shall consider first the development of behavior in the individual, then its development in the race. In unicellular organisms the first, perhaps, includes the second.

The primary facts for development in behavior are two principles to which our analysis of the chief factors in behavior have led us. One of these is that behavior is based fundamentally on the selection of varied movements. The other is the law in accordance with which the resolution of one physiological state into another becomes readier and more rapid through repetition.

In making use of the law of the readier resolution of physiological states after repetition in the study of development, it needs to be kept in mind that this law has been rigidly demonstrated for the lower organisms only in scattered instances. It has been shown to be valid in certain unicellular organisms, but in these cases it has not been shown that the modifications induced are lasting, as must be the case if this law plays a part in the development of behavior. In the lowest metazoa the law has likewise been demonstrated only for a few cases. In the flatworm and the Crustacea we find the law clearly exhibited in the form that is necessary in order that it may play a part in the permanent modification of behavior.

On the other hand, the fact that the law remains undemonstrated for many of the lowest organisms by no means indicates that it is not here valid. We lack proper experiments to show whether it exists or not. It is exceedingly difficult to carry out experiments that shall actually test this matter in the lowest animals. The view that this law is universally valid in organic behavior is thoroughly consistent with all that we know of the behavior of lower organisms, and the fact that it has actually been demonstrated in certain cases favorable for experimentation in unicellular organisms raises a presumption of its general validity. The following discussion of development is based on the assumption that the law is one of general validity. It must be kept in mind that this *is* partly an assumption, but the probability that this will be found true is such that the relation of development to the law is worth setting forth. There is no other need greater in the study of animal behavior than that of a thorough investigation of the validity of this law in the lower organisms.

The question in which we are here interested is then the following: How can behavior develop? That is, how can it change so as to become more effective — more regulatory?

(1) The behavior of any organism may become more effective through an increased tendency for the first weak effects of injurious or beneficial agents to cause the appropriate reaction; in other words, through increased delicacy of perception and discrimination on the part of the organism. Such a change would be brought about through the law of the readier resolution of physiological states after repetition. When the organism is subjected to a slight stimulus, this changes its physiological state, though perhaps not sufficiently to cause a reaction. Such a slight stimulus would be produced by a very weak solution of a chemical, or by a slight increase in temperature. Now, suppose that this weak stimulus, causing no reaction, is regularly followed by a stronger one, as would be the case if the weak chemical or slight warmth were the outer boundary of a strong chemical solution, or of a region of high temperature toward which the organism is moving. This stronger stimulus would produce an intense physiological state, corresponding to a marked negative reaction. That is, the first (weak) physiological state is regularly resolved by the action of the stimulating agent into the second (intense) one, inducing reaction. In time the first state would come to resolve itself into the second one even before the intense stimulus had come into action. As a result, the organism would react now to the weak stimulus, as it had before reacted only to the strong one. It would thus be prevented from entering the region of the chemical or the heat, even before any injury had arisen.

(2) In the same way the organism may come to react positively or negatively to a stimulus that is in itself not beneficial nor injurious, but which serves as a sign of a beneficial or injurious agent, because it regularly precedes such an agent. Suppose that a slight decrease in illumination (a shadow), which is of itself indifferent, regularly precedes the approach of an enemy, as happens in the sea urchin. The slight decrease in light induces a certain physiological state, which is so little marked that in itself it produces no reaction. But through the immediately following attack of the enemy, this indifferent physiological state is

regularly resolved into an intense one, corresponding to a strong negative reaction. Then after many repetitions of this process the indifferent state resolves itself at once into the intense one, and the animal reacts at the change in illumination, before the enemy has reached it. This tendency to react to "representative" factors, rather than to those which are in themselves beneficial or injurious, is, of course, immensely developed in higher animals. All positive or negative reactions to things merely seen or heard, which are not directly beneficial or injurious save when brought into direct contact with the organism, are, of course, reactions to such representative stimuli.

It is clear that neither the tendency to react to faint stimuli, nor that to react to "representative" factors will be increased, save as this is required by the environment. If the indifferent stimulus is not followed with some regularity by the powerful one; that is, if it does not really introduce a powerful agent, then there will be no tendency for the organism to acquire a reaction to this indifferent stimulus, for there will be no regular resolution of the first (faint) physiological change into the second (intense) one. And of course it would be no advantage, but on the contrary a positive disadvantage, for the organism to acquire this tendency to react to *all* weak stimuli. If it reacted negatively to every slight change in the environment, its movements would be seriously impeded; continued locomotion in any one direction would be almost impossible, and its activity would be frittered away in useless and disconnected reactions. The behavior becomes modified, in accordance with the principles above set forth, only as it is to the advantage of the organism that it should be so modified; that is, only as the modification favors the normal current of life activities.

(3) Progress takes place through increase in the complexity and permanence of physiological states, and in the tendency to react to these derived and complex states, instead of to the primitive and simple ones. We may imagine an organism whose physiological state depends entirely on the stimulus now acting upon it, the organism returning completely, as soon as the stimulus ceases, to its original state. Such an organism could react only with relation to the present stimulus, and its reaction to the same stimulus would always be the same. We might even imagine an organism that could change in only one way under the action of stimuli; its reactions to all stimuli would be the same. Such organisms would represent a purely reflex type of behavior. An advance on this condition would be represented by cases where the physiological state induced by a stimulus endures for a short time, influencing the immediately succeeding reactions, and a further advance when the reaction performed by the organism influences its physiological state, and therefore its later

reactions. Other advances would come in the production of different physiological states according to the different organs or parts of the body stimulated; this condition would naturally arise as structural differentiations were developed in the body. As new organs develop and the body becomes more complex, each part will naturally have physiological states peculiar to itself, and will be acted upon by external stimuli, producing changes in its physiological states. This is evidently the case in such organisms as the sea urchin and sea anemone. These partial physiological states of the different organs will then interact, altering each other and combining to form a general state for the entire organism. All the partial physiological states will be regulated, as in the separate organism, by their relation to the normal life current of the organ concerned, and further, their combinations will be regulated by their relation to the general life current of the organism. Whatever interferes with this normal life current will be changed, while that which does not interfere must persist. The partial and general physiological states will be subject to the laws of the combination and regulation of physiological states, just as in simple organisms. They will tend to discharge themselves in action, or by resolution into other states, as in the simple organisms. Thus the behavior of the organism must become in time controlled by these physiological states, derived from many sources besides that of the present stimulus. Behavior is gradually emancipated from its bondage to present external conditions, and depends largely upon the past experience and present needs of the organism. This is the condition we find in higher animals, and especially in man.

The various stages set forth above are merely logical divisions, and probably do not correspond in any close way to actual stages in the development of behavior. There seems to be no reason to suppose that an organism ever existed in which the original state is immediately restored on the cessation of a stimulus. This immediate return to the original state is not what we should expect from analogy even with inorganic substances.[1] Even in unicellular organisms we find a considerable complication of physiological states, depending on past stimuli, past reactions, localization of the stimulus, and present external conditions, as well doubtless as upon other factors.

Progress along the line just set forth will be brought about by the same factors, whatever they may be, that determine the development

---

[1] With relation to colloids, the substances of which organisms are mainly composed, a high authority in physical chemistry remarks as follows: "Their qualities often depend in the clearest way upon the former history of the colloid, its age, its previous temperature, and the time this continued: in short, on the way it has reached its present condition" (Bredig, 1902, p. 183). The facts of behavior in organisms might be cited as illustrations of this statement.

of complexity in structure. Differentiation of structure and of physiological states must go hand in hand. It is not our province to attempt to account for structural differentiations. The problem is the general problem of evolution.

(4) Progress in behavior may take place through increased variety and precision of the movements brought about by stimulation. Certain kinds of movements are much better adapted to relieving an organism from an unfavorable stimulation or securing it a favorable one than are others. This is illustrated by a comparison of the reactions of Amœba and Paramecium, or of the reaction of Bursaria to heat with that of Paramecium, as set forth on page 305. Owing to the difference in the effectiveness of their movements, if an area containing equal numbers of Paramecia and Bursaria is heated at one end, many of the Bursariæ are killed, while all the Paramecia escape.

New and better adapted methods of movement may be acquired through the selection of varied movements, in conjunction with the law of the resolution of physiological states. Under strong stimulation the organism, as it passes from one physiological state to another, tries successively all the movements of which it is capable. One of these movements (the spiral course, in the case of Bursaria) finally removes the organism from the stimulating agent. This happens every time the organism is stimulated in this manner. The result is that each physiological state is resolved into the succeeding one, until that one is reached in which the organism responds by the effectual movement. After a number of repetitions, this resolution takes place immediately, in accordance with the law that after repeated resolutions of one physiological state into another, this resolution takes place spontaneously and rapidly. Thus the organism responds at once with the effectual movement, and escapes.

In the same way the use of new organs might be acquired. Suppose that an Amœba sends forth, as sometimes happens, a long, slender pseudopodium, which may vibrate back and forth, like a flagellum. When stimulated, the overproduced movements of the organism, as it passes from one physiological state to another, include the vibration of this pseudopodium. Suppose that by this vibration the Amœba is at once moved away from the stimulating agent — the pseudopodium acting as does the flagellum in Euglena. If this is repeated, the physiological state inducing other movements will always be resolved finally into that inducing this one, and in time this resolution will take place so rapidly that only this movement will come to actuality. The Amœba will have acquired the habit when stimulated of swimming by means of a flagellum.

Thus the behavior of organisms is of such a character as to pro-

vide for its own development. Through the principle of the production
of varied movements, and that of the resolution of one physiological
state into another, anything that is possible is tried, and anything that
turns out to be advantageous to the organism is held and made permanent.

Thus through development in accordance with the two principles
mentioned, the organism comes to react no longer by trial, — by the over-
production of movements, — but by a single fixed response, appropriate
to the occasion. This is, of course, a great advantage, so long as the
conditions remain such as to make the response appropriate. Such
fixed responses are the general rule in the adult behavior of higher organisms, and are found to a certain extent in all organisms. In the
higher organisms we speak of some of these fixed responses as reflexes,
tropisms, habits, and instincts. The methods which we have discussed
are not the only possibilities for the development of such responses ; other
methods we shall take up later.

After the responses of the organism have become fixed, conditions
may so change that these responses are no longer appropriate. The
organism is then in a less advantageous position than one whose behavior
is determined more purely by trial movements. There will be now a
tendency for the fixed responses to become broken up and for processes
of trial to supplant them, until new fixed responses, appropriate to present
conditions, are produced. But in many cases the fixed responses are
so firmly established as not to give way save after long experience of
their lack of efficiency, and often the organism is destroyed by the new
environment, before it has developed appropriate responses by which
to preserve itself.

(5) We have thus far considered primarily the methods by which
the behavior of a given individual may be modified and made more
effective. It needs to be recalled that differences between the behavior
of different individuals may appear from other reasons. There are
congenital variations among different organisms. Some have naturally
a greater delicacy of perception or discrimination than others. Some
move more rapidly or in more or less varied ways than others, giving
some a more efficient method of reaction without any modification
through experience. These congenital variations play a most important
part in the question next to be considered.

(6) Our discussion thus far has related to individuals. The further
question arises as to how modifications of behavior may arise in the race
as a whole. How does it happen that the behavior of the race becomes
changed in the same way as that of the individual, so that succeeding
generations show the new method of reacting without acquiring it for
themselves?

There seems to be no question but that the power of new individuals to react in certain ways without preliminary trial has been much over-estimated. In most organisms there is in the early stages of development a continued process of trial, through which the habits become established. On the other hand, there is no doubt that individuals do appear with certain ways of reacting which most of their early ancestors did not at the beginning have. The question as to how this happens, therefore, presses for an answer.

The answer formerly given was, that the acquirements of the parent are directly inherited by the offspring. The parent having come to react in a certain way, the condition of the system inducing this reaction is passed on to posterity. In the unicellular organisms there seems to be nothing in the way of this inheritance by the offspring of the reaction methods acquired by the parent. There is no distinction between germ cells and body cells in these organisms; all acquirements pertain to the reproductive cells. Through reproduction by division the offspring *are* the parents, merely divided, and there is no evident reason why they should not retain the characteristics of the parents, however these characteristics were attained. If this is the real state of the case, then in unicellular organisms the life of the race is a direct continuation of the life of the individuals, and any acquirements made by the individuals are preserved to the race.

But in multicellular organisms the facts show that in the immense majority of cases the inheritance of the acquirements of the parents by the offspring does not occur. We know that we do not start with the education acquired by our parents, but must begin at the bottom, and acquire both knowledge and wisdom of action. In other words, we know that we fail to inherit directly the more efficient methods of reaction acquired through experience by our parents, in at least nine hundred and ninety-nine cases out of a thousand. Moreover, the theoretical difficulties in the way of such inheritance are great, and no demonstrative evidence seems to exist that it ever occurs. Thus we are certain that in most cases it does not take place, and must doubt whether it is possible.

If we give up, as most students of heredity do, the inheritance of acquired characters, the alternative explanation for progress in the race is by natural selection of congenital variations. The theory of natural selection may be stated briefly as follows: Organisms vary in many ways, through variations affecting the germ cells. Among these variations are some that help the organism, making it more efficient in escaping enemies or in obtaining food. These organisms, therefore, survive, while those without these helpful variations are killed. The surviving organisms transmit their helpful congenital variations to their offspring, so that in time an entire race may show the characteristics which first arose as accidental variations along with many other useless ones.

A great objection to this theory has been that it deals merely with chance variations in all directions, so that progress along a definite line, it is said, could never be brought about through it. The race progresses just as the individuals do; what is first acquired by the individual is later acquired by the race, as if the law of progress were the same in the two cases. This, it is held, could not be brought about through the selection of chance variations in all directions.

In recent years a most successful attempt has been made by J. Mark Baldwin (1902) and others to show that this objection is not a valid one; that the action of natural selection on characters playing a part in the behavior would, in fact, be guided by laws similar to or identical with those controlling the progress of the individual. To this guidance the name *organic selection* has been given. Organic selection would then account for the progress of the race in a continuous manner and in a definite direction. We shall examine briefly, from this point of view, the action of natural selection on behavior in the lower organisms.

Observation and experiment show that there exist such variations in the behavior of lower organisms as would under certain circumstances give opportunity for the action of natural selection. If into an area containing Paramecia a drop of a 10 per cent sugar solution is introduced, most of the animals enter it and are killed, but a few react negatively on coming in contact with it, and escape. If such solutions were a constant feature of the environment, it seems probable that in time there would be produced through selection a race of Paramecia that would always react negatively to them, and would, therefore, not be endangered by their existence. Similar differences exist among different individuals as to sensitiveness to other chemicals, to heat, and to electricity, as we have seen in previous pages. There is thus undoubtedly an opportunity for the action of natural selection to produce a race of organisms more sensitive to weak stimuli than is the average at present, if the environment should require it. But if the environment does not require it, the action of natural selection, like that of individual accommodation, will not bring it about. By either method only that is preserved which is useful.

There is likewise clearly an opportunity for natural selection to produce a race showing increased precision and adaptiveness in the movements brought about by stimulation. As we have seen on page 305, the reactions of Paramecium to heat are so much more effective than those of Bursaria that if locally heated regions were part of the usual environment of the two organisms, the Bursariæ would, for the greater part, soon be killed, while the Paramecia would not suffer. The latter would, therefore, be selected, as compared with the former. But there exist variations of reaction even among individuals of the same species. Some specimens of Bursaria when stimulated by heat show a greater inclination to swim freely, revolving on the long axis, than do the majority, that sink quickly to the bottom and cease to revolve. The former are saved from the heat, while the latter are killed. In time there might thus be developed a race of Bursariæ that were as well protected by their behavior from the action of heat as are Paramecia.

What are the characteristics that would be preserved by natural selection?   First it seems clear that under usual conditions the regulative power would tend to be preserved. So long as the environment is a changing one, those individuals that can alter their behavior to fit the new conditions would live, while any that cannot do so will be killed, so that any variation in the direction of less regulative power will be cut off. But under quite uniform conditions there might be no advantage in this regulative power, and no selection based upon it.

Second, those variations will be preserved that are in line with the general tendency of the behavior. In other words, those variations will persist that tend in the same direction as the adaptation of the individuals, due to selection of overproduced movements and the law of the resolution of physiological states. This will be made clear by an illustration.

Most ciliate infusoria may swim freely through the water, may creep along surfaces, may exude mucus to form a cyst, and may burrow about in the débris at the bottom of the water. Some show one habit in a more marked way, others another. Let us suppose a ciliate infusorian with a cylindrical body covered uniformly with cilia, that may behave in all these ways. It responds to stimulation by trial of the different reactions which it has at command, continuing, in accordance with the principle of the resolution of physiological states, that reaction which proves successful. Suppose that a number of the individuals come thus to react habitually in the first of the four ways mentioned above, others in the second, others in the third, and still others in the fourth. All these different methods have advantages for meeting unfavorable conditions, and all are found as a prevailing reaction in different ciliates.

We have then four groups of ciliate organisms, all alike structurally, but with different habits. How will natural selection act on these?

(1) In the first group, that swim freely through the water, like Paramecium, all variations that favor quickness of reaction, rapidity of movement, and precision of direction will be advantageous, and the individuals possessing them will tend to be selected. Specimens with body ill-shaped for rapid movement, with cilia weak or unequally distributed,

or with awkward methods of moving, will be killed by their inability to escape with sufficient rapidity from powerful agents. There will thus be a tendency to develop a fishlike form, adapted for rapid movements through the water; close-set, uniform cilia, and a tendency to revolve on the long axis; in other words, such characteristics as we find in Paramecium.

(2) In the second group, which reacts, like Oxytricha, by running along the bottom, variations of an entirely different character will be advantageous. The original cylindrical form can bring but few of its cilia against a surface, and presents much resistance to the water. Variations in the direction of a flat form, bringing many cilia against the surface, and presenting little resistance to the water as it runs along, will be advantageous, and individuals with such variations will be selected. The cilia on the surface kept against the bottom will be the all-important ones, so variations in the direction of increased size, strength, and rapidity of these cilia will be preserved; they will develop into "cirri" and other leglike structures. The cilia on the upper side of the body will be not merely useless, but a hindrance; hence they will tend to be lost. The tendency to revolve on the long axis will be injurious and will likewise tend to disappear by selection of those that do not thus revolve. In this way, under the action of natural selection, an organism will be developed having totally different characteristics from the organisms of the first set, that react by swimming freely. It will naturally approach the characteristics shown by Stylonychia, rather than those of Paramecium.

(3) On the third organism, which reacts to intense agents by secreting a layer of mucus about itself, natural selection will act in a still different manner. There will be no tendency to select rapidly moving individuals, nor those having larger or more numerous cilia, nor those having cilia distributed in any special way; all these characteristics will indeed be disadvantageous. Spiral swimming will not be developed. Those organisms that produce a thicker layer of mucus, of a more resistant character, and do this the more rapidly, will be selected.

(4) The fourth organism, which habitually reacts by burrowing into the detritus at the bottom of the water, will be acted upon by natural selection in a still different way. Only those characteristics which aid the burrowing will be useful and therefore selected. There will be no tendency to produce a swiftly swimming organism, nor one adapted to running along the bottom, nor one secreting a thick and resistant layer of mucus.

To sum up, it appears that only those variations are of advantage that are used, and only such variations can be preserved by the action of natural selection. Only such characteristics can be selected as are in line with the efforts of the organism. A variation which might be of inestimable advantage to an organism that reacts by swimming would be entirely lost on one that burrows in the earth. The organism determines by its own actions the direction of its development under the action of natural selection. When it adopts a certain line of behavior, it decides to a large degree the future career of the race. Development through the action of natural selection must then follow as definite a trend as does the behavior of the individual and indeed the same trend, for it is guided by this behavior. Individual selection guides natural selection.

Individual selection, with its production of definite adaptive reactions, is due, of course, to selection from varied movements, later fixed by the law of the readier resolution of physiological states.[1] With this in mind, we may express what we have just brought out as follows: Individual selection (intelligence) and natural selection are merely different methods of selecting adaptive ways of reacting. The former selects the adaptive response from among diverse reactions of the same individual; while natural selection selects the adaptive response from among diverse reactions of different individuals.

This may be illustrated as follows: Let us suppose an organism whose action system includes the different acts 1, 2, 3, 4, 5, 6, 7, 8, 9. When the physiological processes of this animal are interfered with by external agents, it tends to run through these nine reactions, in the order given above — as Stentor runs through its four or five reactions. Suppose that under a certain frequently recurring injurious condition the reaction 7 is the adaptive one, relieving the interference with the physiological processes. The organism runs through the series to 7, then stops (since the cause for further reaction has ceased). It now retains this reaction as the immediate response to the given condition, through the law of the readier resolution of physiological states. Many of the individuals are killed before 7 is reached, but after this adaptive reaction has become fixed, no others are killed. The young of these individuals must, however, begin at the beginning of the series, so that many will be destroyed.

Let us suppose that in another group there are, among many different individuals, congenital variations in the order in which the nine responses are given. Some respond by the series 2, 3, 7, 1, 4, 5, 6, 8, 9. These reach the adaptive reaction 7 sooner than do those following the usual order, hence fewer are killed by the injurious condition. Others react in the order 7, 4, 3, 5, 1, 2, 6, 8, 9. The first reaction is here the

[1] This is the process known as intelligence, in higher animals. See Chapter XX.

adaptive one. Hence the series goes no farther (since the cause for reaction ceases at once), and these organisms are not killed at all by the injurious condition. They are thus selected, as compared with those reacting in the usual way, and their method of reacting, being congenital, is inherited by posterity. In the course of time all the remaining individuals of this group will respond at once, like those of the previous group, by the reaction 7.

Thus individual selection and natural selection necessarily work to the same result. One selects from among the different acts of the same individual, the other from among those of different individuals. The thing selected is the same in each case, — namely, the adaptive reaction.

If there exist at the same time the power of individual modification and the variations on which natural selection acts, then under uniform conditions the latter will be more effective, since it results in immediate response by the adaptive reaction, while the former requires that every new individual should go through the trial series, with its attendant dangers of destruction. If the conditions are very severe, in time only the individuals which have inherited the immediate adaptive response will survive. Thus, through the action of natural selection these organisms will have an inborn tendency to react directly in an adaptive way, whereas in previous generations most of the individuals of the race acted in this manner only as a result of individual modification through experience.

Furthermore, it may be pointed out that in the course of time an organism which had adopted some special type of behavior, as burrowing, would become quite unadapted to other behavior, as running along the bottom or swimming through the water. It develops structures, under the influence of its adaptive behavior, that make it difficult or perhaps impossible for the organism to react in any other way than by burrowing. After a time, then, it will lose all tendency to react in other ways, because it *cannot* react in other ways, owing to the structural changes it has undergone. In most cases the specialization will not go so far as this, and the organism will retain the power of attempting other methods of reaction; that is, of performing other movements. But these movements will be ineffectual, because the structures of the organism are not adapted to their performance. They will therefore not relieve the organism from stimuli; hence they will be quickly exchanged for the movements which are effective. Thereafter the organism will always react by these movements on which its structure is based. If these first few ineffectual movements are not observed, it will appear that the organism has been rigidly limited from the beginning to this one type of behavior. Apparently there exist few if any organisms which do not show, in their younger stages at least, a few such ineffectual movements.

Baldwin suggests that the same process may go farther than this, in the following way: After the development, under the influence of a certain reaction method, of structures fitted to carry out that method, another congenital variation may occur, by which energy will be discharged directly into this apparatus, in the way necessary for performing the accustomed reaction, without any previous trial. It is urged that after the apparatus has been developed, the further variation required would probably be slight and not unlikely to occur. The organisms having this variation must react more readily and rapidly than those in which a trial is required, hence they might be selected. Thus in time in the entire race the reaction would be limited to this particular method. There seems to be no theoretical difficulty as to the occurrence of such a variation; if it occurs, development would doubtless take place in the way set forth, provided the environment remain sufficiently constant. But perhaps there would be little difference in reality between the behavior of such an organism, and one which had merely developed such structures as to make difficult any kind of reaction save one. The latter would still reserve the capability of developing other reactions, under changed circumstances, while the former would not.

The guidance of natural selection by the actions of the individuals that we have illustrated above, is what has been called "organic selection." The latter is evidently merely an exposition of how natural selection acts, not anything additional to natural selection, or differing from it in principle. For a general discussion of the questions which it involves, reference should be made to J. Mark Baldwin's "Development and Evolution."

Is natural selection, thus guided by individual accommodation, sufficient to account for the progress of the race in behavior? It is clear that natural selection cannot account for the origin of anything; only that can be selected which already exists. All the potency of behavior and of everything else that exists must lie in the laws of matter and energy, — physical and chemical, and possibly vital laws. Whatever the part assigned to natural selection, the superlative importance of these laws remains; they must continue the chief field for scientific investigation. All that natural selection is called upon to explain is the fact that at a given time such and such particular manifestations of these general laws exist, rather than certain other manifestations. In the field of behavior it is called to explain only the fact that this particular organism now behaves in this particular way, rather than in some other one of the infinite number of possible ways. Can it explain this?

The fact is established that organisms which vary in such a way as to make them unfitted to carry out the functions which they undertake are destroyed. The correlative fact that organisms which vary in such a way as to perform their functions better than the average are *not* so usually destroyed, is likewise established. The further fact is established that such congenital variations occur and are often handed on to the offspring. These three facts show that natural selection is beyond question a factor in the development of behavior. The only question is as to the extent of its agency. This depends on the number and extent of the congenital variations that occur. If these are sufficiently numerous and sufficiently varied, then it seems clear that natural selection guided by individual accommodation, would produce the results which we see. Its method of action is exactly what is needed to produce the observed results; the only question is whether the material presented to it in congenital variations is sufficient. The answer to this question must come, if it ever comes, from that study of variations which has received such an impulse in recent years. The recent studies of De Vries in mutation seem especially promising from this point of view. If it should appear that the material presented by congenital variations is not sufficient to account for the observed development, we should be forced apparently to turn once more to the possibility of the inheritance of the characteristics developed during the lifetime of the organism. The question of the inheritance of acquired characters cannot as yet be considered finally settled.

The view that the development of behavior is based largely on selection from among varied movements, with subsequent retention of the selected movements, to which we have come through a study of the behavior of the lower organisms, is of course not a new one. A theory to this effect has been set forth by Spencer and Bain, and has been especially developed in recent years by J. Mark Baldwin. The observations set forth in the present work lead to views differing in some important respects from these developed by Baldwin and Bain, particularly as to the nature of the causes which produce the varied movements. Space will not permit our entering here into a discussion of these differences. The reader may be referred for a discussion of some of the general bearings of this theory to the two volumes of Baldwin (1897, 1902). Possibly the most lucid statement of this theory, in its general bearings, is that recently given by Hobhouse (1901).

LITERATURE XIX

BALDWIN, 1897, 1902; HOBHOUSE, 1901; SPENCER, 1894 (Section 236, pp. 244-245); BAIN, 1888 (p. 315); 1894 (pp. 323, 324).

CHAPTER XX

RELATION OF BEHAVIOR IN LOWER ORGANISMS TO PSYCHIC BEHAVIOR

In describing the behavior of lower organisms we have used in the present work, so far as possible, objective terms — those having no implication of psychic or subjective qualities. We have looked at organisms as masses of matter, and have attempted to determine the laws of their movements. In ourselves we find movements and reactions resembling in some respects those of the lower organisms. We draw away from heat and cold and injurious chemicals, just as Paramecium does. Our behavior depends on physiological states, as does that of Stentor. But in ourselves there is the very interesting additional fact that these movements, reactions, and physiological states are often accompanied by subjective states, — states of consciousness. Different states of consciousness are as varied as the different possibilities of reaction; indeed, more varied. In speaking of behavior in ourselves, and as a rule in higher animals, we use terms based on these subjective states, as pleasure and pain, sensation, memory, fear, anger, reason, and the like.

The peculiarity of subjective states is that they can be perceived only by the one person directly experiencing them, — by the subject. Each of us knows directly states of consciousness only in himself. We cannot by observation and experiment detect such states in organisms outside of ourselves. But observation and experiment are the only direct means of studying behavior in the lower organisms. We can reason concerning their behavior, and through reasoning by analogy we may perhaps conclude that they also have conscious states. But reasoning by analogy, when it is afterward tested by observation and experiment, has often shown itself fallacious, so that where it cannot be tested, we must distrust its conclusiveness. Moreover, in different men it leads to different conclusions, so that it does not result in admitted certainty. Hence it seems important to keep the results of observation and experiment distinct from those of reasoning by analogy, so that we may know what is really established. On this account it is customary among most physiologists not to use, in discussing the behavior of the lower organisms, psychic terms, or those implying subjective states. This has the additional ground that the ideal of most scientific men is to explain behavior in terms of matter and energy, so that the introduction of psychic implications is considered superfluous. While this exclusive use of objective terms has great advantages, it has one possible disadvantage. It seems to make an absolute gulf between the behavior of the lower organisms on the one hand, and that of man and higher animals on the other. From a discussion of the behavior of the lower organisms in objective terms, compared with a discussion of the behavior of man in subjective terms, we get the impression of complete discontinuity between the two.

Does such a gulf actually exist, or does it lie only in our manner of speech? We can best get evidence on this question by comparing the objective features of behavior in lower and in higher organisms. In any animal outside of man, and even in man outside of the self, the existence of perception, choice, desire, memory, emotion, intelligence, reasoning, etc., is judged from certain objective facts — certain things which the organisms do. Do we find in the lower organisms objective phenomena of a similar character, so that the same psychic names would be applied to them if found in higher organisms? Do the objective factors in the behavior of lower organisms follow laws that are similar to the laws of psychic states? Only by comparing the objective factors can we determine whether there is continuity or a gulf between the behavior of lower and higher organisms (including man), for it is only these factors that we know.

Let us then examine some of the concepts employed in discussions of the behavior of higher animals and man, determining whether there exist any corresponding phenomena in lower organisms. We shall not attempt to take into consideration the scholastic definitions of the terms used, but shall judge of them merely from the objective phenomena on which they are based.

When we say that an animal *perceives* something, or that it shows *perception* of something, we base this statement on the observation that it reacts in some way to this thing. On the same basis we could make the statement that Amœba perceives all classes of stimuli which we ourselves perceive, save sound (which is, however, essentially one form of mechanical stimulation). Perception as judged from our subjective experiences means much more: how much of this may be present in animals outside the self we cannot know.

Discrimination is a term based, so far as objective evidence goes, upon the observed fact that organisms react differently to different stimuli. In this sense Paramecium, as we have seen, discriminates acids from alkalies; Amœba discriminates a Euglena cyst from a grain of sand, and in general all lower organisms show discrimination in many phases of their behavior.

Choice is a term based objectively on the fact that the organism accepts or reacts positively to some things, while it rejects or reacts negatively or not at all to others. In this sense all lower organisms show choice, and at this we need not be surprised, for inorganic substances show a similar selectiveness. The distinctive thing about the choice of organisms is that it is regulatory; organisms on the whole choose those things which aid their normal life processes and reject those that do not. This is what justifies the use of the term "choice," as contrasted with the mere selectiveness of inorganic reactions. Choice in this regulatory sense is shown by lower organisms, as we have seen in detail in previous chapters. Choice is not perfect, from this point of view, in either lower or higher organisms. Paramecium at times accepts things that are useless or harmful to it, but perhaps on the whole less often than does man.

The methods by which choice is shown in particular organisms have been set forth in our descriptive chapters. We may refer particularly to the account of choice in the infusoria, given on page 183. The free-swimming infusoria as they move about are continually rejecting certain things and accepting others, and this choice is regulatory. Their behavior is based throughout on the method of trial, and this involves an act comparable to choice in almost every detail. Whatever the condition met, the infusorian must either accept it by going ahead, or reject it by backing and giving the avoiding reaction. We can almost say that its whole behavior is a process of choice; that choice is the essential feature of its behavior. For the other lower organisms that we have taken up, a consideration of details would discover activities involving regulatory choice almost as continuously as in the infusoria.

Is not what we call *attention* in higher organisms, when considered objectively, the same phenomenon that we have called the interference of one stimulus with the reaction to another? At the basis of attention lies objectively the phenomenon that the organism may react to only one stimulus even though other stimuli are present which would, if acting alone, likewise produce a response. The organism is then said to attend to the particular stimulus to which it responds. This fundamental phenomenon is clearly present in unicellular organisms. Stentor and Paramecium when reacting to contact with a solid "pay no attention" to a degree of heat or a chemical or an electric current that would produce an immediate reaction in a free individual. On the other hand, individuals reacting to heat or a chemical may not respond to contact with a mass of bacteria, to which they would under other conditions react positively. In our chapter on reaction under two or more stimuli in the infusoria, many examples of this character are given.

Indeed, attention in this objective sense seems a logical necessity for the behavior of any organism having at its command more than a single action. The characteristic responses to two present stimuli may be incompatible with each other. The organism must then react to one or the other, since it cannot react to both; it thus *attends* (objectively) to one, and not to the other. Only in case there is no reaction at all in the presence of two stimuli, or in case its reaction is precisely intermediate between those required by the two, could the basis of attention be considered lacking. An organism behaving in this way would be quickly destroyed as a result of its indecisive and ineffective behavior.

In higher animals and man we distinguish certain different conditions, — "states of feeling," "emotions," "appetites," "desires," and the like. In all cases except the self, these various states are distinguished through the fact that the organism behaves differently in the different conditions, even though the external stimuli may be the same. We find a parallel condition of affairs in the lower organisms. Here, as we have seen, the behavior under given external conditions depends largely on the physiological condition of the individual. Many illustrations of this fact are given in preceding chapters, so that we need not dwell upon it here.

In the lower organisms we can even distinguish a number of states that are parallel, so far as observation can show, with those distinguished and named in higher animals and man. To begin with some of the simpler ones, the objective correlate of hunger can be distinguished at least as low in the scale as Hydra and the sea anemone. These animals, as we have seen, take food only when hungry, and if *very* hungry, will take substances as food which they otherwise reject. Doubtless hunger could be detected in still lower organisms by proper experiments. A resting condition comparable to sleep is found, as we have seen, in the flatworm (p. 253), while there seems to be no indication of such a state in the infusoria (p. 181). *Fatigue* can of course be distinguished in all living things, including separated muscles.

Correlative with hunger, there exists a state which corresponds so far as objective evidence goes with what we should call in higher animals a *desire* for food. Hydra when hungry opens its mouth widely when immersed in a nutritive liquid. In the flatworm, we can distinguish a certain physiological condition in which the animal moves about in an eager, searching way, as if hunting for food. Even in Amœba we find a pertinacity in the pursuit of food (p. 14 and Fig. 21) such as we would attribute in a higher animal to a desire for it.

All the way up the scale, from Amœba and bacteria to man, we find that organisms react negatively to powerful and injurious agents. In man and higher animals such reactions are usually said to be due to *pain*. In the lower organisms the objective facts are parallel, and naturally lead to the assumption of a physiological state similar to what we have in the higher forms. As to subjective accompaniments of such a state we of course know nothing in animals other than ourselves. The essential cause of the states corresponding to pain is interference with any of the processes of which the organism is the seat, and the correlate in action of these states is a change in movement. This point will be developed in our final chapter.

A similar basis exists for distinguishing throughout the organic series a physiological state corresponding to that accompanying pleasure in man. This is correlated with a relief from interference with the life processes, or with the uninterrupted progression of these processes.

In man and higher animals we often find a negative reaction to that which is not in itself injurious, but which is usually followed by something injurious. The sight of a wild beast is not injurious, considered by itself, but as preceding actual and injurious contact with this beast, it leads to powerful negative reactions. Such reactions are said to be due to *fear*. In fear there is then a negative reaction to a representative stimulus — one that *stands* for a really injurious stimulation. In lower organisms we find the objective indications of a parallel state of affairs. The infusoria react negatively to solutions of chemicals that are not, so far as we can determine, injurious, though they would naturally, under ordinary circumstances, be immediately followed by a solution so strong as to be injurious. Euglena reacts negatively when darkness affects only its colorless anterior end, though we have reason to believe that it is only the green part of the body which requires the light for the proper discharge of its functions. A much clearer case is seen in the sea urchin, which reacts by defensive movements when a shadow falls upon it, though shade is favorable to its normal functions. Objectively, fear has at its basis the fact that a negative reaction may be produced by a stimulus which is not in itself injurious, provided it leads to an injurious stimulation; this basis we find throughout organisms.

Sometimes higher animals and man are thrown into a "state of fear," such that they react negatively to all sorts of stimuli, that under ordinary circumstances would not cause such a reaction. A similar condition of affairs we have seen in Stentor and the flatworm. After repeated stimulation, they react negatively to all stimuli to which they react at all.

The general fact of which the reactions through fear are only a special example is the following: Organisms react appropriately to *representative* stimuli. That is, they react, not merely to stimuli that are in themselves beneficial or injurious, but to stimuli which lead to beneficial or injurious conditions. This is as true of positive as of negative reactions. It is true of Amœba when it moves toward a solid body that will give it an opportunity to creep about and obtain food. It is true of Paramecium when it settles against solids (even bits of filter paper), because usually such solids furnish a supply of bacteria. It is true of the colorless flagellate Chytridium and the white Hydra, when they move toward a source of light and thus come into the region where their prey congregate. There seems to be no general name for this positive reaction to a representative stimulus. In man we call various subjective aspects of it by different names, — foresight, anticipation, prudence, hope, etc.

The fact that lower as well as higher organisms thus react to representative stimuli is of the greatest significance. It provides the chief condition for the advance of behavior to higher planes. At the basis of reaction of this character lies the simple fact that a *change*, even though neutral in its effect, may cause reaction (p. 294). This taken in connection with the law of the resolution of physiological states (p. 291) permits the establishment of a negative or positive reaction, as the case may require, as a response to a given change. The way in which this may take place we have attempted to set forth on page 316.

Related to these reactions to representative stimuli are certain other characteristics distinguished in the behavior of man and higher animals. The objective side of *memory* and what is called *habit* is shown when the behavior of an organism is modified in accordance with past stimuli received or past reactions given. If the behavior is merely changed in a way that is not regulatory, as by fatigue, we do not call this memory. In memory the reaction is modified in such a way that it is now more adequate to the conditions to be met. Habit and memory in this objective sense are clearly seen in the Crustacea, and in the low acœlous flatworm Convoluta (p. 255). Something of a similar character is seen even in the protozoan Stentor. After reacting to a weak stimulus which does not lead to an injurious one it ceases to react when this stimulus is repeated, while if the weak stimulus does lead to an injurious one, the animal changes its behavior so as to react next time in a more effective way; and it repeats this more effective reaction at the next incidence of the stimulus. Habit and memory, objectively considered, are based on the law of the resolution of physiological states (p. 291), which may be set forth in application to the present subject as follows: If a given physiological state, induced by a stimulus, is repeatedly resolved into a succeeding state, this resolution becomes easier, and may take place spontaneously, so that the reaction induced is that due primarily to the second physiological state reached. Wherever we find this law in operation, we have the ultimate basis from which habit and memory (objectively considered) are developed.

From memory in the general sense it is customary to distinguish associative memory. This is characterized objectively by the fact that the response at first given to one stimulus comes, after a time, to be transferred to another one. Examples of associative memory are seen in the experiments of Yerkes and Spaulding on crustaceans, described in Chapter XII. It may be pointed out that the essential basis for associative memory is the same law of the resolution of physiological states which we have set forth in the last paragraph as underlying ordinary memory. The physiological condition induced by the first stimulus (sight of the screen, in Spaulding's experiments) is regularly resolved into that due to the second stimulus (food, in the experiments just mentioned). After a time the resolution becomes spontaneous, so that the physiological state primarily due to the food is reached immediately after the introduction of the screen, even though no food is given. There seems to be no difference in kind, therefore, between associative memory and other sorts; they are based on the same fundamental law. The existence of associative memory has often been considered a criterion of the existence of consciousness, but it is clear that the process underlying it is as readily conceivable in terms of matter and energy as are other physiological processes. Even in inorganic colloids, as we have seen (p. 317), the properties depend on the past history of the colloid, and the way in which it has reached the condition in which it is now found. If this is conceivable in terms of matter and energy, it is difficult to see why the law of the readier resolution of physiological states is not equally so.

Intelligence is commonly held to consist essentially in the modification of behavior in accordance with experience. If an organism reacts in a certain way under certain conditions, and continues this reaction no matter how disastrous the effects, we say that its behavior is unintelligent. If on the other hand it modifies its behavior in such a way as to make it more adequate, we consider the behavior as in so far intelligent. It is the "correlation of experiences and actions" that constitutes, as Hobhouse (1901) has put it, "the precise work of intelligence."

It appears clear that we find the beginnings of such adaptive changes of behavior even in the Protozoa. They are brought about through the law in accordance with which the resolution of one physiological state into another takes place more readily after repetition, — in connection with the other principle that interference with the life processes causes a change of behavior. These laws apparently form the fundamental basis of intelligent action. This fundamental basis then clearly exists even in the Protozoa; it is apparently coextensive with life. It is difficult if not impossible to draw a line separating the regulatory behavior of lower organisms from the so-called intelligent behavior of higher ones; the one grades insensibly into the other. From the lowest organisms up to man behavior is essentially regulatory in character, and what we call intelligence in higher animals is a direct outgrowth of the same laws that give behavior its regulatory character in the Protozoa.

Thus it seems possible to trace back to the lowest organisms some of the phenomena which we know, from objective evidence, to exist in the behavior of man and the higher animals, and which have received special names. It would doubtless be possible to extend this to many other phenomena. Many conditions which we can clearly distinguish in man must be followed back to a single common condition in the lower organism. But this is what we should expect. Differentiation takes place as we pass upward in the scale in these matters as in others. Because we can trace these phenomena back to conditions found in unicellular forms, it does not follow that the behavior of these organisms has as many factors and is as complex as that of higher animals. The facts are precisely parallel with what we find to be true for other functions. Amœba shows respiration, and all the essential features of respiration in man can be traced back to the condition in such an organism. Yet in man respiration is an enormously complex operation, while in Amœba it is of the simplest character possible — apparently little more than a mere interdiffusion of gases. In the case of behavior there is the same possibility of tracing all essential features back to the lower organisms, with the same great simplification as we go back.

THE QUESTION OF CONSCIOUSNESS

All that we have said thus far in the present chapter is independent of the question whether there exist in the lower organisms such subjective accompaniments of behavior as we find in ourselves, and which we call consciousness. We have asked merely whether there exist in the lower organisms objective phenomena of a character similar to what we find in the behavior of man. To this question we have been compelled to give an affirmative answer. So far as objective evidence goes, there is no difference in kind, but a complete continuity between the behavior of lower and of higher organisms.

Has this any bearing on the question of the existence of consciousness in lower animals? It is clear that objective evidence cannot give

a demonstration either of the existence or of the non-existence of consciousness, for consciousness is precisely that which cannot be perceived
objectively. No statement concerning consciousness in animals is open
to verification or refutation by observation and experiment. There
are no processes in the behavior of organisms that are not as readily
conceivable without supposing them to be accompanied by consciousness as with it.

But the question is sometimes proposed: Is the behavior of lower
organisms of the character which we should "naturally" expect and
appreciate if they did have conscious states, of undifferentiated character,
and acted under similar conscious states in a parallel way to man? Or
is their behavior of such a character that it does not suggest to the
observer the existence of consciousness?

If one thinks these questions through for such an organism as Paramecium, with all its limitations of sensitiveness and movement, it appears
to the writer that an affirmative answer must be given to the first of the
above questions, and a negative one to the second. Suppose that this
animal *were* conscious to such an extent as its limitations seem to permit.
Suppose that it could feel a certain degree of pain when injured; that
it received certain sensations from alkali, others from acids, others from
solid bodies, etc., — would it not be natural for it to act as it does?
That is, can we not, through our consciousness, *appreciate* its drawing
away from things that hurt it, its trial of the environment when the
conditions are bad, its attempting to move forward in various directions,
till it finds one where the conditions are not bad, and the like? To
the writer it seems that we can; that Paramecium in this behavior
makes such an impression that one involuntarily recognizes it as a little
subject acting in ways analogous to our own. Still stronger, perhaps,
is this impression when observing an Amœba obtaining food as shown
in Figs. 19 and 21. The writer is thoroughly convinced, after long study
of the behavior of this organism, that if Amœba were a large animal, so
as to come within the everyday experience of human beings, its behavior would at once call forth the attribution to it of states of pleasure
and pain, of hunger, desire, and the like, on precisely the same basis
as we attribute these things to the dog. This natural recognition is
exactly what Münsterberg (1900) has emphasized as the test of a
subject. In conducting objective investigations we train ourselves to
suppress this impression, but thorough investigation tends to restore it
stronger than at first.

Of a character somewhat similar to that last mentioned is another
test that has been proposed as a basis for deciding as to the consciousness of animals. This is the satisfactoriness or usefulness of the concept

of consciousness in the given case. We do not usually attribute consciousness to a stone, because this would not assist us in understanding or controlling the behavior of the stone. Practically indeed it would lead us much astray in dealing with such an object. On the other hand, we usually do attribute consciousness to the dog, because this is useful; it enables us practically to appreciate, foresee, and control its actions much more readily than we could otherwise do so. If Amœba were so large as to come within our everyday ken, I believe it beyond question that we should find similar attribution to it of certain states of consciousness a practical assistance in foreseeing and controlling its behavior. Amœba is a beast of prey, and gives the impression of being controlled by the same elemental impulses as higher beasts of prey. If it were as large as a whale, it is quite conceivable that occasions might arise when the attribution to it of the elemental states of consciousness might save the unsophisticated human being from the destruction that would result from the lack of such attribution. In such a case, then, the attribution of consciousness would be satisfactory and useful. In a small way this is still true for the investigator who wishes to appreciate and predict the behavior of Amœba under his microscope.

But such impressions and suggestions of course do not demonstrate the existence of consciousness in lower organisms. Any belief on this matter can be held without conflict with the objective facts. All that experiment and observation can do is to show us whether the behavior of lower organisms is objectively similar to the behavior that in man is accompanied by consciousness. If this question is answered in the affirmative, as the facts seem to require, and if we further hold, as is commonly held, that man and the lower organisms are subdivisions of the same substance, then it may perhaps be said that objective investigation is as favorable to the view of the general distribution of consciousness throughout animals as it could well be. But the problem as to the actual existence of consciousness outside of the self is an indeterminate one; no increase of objective knowledge can ever solve it. Opinions on this subject must then be largely dominated by general philosophical considerations, drawn from other fields.

LITERATURE XX

CONSCIOUSNESS IN LOWER ANIMALS

CLAPARÈDE, 1901, 1905; TITCHENER, 1902; MINOT, 1902; MÜNSTERBERG, 1900; VERWORN, 1889; BETHE, 1898; YERKES, 1905, 1905 *a*; JORDAN, 1905; v. UEXKULL, 1900 *b*, 1902; WASMANN, 1901, 1905; LUKAS, 1905.

CHAPTER XXI

BEHAVIOR AS REGULATION, AND REGULATION IN OTHER FIELDS

1. INTRODUCTORY

EVERYWHERE in the study of life processes we meet the puzzle of regulation. Organisms do those things that advance their welfare. If the environment changes, the organism changes to meet the new conditions. If the mammal is heated from without, it cools from within; if it is cooled from without, it heats from within, maintaining the temperature that is to its advantage. The dog which is fed a starchy diet produces digestive juices rich in enzymes that digest starch; while under a diet of meat it produces juices rich in proteid-digesting substances. When a poison is injected into a mouse, the mouse produces substances which neutralize this poison. If a part of the organism is injured, a rearrangement of material follows till the injury is repaired. If a part is removed, it is restored, or the wound is at least closed up and healed, so that the life processes may continue without disturbance. Regulation constitutes perhaps the greatest problem of life. How can the organism thus provide for its own needs? To put the question in the popular form, How does it know what to do when a difficulty arises? It seems to work toward a definite purpose. In other words, the final result of its action seems to be present in some way at the beginning, determining what the action shall be. In this the action of living things appears to contrast with that of things inorganic. It is regulation of this character that has given rise to theories of vitalism. The principles controlling the life processes are held by these theories to be of a character essentially different from anything found in the inorganic world. This view has found recent expression in the works of Driesch (1901, 1903).

2. REGULATION IN BEHAVIOR

Nowhere is regulation more striking than in behavior. Indeed, the processes in this field have long served as the prototype for regulatory action. The organism moves and reacts in ways that are advantageous to it. If it gets into hot water, it takes measures to get out again, and the same is true if it gets into excessively cold water. If it enters an injurious chemical solution, it at once changes its behavior and escapes. If it lacks material for its metabolic processes, it sets in operation movements which secure such material. If it lacks oxygen for respiration, it moves to a region where oxygen is found. If it is injured, it flees to safer regions. In innumerable details it does those things that are good for it. It is plain that behavior depends largely on the needs of the organism, and is of such a nature as to satisfy these needs. In other words, it is regulatory.

Behavior is merely a collective name for the most obvious and most easily studied of the processes of the organism, and it is clear that these processes are closely connected with, and are indeed outgrowths from, the more recondite internal processes. There is no reason for supposing them to follow laws different from those of the other life processes, or for holding that regulation in behavior is of a different character from that found elsewhere. But nowhere else is it possible to perceive so clearly how regulation occurs. In the behavior of the lowest organisms we can see not only what the animal does, but precisely how this happens to be regulatory. The method of regulation lies open before us. This method is of such a character as to suggest the possibility of its general applicability to life processes. In the present chapter we shall attempt to sum up the essential points in regulation as shown in behavior, and to make some suggestions as to its possible application to other fields.

A. Factors in Regulation in the Behavior of Lower Organisms

In the lower organisms, where we can see just how regulation occurs, the process is as follows: Anything injurious to the organism causes changes in its behavior. These changes subject the organism to new conditions. As long as the injurious condition continues, the changes of behavior continue. The first change of behavior may not be regulatory, nor the second, nor the third, nor the tenth. But if the changes continue, subjecting the organism successively to all possible different conditions, a condition will finally be reached that relieves the organism from the injurious action, provided such a condition exists. Thereupon the changes in behavior cease, and the organism remains in the favorable condition. The movements of the organism when stimulated are such as to subject it to various conditions, one of which is selected.

This method of regulation is found in its purest form in unicellular organisms. But, as we have seen in preceding pages, it occurs also in higher organisms, and indeed is found in a less primitive form throughout the animal series, up to and including man. It is commonly spoken of as behavior by "trial and error." In connection with this method of behavior, three questions arise, which are fundamental for the theory of regulation. The first is as follows: How is it determined what shall cause the changes in behavior resulting in new conditions? Why does the organism change its behavior under certain conditions, not under others? Second, how does it happen that such movements are produced as result in more favorable conditions? Third, how is the more favorable condition selected? What it this selection and what does it imply?

Our first and third questions may indeed be condensed into one, which involves the essence of regulation. Why does the organism choose certain conditions and reject others? This selection of the favorable conditions and rejection of the unfavorable ones presented by the movements is perhaps the fundamental point in regulation.

It is often maintained that this selection is precisely personal or conscious choice, and that the behavior cannot be explained without this factor. Personal choice it evidently is, and in man it is often conscious choice; whether it is conscious in other animals we do not know. But in any case this does not remove it from the necessity for analysis. Whether conscious or unconscious, choice must be determined in some way, and it is the province of science to inquire as to how this determination occurs. To say that rejection is due to pain, acceptance to pleasure or to other conscious states, does not help us, for we are then forced to inquire why pain occurs under certain circumstances, pleasure under others. Surely this is not a mere haphazard matter. There must be some difference in the conditions to induce these differences in the conscious states (if they exist), and at the same time to determine the differences in behavior. We are therefore thrown back upon the objective processes occurring. Why are certain conditions accepted, others rejected?

Let us examine one or two of the simplest cases of such regulatory selection. The green infusorian *Paramecium bursaria* requires oxygen for its metabolic processes. While swimming about it comes to a region where oxygen is lacking. Thereupon it changes its behavior, turns away, and goes in some other direction. The white *Paramecium caudatum* does the same, and so also do many bacteria; they likewise require oxygen for their metabolic processes. All reject a region without oxygen. The green *Paramecium bursaria* comes to a dark region. The water contains plenty of oxygen, hence the metabolic processes are proceeding uninterruptedly, and passing into darkness does not interfere with them. The animal does not change its behavior, but enters the dark region without hesitation. Later the oxygen in the water has become nearly exhausted. The animal is again swimming about in the light, and the green chlorophyll bodies which it contains are producing a little oxygen which the infusorian uses in its metabolic processes. Now it comes again to a dark region. In the darkness the production of oxygen by the green bodies ceases; they no longer supply the metabolic processes with this necessary factor. Now we find that the infusorian rejects the darkness and turns in another direction. The white *Paramecium caudatum* does not do this, nor do the colorless bacteria. Possessing no chlorophyll, they receive no more oxygen in the light than in the darkness, and they pass into darkness as readily as into light. But many colored bacteria do reject the darkness. They require light in certain other metabolic processes, — in their assimilation of inorganic compounds, — and when they come to the boundary between light and darkness, they return into the light. Most bacteria reject regions containing no oxygen, as we have seen. But in certain bacteria, oxygen is not required for the metabolic processes; on the contrary, it impedes them. These bacteria reject regions containing oxygen, swimming back into the light. In some cases among unicellular organisms the relation of behavior to the metabolic processes is exceedingly precise. Thus, Engelmann (1882 *a*) proved that in *Bacterium* (or *Chromatium*) *photometricum* the ultra-red and the yellow-orange rays are those most favorable to the metabolic processes (assimilation of carbon dioxide, etc.). When a microspectrum is thrown on these bacteria, they are found to react in such a way as to collect in precisely the ultra-red and the yellow-orange. The reaction consists in a change of behavior, — a reversal of movement, — at the moment of passing from the ultra-red or the yellow-orange to any other part of the spectrum. At that same instant the metabolic processes of course suffer interference. Bacteria are not in nature subjected to pure spectral colors in bands, so that there has been no opportunity for the production of this correspondence between behavior and favorable conditions, through the natural selection of varying individuals.

In all these cases the behavior depends upon the metabolic processes, and is of such a character as to favor them. Throughout the present volume we have found similar relations to hold for all sorts of organisms. We find even that when the metabolic processes of a given individual change, the behavior changes in a corresponding way.

Why does the bacterium or infusorian change its behavior and shrink back from the darkness or the region containing no oxgyen? As a matter of fact, it needs the light or the oxygen in its metabolic processes, and it does not shrink back from their absence unless it does need them. But we have no reason to attribute to the bacterium anything like a knowledge or idea of that relation. We do not need any purpose or idea in the mind of the organism, or any "psychoid" or entelechy, to account for the change of behavior, for an adequate objective cause exists. We know experimentally that the darkness or the lack of oxygen interferes with the metabolic processes. This very interference is then evidently the cause of the change of behavior. The organism is known to be the seat of varied processes, proceeding with a certain energy. When there is interference with these processes, the energy overflows into other channels, resulting in changes in behavior. This statement is a formulation of the facts determined by observation and experiment in the most diverse organisms. It is illustrated on almost every page of the present work.

In the lower organisms the processes of metabolism are the chief ones occurring, and behavior is largely determined with reference to them. In higher organisms these usually retain their commanding rôle, but an immense number of coördinated and subsidiary processes also occur, and changes in behavior may be induced by interference with any of these.

The answer to our first question is then as follows: The organism changes its behavior as a result of interference or disturbance in its physiological processes.

Our second question was: How does it happen that such movements are produced as bring about more favorable conditions? This question we have already answered, so far as lower organisms are concerned, in our general statement on page 339. The organism does not go straight for a final end. It merely acts, — in all sorts of ways possible to it, — resulting in repeated changes of the environmental conditions. The fundamental fact must be remembered that the life processes depend upon internal and external conditions, and are favored by conditions that are rather generally distributed throughout the environment of organisms. If there were no favorable conditions attainable, of course no change of behavior could attain them. But the favorable conditions actually exist, and if the changes of behavior continue, subjecting the organism to all possible different conditions, a condition will finally be reached that is favorable to the life processes. Often only a slight change of behavior is required in order to bring about favorable conditions. If an organism swims suddenly into a heated area, almost any change in the direction of movement is likely to restore the conditions previously existing. Adjustment, then, is reached by repeated changes of movement.

Our third question was: How does the organism select the more favorable condition thus reached? This question now answers itself.

It was the interference with the physiological processes that caused the changes in behavior. As soon therefore as this interference ceases, there is no further cause for change. The organism selects and retains the favorable condition reached, merely by ceasing to change its behavior when interference ceases.

Thus in the lowest organisms we find regulation occurring on the basis of the three following facts: —

1. Definite internal processes are occurring in organisms.

2. Interference with these processes causes a change of behavior and varied movements, subjecting the organism to many different conditions.

3. One of these conditions relieves the interference with the internal processes, so that the changes in behavior cease.

It is clear that regulation taking place in this way does not require that the end or purpose of the action shall function in any way as part of its cause, as is held in various vitalistic theories. There is no evidence that a final aim is guiding the organism. None of the factors above mentioned appear to include anything differing in essential principle from such methods of action as we find in the inorganic world.

Now an additional factor enters the problem. By the process which we have just considered, the organism reaches in time a movement that brings relief from the interfering conditions. This relieving response becomes fixed through the operation of the law of the readier resolution of physiological states as a result of repetition (Chapter XVI, Section 10). After reaching the relieving response a number of times by a repeated succession of movements, a recurrence of the interfering condition induces more quickly the relieving response, and in time this becomes the immediate reaction to this interfering condition.

It is in this second stage of the process, when the relieving response has become set through the law of the readier resolution of physiological states by repetition, that an end or purpose seems to dominate the behavior. This end or purpose of course actually exists, as a subjective state called an idea, in man. Whether any such subjective state exists in the lower organism that has gone through the process just sketched, of course we do not know. But some objective phenomenon, as a transient physiological state, would seem to be required in the lower animal, corresponding to the objective physiological accompaniment of the idea in man. The behavior in this stage is that which, in its higher reaches at least, has been called intelligent.

But so far as the objective occurrences are concerned, there would seem to be nothing in this later stage of the behavior involving anything different in essential principle from what we find in the inorganic world. The only additional factor is the law of the readier resolution of physiological states after repetition. While possibly our statement of this law may not be entirely adequate, there would seem to be nothing implied by it that is specifically vital, in the sense that it differs in essential principle from the methods of action seen in the inorganic world. This law of the readier resolution of physiological states after repetition presents indeed many analogies with various chains of physical and chemical action.[1] It certainly by no means requires in itself the action of any "final cause," — that is, of an entity that is at the same time purpose and cause. On the other hand, it undoubtedly does produce that type of behavior which has given rise to the conception of the purpose acting as cause. This conception is in itself of course a correct one, so far as we mean by a purpose an actual physiological state of the organism, determining behavior in the same manner as other factors determine it. But such a physiological state (subjectively a purpose) is a result of a foregoing objective cause, and acts to produce an effect in the same way as any other link in the causal chain. It would seem therefore to present no basis for theories of vitalism, so far as these depend on anything like the action of final causes.

That regulation takes place in the behavior of many animals in the manner above sketched may be affirmed as a clearly established fact, and it seems to be perhaps the only intelligible way in which regulatory behavior could be developed in a given individual.

But we are, of course, confronted by the fact that many individuals are provided at birth with definite regulatory methods of reaction to certain stimuli. In these cases the animal is not compelled to go through the process of performing varied movements, with subsequent fixation of the successful movement. How are such cases to be accounted for?

If the regulatory method of reaction acquired through the process sketched in the preceding paragraphs could be inherited, there would of course be no difficulty in accounting for such congenital regulatory reactions. In Protozoa this is apparently the real state of the case; there appears to be no reason why the products of reproduction by division should not inherit the properties of the individual that divides, however these properties were attained. But in the Metazoa such inheritance of acquirements presents great theoretical difficulties, and has not been experimentally demonstrated to occur, though it is perhaps too early to consider the matter as yet out of court. If such inheritance does not occur, the existence of congenital definite regulatory reactions would seem explicable only on the basis of the natural selection of individuals having varying methods of reaction, unless we are to adopt the theories of vitalism. In the method we have sketched above, a certain reaction that is

[1] See note, page 317.

regulatory is selected, through the operation of physiological laws, from among many performed by the same individual. In natural selection the same reaction is selected from among many performed by *different* individuals — in both cases because it is regulatory — because it assists the life processes of the organism. The two factors must then work together and produce similar results.[1] In both, the essential point is a selection from among varied activities.

We must here notice the fact that we often find in organisms behavior that is not regulatory. How are we to account for this? Without going into details, it is clear that there are a number of factors that would produce this result. First, interference with the life processes is not the only cause of reaction. The organism is composed of matter that is subject to the usual laws of physics and chemistry. External agents may of course act on this matter directly, causing changes in movement that are not regulatory. Second, the organism can perform only those movements which its structure permits. Often none of these movements can produce conditions that relieve the existing interference with the life processes. Then the organism can only try them, without regulatory results, and die (see, for example, such a case in the flatworm, p. 244). Further, certain responses may have become fixed, in the way described above, because under usual conditions they produce adjustment. Now if the conditions change, the organism still responds by the fixed reaction, and this may no longer be regulatory. The organism may then be destroyed before a new regulatory reaction can be developed by selection from varied movements. This condition of affairs is of course often observed.

All together, the regulatory character of behavior as found in many animals seems intelligible in a perfectly natural, directly causal way, on the basis of the principles brought out above. We may summarize these principles as (1) the selection through varied movements of conditions not interfering with the physiological processes of the organism ("trial and error"); (2) the fixation of the adaptive movements through the law of the readier resolution of physiological states after repetition.

3. REGULATION IN OTHER FIELDS

Is it possible that individual regulation in other fields is based on the same principles that we have set forth above for behavior? Bodily movement is only one of the many activities that vary, and variations of any of the organic activities may impede or assist the physiological

---

[1] For a discussion of the relation of these two factors, see Chapter XIX.

processes of the organism.   Is it possible that interference with the physiological processes may induce changes in other activities, — in chemical processes, in growth, and the like, — and that one of these activities is selected, as in behavior, through the fact that it relieves the interference that caused the change?

There is some evidence for this possibility.   Let us look, for example, at regulative changes in the chemical activity of the organism, such as we see in the acclimatization to poisons, in the responses to changes in temperature, or in the adaptation of the digestive juices to the food. What is the material from which the regulative conditions may be selected?   One of the general results of modern physical chemistry is expressed by Ostwald (1902, p. 366) as follows: "In a given chemical structure all processes that are so much as possible, are really taking place, and they lead to the formation of all substances that can occur at all."   Some of these processes are taking place so slowly that they escape usual observation; we notice only those that are conspicuous. But in its enzymes the body possesses the means (as Ostwald sets forth) of hastening any of these processes and delaying others, so that the general character of the action shall be determined by the more rapid process.   Such enzymes are usually present in the body in inactive forms (zymogens), which may be transformed into active enzymes by slight chemical changes, thus altering fundamentally the course of the chemical processes in the organisms.

It is evident, then, that the organism has presented to it, by the condition just sketched, unlimited possibilities for the selection of different chemical processes.   The body is a great mass of the most varied chemicals, and in this mass thousands of chemical processes, in every direction, — all those indeed that are possible, — are occurring at all times.   There is then no difficulty as to the sufficiency of the material presented for selection, if some means may be found for selecting it.

Further, it is known that interference with the physiological processes does result in many changes in the internal activities of the organism, as well as in its external movements.   Intense injurious stimulation causes not merely excess movements of the body as a whole, but induces marked changes in circulation, in respiration, in temperature, in digestive processes, in excretion, and in other ways.   Such marked internal changes involve, and indeed are constituted by, alterations of profound character in the chemical processes of the organism.   These chemical changes are sometimes demonstrated by the production of new chemicals under such circumstances.   Furthermore, it is clear that the internal changes due to interference with the physiological processes are not stereotyped in character, but varied.   Under violent injurious stimulation, respiration becomes for a time rapid, then is almost suspended. The heart beats for a time furiously, then feebly, and there is similar variation in other internal symptoms.

Thus it seems clear that interference with the life processes does produce varied activities in other ways than in bodily movements; and that among these it results in varied chemical processes. There is then presented opportunity for regulation to occur in the same way as in behavior. Certain of the processes occurring relieve the disturbance of the physiological functions. There results a cessation of the changes. In other words, a certain process is selected through the fact that it does relieve. It is well known, through the work of Pawlow (1898), that the adaptive changes in the activities of the digestive glands, fitting the digestive juices to the food taken, do not occur at once and completely under a given diet, but are brought about gradually. As the dog is continued on a diet of bread, the pancreatic juice becomes more and more adapted to the digestion of starch. This slow adaptation is of course what should be expected if the process occurs in anything like the manner we have sketched.

At a later stage, if the laws of these processes are the same as those for behavior, there will be present certain fixed methods of chemical response, by which the organism reacts to certain sorts of stimulation. That the law of the readier resolution of physiological states after repetition holds in this field, is clearly indicated by the work of Pawlow. He found that the pancreas under a uniform diet does tend to acquire a fixed method of reaction to the introduction of the food, that is not easily changed. In the dog which has digested starch for a month, the pancreatic juice is not readily changed back to that adapted to the digestion of meat. As a result, definite organs will in the course of time have left open to them only certain limited possibilities of variation — due to the development of something corresponding to the "action system" in behavior. Thus, in the pancreas, there will not exist unlimited possibilities as to the chemical changes that may occur. Its "action system" will be limited perhaps to the production of varied quantities of a certain set of enzymes, — amylopsin, trypsin, etc. The proper selection of these few possibilities will then occur by the method sketched. When digestion is disturbed by food that is not well digested, variations in the production of the different enzymes will be set in train, and one of these will in time relieve the difficulty, through the more complete digestion of the food. Thereupon the variations will cease, since their cause has disappeared. By still more complete fixation of the chemical response, through the law of the readier resolution of physiological states after repetition, or the analogue of this law, an organ or organism may largely lose its power of varying its chemical behavior and thus be unable to meet new conditions in a regulative way. A condition comparable to the production of a fixed reflex in behavior will result.

It is perhaps more difficult to apply the method of regulation above set forth to processes of growth and regeneration. Yet there is no logical difficulty in the way. The only question would be that of fact, — whether the varied growth processes necessary do, primitively, occur under conditions that interfere with the physiological processes. When a wound is made or an organ removed, is the growth process which follows always of a certain stereotyped character, or are there variations? It is well known, of course, that the latter is the case. In the regeneration of the earthworm, Morgan (1897) finds great variation; he says that in trying many experiments, one finds that what ninety-nine worms cannot do in the way of regeneration, the one hundredth can. The very great variations in the results of operations on eggs and young stages of animals are well known. Removal of an organ is known to produce great disturbance of most of the processes in the organism, and among others in the process of growth.

It appears not impossible then that regulation may be brought about in growth processes in accordance with the same principles as in behavior. A disturbance of the physiological processes results in varied activities, and among these are varied growth activities. Some of these relieve the disturbance; the variation then ceases and these processes are continued. In any given highly organized animal or plant the different possibilities of growth will have become decidedly limited; and it is only from this limited number of possibilities that selections can be made. In some cases, by the fixation of certain processes through the analogue of the law of the readier resolution of physiological states, the organism or a certain part thereof will have lost the power of responding to injury save in one definite way. Under new conditions this one way may not be regulatory, yet it may be the only response possible. Thus may result the formation under certain conditions of heteromorphic structures, — a tail in place of a head, or the like, from a part of the body that (in normal development perhaps) is accustomed to produce such an organ. This would again correspond to the production of a fixed reflex action in behavior, even under circumstances where this action is not regulatory.

It appears to the writer that the method of form regulation recently set forth in a most suggestive paper by Holmes (1904) is in agreement with the general method of regulation here set forth, and may be considered a working out of the details of the way in which growth regulation might take place along these lines. Holmes has of course emphasized other features of the process in a way that is not called for in the present work.

Some suggestions as to the possibility of regulation along the line of the selection of overproduced activities are found in J. Mark Baldwin's valuable collection of essays entitled "Evolution and Development."

It may be noted that regulation in the manner we have set forth is what in behavior is commonly called intelligence. If the same method of regulation is found in other fields, then there is no reason for refusing to compare the action there to intelligence. Comparison of the regulatory processes that are shown in internal physiological changes and in regeneration, to intelligence seems to be looked upon sometimes as unscientific and heretical. Yet intelligence is a name applied to processes that actually exist in the regulation of movement, and there is no *a priori* reason why similar processes should not occur in regulation in other fields. Movement is after all only the general result of the more recondite chemical and physical changes occurring in organisms, and therefore cannot follow laws differing in essential character from the latter. We are dealing in other fields with the same substance that is capable of performing the processes seen in intelligent action, and these could not occur as they do if the underlying physical and chemical processes did not obey the same laws. In a purely objective consideration there seems no reason to suppose that regulation in behavior (intelligence) is of a fundamentally different character from regulation elsewhere.

4. SUMMARY

We may sum up the fundamental features in the method of individual regulation above set forth as follows: —

The organism is a complex of many processes, of chemical change, of growth, and of movement; these are proceeding with a certain energy. These processes depend for their unimpeded course on their relations to each other and on the relations to the environment which the processes themselves bring about. When any of these processes are blocked or disturbed, through a change in the relations to each other or the environment, the energy overflows in other directions, producing varied changes, — in movement, and apparently also in chemical and growth processes. These changes of course vary the relations of the processes to each other and to the environment; some of the conditions thus reached relieve the interference which was the cause of the change. Thereupon the changes cease, since there is no further cause for them; the relieving condition is therefore maintained. After repetition of this course of events, the process which leads to relief is reached more directly, as a result of the law of the readier resolution of physiological states after repetition. Thus are produced finally the stereotyped changes often resulting from stimulation.

This method of regulation is clearly seen in behavior, where its operation is, in the later stages, what is called intelligence. Its application to chemical and form regulation is at present hypothetical, but appears possible.

BIBLIOGRAPHY

THE following is a list of the works cited in the text. It is not a complete bibliography of behavior in lower animals, but will be found to contain most of the more important papers on the lowest groups. The authors' names are given in alphabetical order, with their works arranged according to the date of their appearance. In the text the works are cited by the name of the author accompanied by the date; for the complete title reference is to be made to the present list.

ADAMS, G. P., 1903. On the negative and positive phototropism of the earthworm Allolobophora fœtida (Sav.), as determined by light of different intensities: Amer. Journ. Physiol., IX, 26–34. — ALLABACH, L. F., 1905. Some points regarding the behavior of Metridium: Biol. Bul., X, 35–43.

BAIN, ALEXANDER, 1888. The emotions and the will. 604 pp. London. — ID., 1894. The senses and the intellect. 3d ed. New York. — BALBIANI, E. G., 1861. Recherches sur les phénomènes sexuels des infusoires: Journ. Physiol. (Brown-Séquard), IV, 102–130; 194–220; 431–448; 465–520 (also separate, Paris, 1862). — ID., 1873. Observations sur le Didinium nasutum: Arch. d. Zool. Exp., II, 363–394. — BALDWIN, J. MARK, 1897. Mental development in the child and in the race. Methods and processes. 2d ed. 496 pp. New York. — ID., 1902. Development and evolution. 395 pp. New York. — BANCROFT, F. W., 1904. Note on the galvanotropic reactions of the medusa Polyorchis penicillata A. Agassiz: Journ. Exp. Zool., I, 289–292. — ID., 1905. Ueber die Gültigkeit des Pflüger'schen Gesetzes für die galvanotropischen Reaktionen von Paramæcium: Arch. f. d. ges. Physiol., CVII, 535–556. — BARRATT, J. O. W., 1905. Der Einfluss der Konzentration auf die Chemotaxis: Zeitschr. f. allg. Physiol., V, 73–94. — BEER, BETHE, und v. UEXKÜLL, 1899. Vorschläge zu einer objectivirender Nomenclatur in der Physiologie des Nervensystems: Centralb. f. Physiol., June 10, 5 pp. — BERNSTEIN, J., 1900. Chemotropische Bewegung eines Quecksilbertropfens: Arch. f. d. ges. Physiol., LXXX, 628–637. — BETHE, ALB., 1898. Dürfen wir den Ameisen und den Bienen psychische Qualitäten zuschreiben?: Arch. f. d. ges. Physiol., LXX, 15–100. — BIRUKOFF, B., 1899. Untersuchungen über Galvanotaxis: Arch. f. d. ges. Physiol., LXXVII, 555–585. — ID., 1904. Zur Theorie der Galvanotaxis: Arch. f. Anat. u. Physiol., Physiol. Abth., 271–296. — BOHN, G., 1903. De l'évolution des connaissances chez les animaux marins littoraux: Bul. Institut Gén. Psychol., no. 6. 67 pp. — ID., 1903 a. Les Convoluta roscoffensis et la theorie des causes actuelles: Bul. Mus. d. Hist. Nat., 352–364. — ID., 1905. Attractions et Oscillations des animaux marins sous l'influence de la lumière: Inst. Gén. Psychol., Mémoires, I., 110 pp. — BREDIG, 1902. Die Elemente der chemischen Kinetik, mit besonderer Berücksichtigung der Katalyse und der Fermentwirkung: Ergebnisse der Physiologie, I Abth. 1, pp. 134–212. — BÜRGER, O., 1903. Ueber das Zusammenleben von

Antholoba reticulata Couth. und Hepatus chilensis M. E.: Biol. Centralbl., XXIII, 677-678. — BÜTSCHLI, O., 1880. Protozoa, I Abth. Bronn's Klassen und Ordnungen des Thierreichs. Leipzig. — ID., 1889. Protozoa, III Abth. Infusoria. Bronn's Klassen und Ordnungen des Thierreichs, I Bd. Leipzig. — ID., 1892. Untersuchungen über mikroskopische Schäume und das Protoplasma. 234 pp. Leipzig.

CARLGREN, O., 1899. Ueber die Einwirkung des constanten galvanischen Stromes auf niedere Organismen: Arch. f. Anat. u. Physiol. Physiol. Abth., 49-76. — ID., 1905. Ueber die Bedeutung der Flimmerbewegung für die Nahrungstransport bei den Actiniarien und Madreporarien: Biol. Centralbl., XXV, 308-322. — ID., 1905 a. Der Galvanotropismus und die innere Kataphorese: Zeitschr. f. allg. Physiol., V, 123-130. — CLAPARÈDE, ED., 1901. Les animaux sont-ils conscients?: Revue Philosophique, LI, 24 pp. (Translation in International Quarterly, VIII, 296-315.) — ID., 1905. La psychologie comparée est-elle légitime?: Archives d. Psychol., V, 13-35. — COEHN, A. and BARRATT, W., 1905. Ueber Galvanotaxis vom Standpunkte der physikalische Chemie: Zeitschr. f. allg. Physiol., V, 1-9.

DALE, H. H., 1901. Galvanotaxis and chemotaxis of ciliate infusoria. Part 1. Journ. of Physiol., XXVI, 291-361. — DAVENPORT, C. B., 1897. Experimental morphology, Vol. 1, 280 pp. — DRIESCH, H., 1901. Die organischen Regulationen. Vorbereitungen zu einer Theorie des Lebens. 228 pp. Leipzig. — ID., 1903. Die "Seele" als elementarer Naturfaktor. 97 pp. Leipzig.

ENGELMANN, T. W., 1879. Ueber Reizung contraktilen Protoplasmas durch plötzliche Beleuchtung: Arch. f. d. ges. Physiol., XIX, 1-7. — ID., 1881. Zur Biologie der Schizomyceten: Arch. f. d. ges. Physiol., XXVI, 537-545. — ID., 1882. Ueber Licht- und Farbenperception niederster Organismen: Arch. f. d. ges. Physiol., XXIX, 387-400. — ID., 1882 a. Bacterium photometricum, ein Beitrag zur vergleichenden Physiologie des Licht- und Farbensinnes: Arch. f. d. ges. Physiol., XXX, 95-124. — ID., 1888. Die Purpurbacterien und ihre Beziehungen zum Licht: Bot. Zeitung, XLVI, 661-669; 677-689; 693-701; 709-720. — ID., 1894. Die Erscheinung der Sauerstoffausscheidung chromophyllhaltiger Zellen im Licht bei Anwendung der Bacterienmethode: Arch. f. d. ges. Physiol., LVII, 375-386.

FAMINTZIN, A., 1867. Die Wirkung des Lichtes auf Algen und einige andere nahe verwandte Organismen: Jahrb. f. wiss. Bot., VI, 1-44.

GAMBLE, F. W. and KEEBLE, F., 1903. The bionomics of Convoluta roscoffensis, with special reference to its green cells: Quart. Journ. Micr. Sci., XLVII, 363-431. — GARREY, W. E., 1900. The effect of ions upon the aggregation of flagellated infusoria: Amer. Journ. Physiol., III, 291-315. — GREELEY, A. W., 1904. Experiments on the physical structure of the protoplasm of Paramecium and its relation to the reactions of the organism to thermal, chemical, and electrical stimuli: Biol. Bul. VII, 3-32.

HABERLANDT, G., 1905. Ueber den Begriff "Sinnesorgane" in der Tier- und Pflanzenphysiologie: Biol. Centralbl., XXV, 446-451. — HARPER, E. H., 1905. Reactions to light and mechanical stimuli in the earthworm, Perichæta bermudensis (Beddard): Biol. Bul., X, 17-34. — HARRINGTON, N. R. and LEAMING, E., 1900. The reaction of Amœba to light of different colors: Amer. Journ. Physiol., III, 9-16. — HERTEL, E., 1904. Ueber Beeinflussung des Organismus durch Licht, speziell durch die chemisch wirksamen Strahlen: Zeitschr. f. allg. Physiol., IV, 1-43. — HOBHOUSE, L. T., 1901. Mind in evolution. 415 pp. London. — HODGE, C. F. and AIKINS, H. A., 1895. The daily life of a protozoan; a study in comparative psycho-physiology: Amer. Journ. Psychol., VI, 524-533. — HOLMES, S. J., 1901. Phototaxis in the amphipoda: Amer. Journ. Physiol., V, 211-234. — ID., 1903. Phototaxis in Volvox: Biol. Bul., IV, 319-326. — ID., 1904. The problem of form regulation: Arch. f. Entw.-mech., XVIII, 265-305. — ID., 1905. The selection of random movements as a factor in phototaxis: Journ. Comp Neurol. and Psych., XV, 98-112. — HOLT, E. B. and LEE, F. S., 1901. The theory of phototactic response: Amer. Journ. Physiol., IV, 460-481.

JAMES, W., 1901. Principles of Psychology. 2 vols. New York. — JEN-NINGS, H. S., 1897. Studies on reactions to stimuli in unicellular organisms. I. Reactions to chemical, osmotic, and mechanical stimuli in the ciliate infusoria: Journ. of Physiol., XXI, 258-322. — ID., 1899. Studies, etc. II. The mechanism of the motor reactions of Paramecium: Amer. Journ. Physiol., II, 311-341. — ID., 1899 a. The psychology of a protozoan: Amer. Journ. Psychol., X, 503-515. — ID., 1899 b. Studies, etc. III. Reactions to localized stimuli in Spirostomum and Stentor: Amer. Naturalist, XXXIII, 373-389. — ID., 1899 c. Studies, etc. IV. Laws of chemotaxis in Paramecium: Amer. Journ. Physiol., II, 355-379 — ID., 1899 d. The behavior of unicellular organisms: Biological Lectures delivered at the Marine Biological Laboratory at Woods Hole in the summer of 1899, 93-112. — ID., 1900. Studies, etc. V. On the movements and motor reflexes of the Flagellata and Ciliata: Amer. Journ. Physiol., III, 229-260. — ID., 1900 a. Studies, etc. VI. On the reactions of Chilomonas to organic acids: Amer. Journ. Physiol., III, 397-403. — ID., 1900 b. Reactions of infusoria to chemicals: a criticism: Amer. Naturalist, XXXIV, 259-265. — ID., 1901. On the significance of the spiral swimming of organisms: Amer. Naturalist, XXXV, 369-378. — ID., 1902. Studies, etc. IX. On the behavior of fixed infusoria (Stentor and Vorticella), with special reference to the modifiability of protozoan reactions: Amer. Journ. Physiol., VIII, 23-60. — ID., 1902 a. Artificial imitations of protoplasmic activities and methods of demonstrating them: Journ. Appl. Micr. and Lab. Methods, V, 1597-1602. — ID., 1904. Reactions to heat and cold in the ciliate infusoria. Contributions to the study of the behavior of lower organisms: Carnegie Institution of Washington, Publication 16, pp. 5-28. — ID., 1904 a. Reactions to light in ciliates and flagellates: ibid., pp. 29-71. — ID., 1904 b. Reactions to stimuli in certain Rotifera: ibid., pp. 73-88. — ID., 1904 c. The theory of tropisms: ibid., pp. 89-107. — ID., 1904 d. Physiological states as determining factors in the behavior of lower organisms: ibid., pp. 109-127. — ID., 1904 e. The movements and reactions of Amœba: ibid., pp. 129-234. — ID., 1904 f. The method of trial and error in the behavior of lower organisms: ibid., pp. 235-252. — ID., 1904 g. Physical imitations of the activities of Amœba: Amer. Naturalist, XXXVIII, 625-642. — ID., 1904 h. The behavior of Paramecium. Additional features and general relations: Journ. Comp. Neurol. and Psychol., XIV, 441-510. — ID., 1905. The basis for taxis and certain other terms in the behavior of infusoria: Journ. Comp. Neurol. and Psychol., XV, 138-143. — ID., 1905 a. Modifiability in behavior. I. Behavior of sea anemones: Journ. Exp. Zool., II, 447-472. — ID., 1905 b. The method of regulation in behavior and in other fields:

Journ. Exp. Zool., II, 448–494. — JENNINGS, H. S. and CROSBY, J. H., 1901. The manner in which bacteria react to stimuli, especially to chemical stimuli: Amer. Journ. Physiol., VI, 31–37. — JENNINGS, H. S. and JAMIESON, CLARA, 1902. The movements and reactions of pieces of infusoria: Biol. Bul., III, 225–234. — JENNINGS, H. S. and MOORE, E. M., 1902. On the reactions of infusoria to carbonic and other acids, with especial reference to the causes of the gatherings spontaneously formed: Amer. Journ. Physiol., VI, 233–250. — JENSEN, P., 1893. Ueber den Geotropismus niederer Organismen: Arch. f. d. ges. Physiol., LIII, 428–480. — ID., 1901. Untersuchungen über Protoplasmamechanik: Arch. f. d. ges. Physiol., LXXXVII, 361–417. — ID., 1902. Die Protoplasmabewegung: Asher und Spiro's Ergebnisse der Physiologie, I. 42 pp. — JORDAN, H., 1905. Einige neuere Arbeiten auf dem Gebiete der "Psychologie" wirbelloser Tiere: Biol. Centralbl., XXV, 451–464, 473–479.

KÜHNE, W., 1864. Untersuchungen über das Protoplasma und die Contractilität. 158 pp. Leipzig.

LE DANTEC, F., 1895. La matière vivante. 191 pp. Paris. — LEIDY, J., 1879. Fresh-water rhizopods of North America: Rep. U. S. Geol. Survey of the Territories, XII, 334 pp., 48 pl. — LOEB, J., 1891. Untersuchungen zur physiologischen Morphologie der Thiere. I. Ueber Heteromorphose. 80 pp. Würzburg. — ID., 1893. Ueber künstliche Umwandlung positiv heliotropischer Thiere in negativ heliotropische und umgekehrt: Arch. f. d. ges. Physiol., LIV, 81–107. — ID., 1894. Beiträge zur Gehirnphysiologie der Würmer: Arch. f. d. ges. Physiol., LVI, 247–269. — ID., 1895. Zur Physiologie und Psychologie der Actinien: Arch. f. d. ges. Physiol., LIX, 415–420. — ID., 1897. Zur Theorie der physiologischen Licht und Schwerkraftwirkungen: Arch. f. d. ges. Physiol., LXVI, 439–466. — ID., 1900. Comparative physiology of the brain and comparative psychology. 309 pp. New York. — ID., 1900 a. On the different effects of ions upon myogenic and neurogenic rhythmical contractions and upon embryonic and muscular tissue: Amer. Journ. Physiol., III, 384–396. — LOEB, J. and BUDGETT, S. P., 1897. Zur Theorie der Galvanotropismus. IV. Mittheilung. Ueber die Ausscheidung electropositiver Ionen an den äusseren Anodenfläche protoplasmatischer Gebilde als Ursache der Abweichungen vom Pflüger'schen Erregungsgesetz: Arch. f. d. ges. Physiol., LXVI, 518–534. — LUDLOFF, K., 1895. Untersuchungen über den Galvanotropismus: Arch. f. d. ges. Physiol., LIX, 525–554. — LUKAS, F., 1905. Psychologie der niedersten Tiere. 276 pp. Wien und Leipzig. — LYON, E. P., 1904. On rheotropism. I. Rheotropism in fishes: Amer. Journ. Physiol., XII, 149–161. — ID., 1905. On the theory of geotropism in Paramœcium: Amer. Journ. Physiol., XIV, 421–432.

MARSHALL, W., 1882. Ueber einige Lebenserscheinungen der Süsswasserpolypen und über eine neue Form von Hydra viridis: Zeitschr. f. wiss. Zool., XXXVII, 664–702. — MASSART, J., 1889. Sensibilité et adaptation des organismes à la concentration des solutions salines: Arch. de Biol., IX, 515–570. — ID., 1891. Recherches sur les organismes inférieurs. II. La sensibilité à la concentration chez les êtres unicellulaires marins: Bul. Acad. roy. Sci. Belgique, (3), XXII, 158–167. — ID., 1891 a. Recherches sur les organismes inférieurs. III. La sensibilité à la gravitation: Bul. Acad. roy. Belgique (3), XXII, 158–167. — ID., 1901. Essai de classification des réflexes non-nerveux: Annales de l'Inst. Pasteur. 39 pp. (German translation in: Biol. Centralbl., XXII, 9–23, 41–52, 65–79.) — ID., 1901 a. Recherches sur les organismes inférieurs. IV. Le lancement des trichocystes (chez Paramecium aurelia): Bul. Acad. roy. Belgique (Classe des Sci.), no. 2, pp. 91–106. — MAST, S. O., 1903. Reactions to temperature changes in Spirillum, Hydra, and fresh-water planarians: Amer. Journ. Physiol., X, 165–190. — ID., 1906. Light reactions in lower organisms. I. Stentor cæruleus. To appear in : Journ. Exp. Zool. MAUPAS, E., in: Binet, 1889. The psychic life of micro-organisms, pp. 48, 49. Chicago. — MENDELSSOHN, M., 1895 Ueber den Thermotropismus einzelliger Organismen: Arch. f. d. ges. Physiol., LX, 1–27. — ID., 1902. Recherches sur la thermotaxie des organismes unicellulaires: Journ. de Physiol. et de Path. Gén., IV, 393–410. — ID., 1902 a. Recherches sur la interférence de la thermotaxie avec d'autres tactismes et sur le mécanisme du mouvement thermotactique: Journ. de Physiol. et de Path. Gen., IV, 475–488. — ID., 1902 b. Quelques considérations sur la nature et la rôle biologique de la thermotaxie: Journ. de Physiol. et de Path. Gén., IV, 489–496. — MINOT, C. S., 1902. The problem of consciousness in its biological aspects: Science (N. S.), XVI, 1–12. — MIYOSHI, M., 1897. Studien über die Schwefelrasenbildung und der Schwefelbacterien der Thermen von Yumoto bei Nikko: Journ. Coll. Sci. Imp. Univ. Tokyo, X, 143–173. — MOEBIUS, K., 1873. Die Bewegungen der Thiere und ihr psychischer Horizont. 20 pp. Kiel. — MOORE, ANNE, 1903. Some facts concerning the geotropic gatherings of Paramecium: Amer. Journ. Physiol., IX, 238–244. — MORGAN, C. LLOYD, 1900. Animal behavior. 344 pp. London. — MORGAN, T. H., 1897. Regeneration in Allolobophora fœtida: Arch. f. Entw -mech., V, 570–586. — MÜNSTERBERG, H., 1900. Grundzüge der Psychologie, Bd. I. 565 pp. Leipzig.

NAEGELI, C., 1860. Ortsbewegungen der Pflanzenzellen und ihrer Theile (Strömungen): Naegeli's Beiträge zur wiss. Bot., Hft. II, 59–108. — NAGEL, W., 1892. Das Geschmacksinn der Actinien: Zool. Anz., XV, 334–338. — ID., 1894. Vergleichendphysiologische und anatomische Untersuchungen über den Geruchs- und Geschmacksinn und ihre Organe: Bibliotheca Zoologica, XVIII, 207 pp. — ID., 1894 a. Experimentelle sinnesphysiologische Untersuchungen an Coelenteraten: Arch. f. d. ges. Physiol., LVII, 495–552. — ID., 1899. Ueber neue Nomenclatur in der vergleichenden Sinnesphysiologie: Centralbl. f. Physiol., Hft. XII., 4 pp. — NUEL, 1904. La vision. 376 pp. Paris.

OLTMANNS, F., 1892. Ueber photometrische Bewegungen der Pflanzen: Flora, LXXV, 183–266. — OSTWALD, WILHELM, 1902. Vorlesungen über Naturphilosophie. 457 pp. Leipzig. — OSTWALD, WOLFGANG, 1903. Zur Theorie der Richtungsbewegungen schwimmender niederer Organismen: Arch. f. d. ges. Physiol., XCV, 23–65.

PARKER, G. H., 1896. The reactions of Metridium to food and other substances: Bul. Mus. Comp. Zool. Harvard Coll., XXIX, 102–119. — ID., 1905. The reversal of ciliary movements in metazoans: Amer. Journ. Physiol., XIII, 1–16. — ID., 1905 a. The reversal of the effective stroke of the labial cilia of sea anemones by organic substances: Amer. Journ. Physiol., XIV, 1–5. — PARKER, G. H. and BURNETT, F. L., 1900. The reactions of planarians with and without eyes, to light: Amer. Journ. Physiol., III, 271–284. — PAWLOW, J. P., 1898. Die Arbeit der

Verdauungsdrüsen. 199 pp. Wiesbaden. — PEARL, R., 1900. Studies on electrotaxis. I. On the reactions of certain infusoria to the electric current: Amer. Journ. Physiol., IV, 96–123. — ID., 1901. Studies on the effects of electricity on organisms. II. The reactions of Hydra to the constant current: Amer. Journ. Physiol., V, 301–320. — ID., 1903. The movements and reactions of fresh-water planarians. A study in animal behavior: Quart. Journ. Micr. Sci., XLVI, 509–714. — PENARD, E., 1902. Faune rhizopodique du bassin du Leman. 714 pp. Geneva. — PERKINS, H. F., 1903. The development of Gonionema murbachii: Proc. Philadelphia Acad. Nat. Sci. for 1902, pp. 750–790. — PFEFFER, W., 1884. Locomotorische Richtungsbewegungen durch chemische Reize: Unters. a. d. bot. Inst. Tübingen, I, 364–482. — ID., 1888. Ueber chemotactische Bewegungen von Bacterien, Flagellaten und Volvocineeʀ: Unters. a. d. bot. Inst. Tübingen, II, 582–661. — ID., 1904. Pflanzenphysiologie, zweiter Ausgabe. II. Leipzig. — PREYER, W., 1886. Ueber die Bewegungen der Seesterne: Mitth. a. d. zool. Stat. z. Neapel, VII, 27–127; 191–233. — PÜTTER, A., 1900. Studien über Thigmotaxis bei Protisten: Arch. f. Anat. u. Physiol., Physiol. Abth., Supplementband, 243–302. — ID., 1904. Die Reizbeantwortungen der ciliaten Infusorien. [Versuch einer Symptomatologie]: Zeitschr. f. allg. Physiol., III, 406–454.

RADL, E., 1903. Untersuchungen über die Phototropismus der Thiere. 188 pp. Leipzig. — RHUMBLER, L., 1898. Physikalische Analyse von Lebenserscheinungen der Zelle I. Bewegung, Nahrungsaufnahme, Defäkation, Vacuolen-Pulsationen, und Gehäusebau bei lobosen Rhizopoden: Arch. f. Entw.-mech., VII, 103–350. — ID., 1905. Zur Theorie der Oberflächenkräfte der Amöben: Zeitschr. f. wiss Zool., LXXXIII, 1–52. — ROESLE, E., 1902. Die Reaktion einiger Infusorien auf einzelne Induktionsschläge: Zeitschr. f. allg. Physiol., II, 139–168. — ROMANES, G. J., 1885. Jellyfish, starfish, and sea urchins. 323 pp. New York. — ROTHERT, W., 1901. Beobachtungen und Betrachtungen über tactische Reizerscheinungen: Flora, LXXXVIII, 371–421. — ID., 1903. Ueber die Wirkung des Aethers und Chloroforms auf die Reizbewegungen der Mikroorganismen: Jahrb. f. wiss. Bot., XXXIX, 1–70. — ROUX, W., 1901. Ueber die "morphologische Polarisation" von Eiern und Embryonen durch den electrischen Strom: Sitz.-ber. d. k. Akad. d. Wiss. z. Wien, Math. u. Naturw. Classe, CI, 27–228. (Roux's Gesammelte Abhandlungen, II, 540–765.)

SCHWARZ, F., 1884. Der Einfluss der Schwerkraft auf die Bewegungsrichtung von Chlamidomonas und Euglena: Ber. Bot. Gesellsch., II, 51–72. — SEMON, R., 1904. Die Mneme als erhaltendes Prinzip im Wechsel des organischen Geschehens. 353 pp. Leipzig. — SMITH, AMELIA C., 1902. The influence of temperature, odors, light, and contact on the movements of the earthworm: Amer. Journ. Physiol., VI, 459–486. — SOSNOWSKI, J., 1899. Untersuchungen über die Veränderungen der Geotropismus bei Paramecium aurelia: Bul. Internat. Acad. Sci. Cracovie, 130–136. — SPAULDING, E. G., 1904. An establishment of association in hermit crabs, Eupagurus longicarpus: Journ. Comp. Neurol. and Psychol., XIV, 49–61. — SPENCER, H., 1894. Principles of psychology, 3d ed. 2 vols. New York. — STAHL, E., 1884. Zur Biologie der Myxomyceten: Bot. Zeitung, XL, 146–155; 162–175; 187–191. — STATKEWITSCH, P., 1903. Ueber die Wirkung der Induktionsschläge auf einige Ciliata: Le Physiologiste Russe, III, 55 pp. — ID., 1903 a. Galvanotropism and galvanotaxis of organisms. Part First. Galvanotropism and

galvanotaxis of ciliate infusoria (Russian). Dissertation. 160 pp. Moscow. —
Id., 1904. Galvanotropismus und Galvanotaxis der Ciliata. Erste Mittheilung:
Zeitschr. f. allg. Physiol., IV, 296–332. — Id., 1904 a. Zur Methodik der biologis-
chen Untersuchungen über die Protisten: Arch. f. Protistenkunde, V, 17–39. —
Strasburger, E., 1878. Wirkung des Lichtes und der Wärme auf Schwärmsporen:
Jenaische Zeitschr. f. Naturw. (N. F.), XII, 551–625. Also separate, 75 pp. Jena.
Thorndike, E., 1898. Animal intelligence. An experimental study of the
associative processes in animals: The Psychol. Review, Monograph Suppl., II,
109 pp. — Titchener, E. B., 1902. Were the earliest organic movements con-
scious or unconscious? Pop. Sci. Monthly, LX, 458–469. — Torrey, H. B., 1904.
On the habits and reactions of Sagartia davisi: Biol. Bul., VI, 203–216. — Id., 1904 a.
Biological studies on Corymorpha palma and its environment: Journ. Exp. Zool., I,
395–422. — Trembley, A., 1744. Mémoires pour servir à l'histoire d'un genre de
polypes d'eau douce à bras en forme de cornes. Paris.
Uexküll, J. v., 1897. Ueber Reflexe bei den Seeigeln: Zeitschr. f. Biol.,
XXXIV, 298–318. — Id., 1897 a. Vergleichend sinnesphysiologische Untersuchun-
gen. II. Der Schatten als Reiz für Centrostephanus longispinus: Zeitschr. f. Biol.,
XXXIV, 319–339. — Id., 1899. Die Physiologie der Pedicellarien: Zeitschr. f.
Biol., XXXVII, 334–403. — Id., 1900. Die Physiologie des Seeigelstachels:
Zeitschr. f. Biol., XXXIX, 73–112. — Id., 1900 a. Die Wirkung von Licht und
Schatten auf die Seeigel: Zeitschr. f. Biol., XL, 447–476. — Id., 1900 b. Ueber
die Stellung der vergleichenden Physiologie zur Hypothese der Thierseele: Biol.
Centralbl., XX, 497–502. — Id., 1902. Im Kampf um die Tierseele: Asher und
Spiro's Ergebnisse der Physiologie, I. 24 pp.
Verworn, M., 1889. Psycho-physiologische Protistenstudien. Experimentelle
Untersuchungen. 219 pp. Jena. — Id., 1889 a. Die polare Erregung der Protisten
durch den galvanischen Strom: Arch. f. d. ges. Physiol., XLV, 1–36. — Id.,
1889 b. Die polare Erregung der Protisten durch den galvanischen Strom: Arch.
f. d. ges. Physiol., XLVI, 281–303. — Id., 1892. Die Bewegung der lebendigen
Substanz. 103 pp. Jena. — Id., 1895. Allgemeine Physiologie. 584 pp. Jena.
— Id., 1896. Untersuchungen über die polare Erregung der lebendigen Substanz
durch den constanten Strom. III. Mittheilung: Arch. f. d. ges. Physiol., LXII,
415–450. — Id., 1896 a. Die polare Erregung der lebendigen Substanz durch den
constanten Strom. IV. Mittheilung: Arch. f. d. ges. Physiol., LXV, 47–62. —
Id., 1899. General physiology, transl. by F. S. Lee. 615 pp. New York.
Wager, H., 1900. On the eye-spot and flagellum of Euglena viridis: Journ.
Lin. Soc. London, XXVII, 463–481. — Wagner, G., 1905. On some movements
and reactions of Hydra: Quart. Journ. Micr. Sci., XLVIII, 585–622. — Wallen-
gren, H., 1902. Zur Kenntnis der Galvanotaxis. I. Die anodische Galvanotaxis:
Zeitschr. f. allg. Physiol., II, 341–384. — Id., 1902 a. Inanitionserscheinungen der
Zelle: Zeitschr. f. allg. Physiol., I, 67–128. — Id., 1903. Zur Kenntnis der Galvano-
taxis. II. Eine Analyse der Galvanotaxis bei Spirostomum: Zeitschr. f. allg. Physiol.,
II, 516–555. — Wasmann, E., 1901. Nervenphysiologie und Tierpsychologie:
Biol. Centralbl., XXI, 23–31. — Id., 1905. Instinct und Intelligenz im Tierreich.
Dritte Auflage. 276 pp. Freiburg i. Br. — Wilson, E. B., 1891. The helio-
tropism of Hydra: Amer. Naturalist, XXV, 413–433.

YERKES, R. M., 1902.   Habit formation in the green crab, Carcinus granulatus:
Biol. Bul., III, 241–244. — ID., 1902 *a*.   A contribution to the physiology of the
nervous system of the medusa Gonionemus murbachii.   I.   The sensory reactions
of Gonionemus: Amer. Journ. Physiol., VI, 434–449. — ID., 1902 *b*.   A contribution,
etc.   II.   The physiology of the central nervous system: Amer. Journ. Physiol.,
VII, 181–198. — ID., 1903.   A study of the reactions and reaction time of the medusa
Gonionema murbachii to photic stimuli: Amer. Journ. Physiol., IX, 279–307. —
ID., 1903 *a*.   The instincts, habits, and reactions of the green frog: Harvard Psychol.
Studies, I, 579–597. — ID., 1903 *b*.   Reactions of Daphnia pulex to light and heat:
Mark Anniversary Volume, 361–377. — ID., 1904.   The reaction time of Gonio-
nemus murbachii to electric and photic stimuli: Biol. Bul., VI, 84–95. — ID., 1905.
Concerning the genetic relations of types of action: Journ. Comp. Neurol. and
Psychol., XV, 132–137. — ID., 1905 *a*.   Animal psychology and the criteria of the
psychic: Journ. of Phil., Psychol., and Scient. Meth., II, 141–149. — YERKES, R. M.
and HUGGINS, G. E., 1903.   Habit formation in the crawfish Cambarus affinis:
Harvard Psychol. Studies, I, 565–577.

ZIEGLER, H. E., 1900.   Theoretisches zur Thierpsychologie und vergleichenden
Neurophysiologie: Biol. Centralbl., XX, 1–16.

INDEX

Acclimatization to stimuli, in Amœba, 24; in Paramecium, 52; in sea anemones, 207; general, 294; to heat, 101; to poisons, 346.

Accommodation, see *adaptiveness* and *regulation.*

Acids, collection in, by Paramecium, 65, 67; by other infusoria, 122.

Actinia, taking of food when cut in two, 227.

Actinians, see *sea anemones.*

Action system, Paramecium, 107; of infusoria in general, 110; of Cœlenterata, 189; general, 300.

Activity, cause of, 284, 285.

Adams, behavior of earthworm, 248.

Adaptiveness of behavior, in Amœba, 23; in bacteria, 39; in Paramecium, 45, 79, 109; of changes of behavior in Stentor, 178; in food reactions of Gonionemus, 221; in cœlenterates, 230; in reactions to representative stimuli, 296; general factors, 299, 305, 338–350.

Adjustment, 342 (see *adaptiveness* and *regulation*).

Aiptasia, reaction to local stimulation, 199; setting of reaction by repetition, 206; acclimatization to stimuli, 207; relation to gravity, 211; food reactions, 223–226; rapid contraction, 228.

Allabach, behavior of Metridium, 224.

Allolobophora, testing movements, 247.

Alternating electric currents, reaction to, in Paramecium, 83.

Amœba angulata, 5; proteus, 2, 12, 13; velata, 5, 8; verrucosa, 2, 18.

Amœba, structure, 1; movements, 2; behavior and reactions, 6–25; food taking, 13–19; relation of behavior to tropism theory, 269; relation to reflexes, 279; question of consciousness, 336.

Amylobacter, 30, 32, 38.

Anaërobic bacteria, 31, 341.

Analysis of behavior, 283–313.

Anode, movement toward, in Paramecium, 81, 85, 98; in Flagellata, 152; in Opalina, 152, 159; in infusoria in general, 163.

Antholoba, attachment to crabs, 197.

Antitype, 277.

Anuræa, reaction to electric current, 242.

Association, in hermit crabs, 257, 290; general, 334.

Attached infusoria, reactions, 116; complexity of behavior, 180.

Attention, 330.

Authorities cited, 351.

Avoiding reaction, in Paramecium, 47, 53; adaptiveness of, 79; in Chilomonas, 111; in Euglena, 112; in other flagellates, 113; in other ciliates, 113; in light reactions, 149; relation to localization, 117; relation to reflexes, 279.

Bacteria, structure, 26; movements, 26; behavior and reactions, 27–40; relation of behavior to tropism theory, 271; relation to reflexes, 278; regulation in behavior, 341.

Bacterium chlorinum, 37; megatherium, 34; termo, 30, 32, 33, 34.

Bain, selection of overproduced movements, 302, 327.

Balantidium, reaction to chemicals, 122.

Balbiani, behavior of conjugating Paramecia, 104; use of trichocysts, 186.

Baldwin, law of dynamogenesis, 289; selection of overproduced movements, 302, 327; organic selection, 321, 326; regulation, 349.

Bancroft, reaction of infusoria to electricity, 167; of medusæ to electricity, 208, 210.

Barratt, reaction of Paramecia to chemicals, 64; theory of reaction to electricity (with Coehn), 165.

Bell of medusa, independent contractility, 228.

Beer, Bethe, and v. Uexküll, terminology, 275; reflex and antitype, 277.

Bethe, behavior of ants and bees, 258.

Bibliography, 351.

Bilateral animals, relation of behavior to tropism theory, 271, 273.

Binet, food habits of infusoria, 186.

Birukoff, reaction of Paramecium to induction shocks, 83, 88.

Blowfly larva, behavior, 249.

Bodo, reaction to chemicals, 124.

Bohn, behavior of hermit crabs, 211, 250; of Convoluta, 254; of littoral animals, 255; local action theory of tropisms, 274.

Botrydium, reactions to light, 143, 144.